GEOLOGY AND MINERALOGY RESEARCH DEVELOPMENTS

SANDSTONE

GEOCHEMISTRY, USES AND ENVIRONMENTAL IMPACT

GEOLOGY AND MINERALOGY RESEARCH DEVELOPMENTS

Additional books in this series can be found on Nova's website under the Series tab.

Additional e-books in this series can be found on Nova's website under the e-book tab.

GEOLOGY AND MINERALOGY RESEARCH DEVELOPMENTS

SANDSTONE

GEOCHEMISTRY, USES AND ENVIRONMENTAL IMPACT

AFSOON MOATARI KAZEROUNI
EDITOR

nova publishers
New York

Library of Congress Cataloging-in-Publication Data

Sandstone : geochemistry, uses and environmental impact / [edited by] Afsoon Moatari Kazerouni (Geology Department, Aarhus University, Copenhagen, Denmark).
pages cm
Includes bibliographical references and index.
ISBN: 978-1-62948-382-5 (hardcover)
1. Sandstone. I. Kazerouni, Afsoon Moatari, editor of compilation.
QE471.15.S25S247 2013
552'.5--dc23

2013039447

Published by Nova Science Publishers, Inc. † New York

CONTENTS

PREFACE

The book "Sandstone: Geochemistry, Uses and Environmental Impact" provides a broad perspective of research interests center around environmental geochemistry, and coupling of geologic, and geochemistry in sandstones.

The book describes the pathways by which potentialy minerals can access and concentrarte and reviews the different environmental assessment monitoring samples. Consideration is given to existing remediation methodologies and those being researched. Finally, the book reviews the importance of geochemical modeling in describing and predicting the interaction between the sandstone geochemistry and environment and emphasizes how preplanning during project studies can result in incorporation of technologies which will greatly impact the environment.

Topics are related to geochemistry and the understanding of ground-water systems as geochemistry has contributed significantly to the understanding of ground-water systems and analytical tecnicques used. Water imbibition is fundamental to waterflood performance in low permeability reservoir rocks. Since measuring the thermal properties of rocks and rock-fluid systems is difficult and time consuming and the results from such measurements are of limited value unless complete descriptions of the rock and fluids are given. A need exists for a method of predicting thermal behavior from other more easily measurable properties.

Presented here are correlations developed for predicting the thermal conductivity of consolidated sandstones from knowledge of density, porosity, permeability, and formation resistivity factor are discussed. The breath of the topics covered will be of interest to researchers and practicing engineers alike. This book contains three chapters. A brief description of each chapter follows.

The first chapter is devoted to the problems of *Sandstone Minerals as Indicators of Helium Residence Time in a Rock - Ground Water System.*

Several fundamental and applied problems require quantification of slow rates of solute transport in quasi-stagnant ground and pore water. Helium appears to be a promising tracer for ancient ground waters.

In this review a new method is discussed, which allows quantifying the He inventory in a rock - pore water system based on mineral records and constraining the He residence time in this system. This new method is based on investigations of rock, mineral and ground water samples from the alternating shale - sandstone sequence of the Permo-Carboniferous Trough, Northern Switzerland.

The second chapter presents the state of the art of *Water Imbibition into Sandstones: Influence of Flow Rate on Water Distribution and Acoustic Response.*

Fluid injection into a porous medium in order to displace another fluid is a common procedure while dealing with the recovery of subsurface fluids.

The seismic method has been widely used to image underground structures and, consequently, map reservoirs. More specifically, the time-lapse method consists in acquiring successive 3D seismic data of the same area over a period of time using acoustic sensors arrays.

The aim of the experimental work is to observe changes in seismic signals resulting from changes in the acoustic impedance. Since the acoustic impedance is the product of velocity and density, time-lapse signals are affected by the compressibilities of the reservoir rock and of pore fluids. This acoustic monitoring is also the basis to track saturation front displacements.

The interaction of seismic waves with the fluid-rock system is complex and complicates the interpretation of time-lapse signals. In order to better understand this interaction, experiments at laboratory scale were performed.

The third chapter presents experimental results and assesses the *Effective Thermal Conductivity of Dry- and Fluid-Saturated Sandstones at High Temperatures and High Pressures. Review of the Experimental Methods and Modeling.*

Available sources of data for the effective thermal conductivity of dry and fluid (gas, water, and oil)–saturated sandstones at high temperatures and high pressures have been collected and evaluated.

A comprehensive review of the different contact and contact-free measurement techniques used to accurately measure of the thermal conductivity of sandstones from various regions of the world were provided.

The uncertainties of each methods and their advantages and disadvantages were discussed.

The review also provides a detailed overview of the fundamentals of the authors own results on thermal conductivity measurements of dry and fluid-saturated sandstones from various locations.

The fourth chapter demonstrates *Diagenetic Chlorite: Example of an Intra Sandstone Resevoir Seal – From Palaeocene Tyr Member Sandstone, Siri Canyon, Danish North Sea.*

The study suggests the conditions for formation of an intra-formational chlorite seal and relates its possible formation to high-porosity/high-permeability zones of the reservoir. The study suggests that the formation of an intra-sandstone seal of diagenetic chlorite relates to the distribution of early diagenetic microquartz (which preserve porosity and permeability), and early diagenetic chlorite which reduces the permeability.

The fifth chapter document the diagenetic steps which release silica and to and quantify the possible timing of silica export from the shale into neighboring sandstones.

A major issue in the study is *The Effect of Burial on Clay Mineral Diagenesis and Quartz Cementation in Mudstones-Examples from the Siri Canyon, Danish North Sea.*

This study explores the potential sources of SiO_2 in the shale, and the timing of their active phases. The primary sources of silica during the diagenesis of the Sele Shale are biogenic opal and volcanic ash.

Finally, let me conclude this preface by thanking all the authors who have contributed to the realization of this book, without their unfailing support this project would not have even started. I thank them for their participation and patience during the preparation of this book and I am also grateful that they have entrusted me to edit their contributions. I hope the readers will find this book useful. I am looking forward to receive comments and constructive feedbacks about contents.

Dr. Afsoon Moatari Kazerouni
May 2013

In: Sandstone
Editor: Afsoon Moatari Kazerouni

ISBN: 978-1-62948-382-5

Chapter 1

SANDSTONE MINERALS AS INDICATORS OF HELIUM RESIDENCE TIME IN A ROCK: GROUND WATER SYSTEM

***M. Gannibal*[1] *and I. Tolstikhin*[1,2,*]**
[1]Geological Institute, Kola Scientific Centre, RAS, Apatity, Russia
[2]Space Research Institute, RAS, Moscow, Russia

ABSTRACT

Several fundamental and applied problems require quantification of slow rates of solute transport in quasi-stagnant ground and pore water. Helium appears to be a promising tracer for ancient ground waters. In this review a new method is discussed, which allows (i) quantifying the He inventory in a rock - pore water system based on mineral records and (ii) constraining the He residence time in this system. This new method is based on investigations of rock, mineral and ground water samples from the roughly 1 km thick alternating shale - sandstone sequence of the Permo-Carboniferous Trough (PCT), Northern Switzerland. In the PCT rocks the parent elements of radiogenic He isotopes, U, Th and Li, are mainly concentrated in the shales, where they reside in biotite, muscovite, clay minerals and organic matter. However, almost all ^{3}He and ^{4}He that have been produced in these minerals since their deposition have been lost into the complementary reservoir, i.e., the pore water. This is supported by the similarity between calculated ^{3}He / ^{4}He production ratio of the shales (7.7×10^{-8}) and ^{3}He / ^{4}He ratios observed in PCT ground water samples (9.1×10^{-8}).

In contrast to shales, sandstones contain less U, Th and much less Li; therefore helium released from the shales to the pore water will migrate along the concentration gradient in the water to the sandstone layers. As a U- and Th-poor mineral, quartz separated from sandstones contains excess He with ^{3}He / ^{4}He ratios identical to the shale production ratio.

Hence, He concentrations and ^{3}He / ^{4}He ratios suggest that He produced in the shales saturates the internal helium-accessible volumes (HAV) in quartz and plagioclase in the sandstones.

[*] E-mail: igor.tolstikhin@gmailo.com.

Special laboratory experiments were conducted to confirm the saturocess and to quantify the helium-accessible volume in quartz: (i) saturation of minerals with He under controlled pressure – temperature conditions, (ii) extraction of He from natural and saturated quartz samples by (mainly) isothermal heating under low temperatures (< 400 ^{O}C) thus preserving their HAV, and (iii) analysis of the He concentrations. According to these experiments, the He concentrations in pore water and in the HAV in quartz grains approach equilibrium during $\approx 10^4$ years. This is much less than the He residence time in the PCT system (above 10^6 years), indicating that equilibrium has been achieved. The HAV is estimated from the artificially saturated He concentration and PT conditions during the saturation experiment. In 36 quartz separates from 14 sandstone samples, the HAVs constitute from 0.017 % to 0.16 % of the grain volume. The HAV, the initial (unsaturated) He-concentration in quartz grains and the PCT temperature determine the internal He partial pressure with the average value $P_{He} = 0.49 \pm 0.21$ atm (1 σ). The product of the He pressure and solubility (7.35×10^{-3} cc STP He / (cc H_2O atm) for PCT aquifers) is the mineral-derived He concentration in the pore water, $C_{PW} = 0.0036 \pm 0.0017$ cc STP He / cc H_2O. This value is in full accord with the measured He concentrations in PCT ground waters, varying from 0.0016 to 0.0045 cc STP He / cc H_2O. Using quartz crystals as detectors, the He concentrations in the pore water across the entire PCT sequence have been quantified. These "quartz-derived" concentrations are used to develop a He migration model that gives rather long residence time for the He-atoms in the PCT rock - pore water system of about 10^7 years and that within the aquifers of 10^4 to 10^5 years.

Keywords: Helium, argon, isotope, saturation, residence time, quartz, sandstone, ground water, Northern Switzerland

1. INTRODUCTION

Dating of the stagnant ground waters is of extreme importance for both fundamental and applied problems, e.g., interactions between aquifers and aquitards and their role in the chemical evolution of movable economically-important ground waters; identification of environments with a stagnant ground water regime as potential sites for waste repositories; understanding and modelling of water-rock interactions, etc. Also, any attempt to estimate the capacity of aquifers (e.g., for drinking water supply) depends on a quantitative understanding of the situation underground.

There are several radioactive tracers successfully applied to estimate flow rate and chemical evolution of relatively young ground waters (Pearson et al., 1991; Mazor, 1991; Kipfer et al., 2002). Dating of stagnant porewaters appears to be a more complicated problem: since their residence times can exceed millions of years, most radioactive tracers cannot be used just because they decay on shorter time scales (Lehmann et al., 1992). Also on these long time scales underground production can overshadow the atmospheric input and thus prevent the chronological interpretation (Andrews and Fontes, 1991). Radiogenic helium appears to be a promising tracer for such waters.

In crustal rocks helium isotopes are mainly produced by α-decay of radioactive ^{238}U, ^{235}U, ^{232}Th series (radiogenic ^4He), and by nuclear reactions, ^6Li (n, α) $^3H \rightarrow \beta^- \rightarrow {}^3He$ (nucleogenic ^3He). Both processes are exothermic, so that the released particles produce damage tracks in the host material. These tracks allow radiogenic (nucleogenic) helium atoms to release readily from sedimentary, magmatic and metamorphic rocks, so that generally less than ~10% of the radiogenic or nucleogenic helium has been retained since the time of rock formation. (e.g., Mamyrin and Tolstikhin, 1984).

The released atoms reside in related ground or pore waters from where they are transported either by diffusion or water flow and are eventually discharged into the atmosphere. If pore water is stagnant, helium can only be removed by diffusion through interstitials filled by water. In this case helium is usually being accumulated in the pore water. Observed concentrations therefore depend on U and Th inventory, the rock age and the rate of helium removal. Measurements of U, Th and He concentrations in rock and pore water give both the present-day abundance of helium and its production rate; the helium residence time in a water-rock system can be assessed using these data.

This approach is widely used since the pioneering contributions by V.P. Savchenko (Savchenko, 1935), some examples are available in Stute et al. (1992), Marty et al. (1993), Tolstikhin et al. (1996). However the time estimates, obtained using helium balance, are only valid if helium flux from external sources is negligible.

This condition, i.e., a negligible He flux from a greater depth, is quite crucial, and was considered invalid in a number of publications. For example, Torgersen and Clarke (1985, 1987) proposed a "whole-crustal flux" or a "flux through the crust-mantle boundary" as an explanation for high He concentrations and a slightly enhanced ^{3}He abundance (relative to a typical crustal value $^{3}He/^{4}He \approx 2 \times 10^{-8}$, Mamyrin and Tolstikhin, 1984) in "young" ground waters of the Great Artesian Basin in Australia (GAB, see also Torgersen 1989; Torgersen et al., 1992). Castro et al. (2000) argue along the same line based on observations and modelling of He data in three different aquifer systems with calculated ground water residence times mainly below 10^4 years. These authors postulate that "it has to be expected that radiogenic ^{4}He is transported into the aquifers from the deep crust by advection, diffusion and dispersion". If such transfer of He isotopes from the mantle and deep crust into aquifer systems operated universally, He isotope data could not be used to infer residence times in rock-water systems. In the case of the GAB ground waters, however, Lehmann et al. (2003a) re-considered the proposed residence times using ^{81}Kr/Kr tracer, up to 4×10^5 years. Taking into account this long time scale, Lehmann et al. (2003a) concluded that both helium isotopes and the subsurface-produced portion of ^{36}Cl in the GAB ground waters of the sandstone aquifers were mainly derived from the adjacent shale aquitards. This indicates that at least for old ground water systems the *in-situ* production of He isotopes and their redistribution between aquifers and aquitards cannot be neglected. The assumption of a wide-spread He flux from the mantle and/or deep crust is further brought into question: Ballentine et al. (2002) argue that such a flux exists only in specific tectonic settings (e.g.).

In some cases, *in-situ* production is indeed considered to be a major source of He isotopes in sedimentary basins. The central role of in-situ production follows from experimental studies of aquifer systems with low flow velocities and long residence times of helium atoms (e.g., Mazor, 1972; Marty et al., 1988; Pearson et al., 1991), of aquitards (e.g., Osenbrück et al., 1998; Rübel et al., 2002), and from modelling helium concentration fields in cores of ancient sedimentary basins undisturbed by recent tectonic events (e.g., Ivanov et al., 1978). In such systems the following processes generally control the radiogenic noble gas isotope concentrations. First, the local radiogenic isotope production provides a source for fluxes from the host rocks into porewater. Second, the radiogenic atoms migrate from rocks (especially those characterised by the low retention coefficients, such as shale and claystone) into adjacent aquifers. And finally, the hydrogeologic setting of the system controls fluxes at a larger scale. For example, the flux of radiogenic species from the deep crust or from the mantle towards higher-lying hydrological systems can become partly or completely interrupted by deep-seated aquifers, depending on their prevailing ground water flow velocity. Permeable aquifers near the interface between crystalline basement and sedimentary sequences are often observed (e.g., Pearson et al., 1991). Solomon et al. (1996) and Sheldon et al. (2003) suggested *in situ* production as a major source of radiogenic helium even for young waters (~10 to 100 years) in shallow aquifers.

However, the two helium sources outlined above can co-exist: mixing of water flows (or helium fluxes) from different sources (reservoirs) is certainly a very typical of ground water evolution. The problem is to adequately characterize the sources, quantify mixing proportions, and then derive (if possible) respective time parameters. For example, when helium inventory in ground waters is discussed, its migration from adjacent aquitards should evaluated first of all, and only if this source could not supply enough helium deep crustal or mantle sources should be involved. Difference between these two cases is important: taking the aquitard(s) into account allows intra-basin production to be considered as the major source of radiogenic noble gas isotopes. This in turn permits adequate estimation of their residence time scales and redistribution between the two rock types.

Distinguishing between local production and inflow of helium from an external source generally requires a thorough expensive study. In some cases measurements of helium concentrations in ground waters offer the solution. For example, helium concentrations in both underlying and overlaying aquifers are often lower than in stagnant waters of a central core of a sedimentary basin, thus favouring local production (Ivanov et al., 1978). Also the isotope composition of helium, argon and some specific elemental and isotopic ratios (e.g., Lehmann et al. 2003a) helps to understand the helium source.

Another problem related to (almost) impermeable water-rock systems is the pore water extraction technique. Rübel and co-authors (Rübel et.al., 2002) have established He versus depth profiles in impermeable rocks by placing freshly drilled rock cores into vacuum containers (immediately after core recovery) and measuring, after a few weeks, helium amounts emanated from pore water in these rocks. This method, however, is limited to situations where appropriate rock samples can be sealed into vacuum containers within ca. 30 minutes after core recovery from a borehole. However some unaccounted loss of helium takes place due to change of the rock structure caused both by the drilling process and by the steep pressure decrease during core uplift.

In this contribution a new another approach (inspired by detection of trapped He in minerals) is presented that allows helium concentrations in stagnant water to be constrained. Thus, quartz grains separated from sandstone of the Permo-Carboniferous Trough (PCT) in Northern Switzerland contain "excess helium", i.e., more helium than could have been produced in-situ during the time since the sediments were formed (Tolstikhin et al., 1996). Furthermore, the measured $^{3}He/^{4}He$-ratio in excess helium clearly indicates the adjacent Li-rich shales as the helium source. It was therefore proposed that these mineral records could be used to determine the concentrations of helium in pore waters of impermeable rocks, an important parameter which cannot easily be obtained otherwise.

Laboratory saturation experiments allow modelling of the natural process of helium migration from pore waters into HAV in quartz grains and thus check the above proposal. These experiments involve exposure to controlled P-T – t conditions till the external and internal (in the HAV) helium (or other gas or gas mixture) pressure approaches equilibrium. Modifications of this technique were widely used to measure gas solubility and diffusivity in natural and man-made materials: waters, melts, vitreous and crystalline media, etc. Thus, Gerling (1940) applied this approach to study He solubility in gabbro-diabase melt. Procedures of such investigations may include penetration of the gases through glassy membrane (Shelby, 1971; Jambon and Shelby, 1980); the most widely accepted technique, however, is saturation of samples of interest with certain gas under controlled P-T-t-conditions, as it suits equally for melts and glasses (e.g., Shelby et al., 1976; Carroll and Stolper 1993), crystalline rocks and minerals (e.g., Roselieb et al., 1997) and fluids (e.g., Top et al., 1987).

Conditions (temperature, pressure, composition of gas mixture, exposure time) of experimental treatment were controlled mainly by the purposes of the given experiment, which, apart from determination of diffusivity and/or solubility, include: investigation of sites of NG atoms in the natural minerals (e.g., Gerling et al., 1965; Roselieb et al., 1997), identifying mechanism of excess argon migration into the minerals (Karpinskaya, 1965), or even looking for the source of terrestrial planet atmospheres (Broadhurst et al., 1992). Dunai and Roselieb (1996) used stepwise saturation instead of stepwise release to assess helium retention in garnet.

While the experimental approach applied by Lehmann et al. (2003b), Tolstikhin et al. (2005, 2011) and Gannibal (2012) was quite common, interpretation of the results was original and allowed to develop and apply a new approach in hydrology: using mineral grains as detectors of He concentrations in interstitial pore waters. This review includes: (i) investigation of helium sources within the water - rock system (shale / sandstone sequence, Permo-Carboniferous Trough, Northern Switzerland); (ii) modelling of helium penetration into quartz grains separated from PCT sandstones; (iii) estimates of the helium partial pressure in quartz grains (in gas phase of fluid inclusions) and the helium concentrations in adjacent pore waters; (iv) estimates of the He residence times within the Permo-Carboniferous Trough.

2. Experimental Procedures

The experimental procedures included helium impregnation and extraction with subsequent measurements of its concentrations and isotopic ratios. Samples were saturated with helium in a crucible, which can withstand helium pressures of up to 150 atmospheres and temperatures up to 600 ^{O}C. Samples were loaded into the crucible, then helium was admitted (from a tank, $^3He/^4He = 2 \times 10^{-8}$) up to a certain pressure and the crucible was heated to a selected temperature. Both temperature and pressure were kept constant during given experiment with the accuracy ±0.25 atm and ±4 ^{O}C, respectively. The helium pressure varied from one saturation experiment to another within the range of from 1 to 30 atm. The temperature was almost constant in all experiments and never exceeded 300 ^{O}C.

Helium was extracted from the samples by: (i) vacuum crushing to the average size of particles in the powder 0.01 ± 0.004 mm (Ikorsky and Kusch, 1992); (ii) step-wise heating and/or fusion at a temperature of 1600°C in Mo-crucible in a double-vacuum electric furnace (Kamensky et al., 1984), (iii) isothermal heating. The latter procedure allows preservation of the samples during degassing, so that a sample can be used for repeated saturation/release experiments. In all cases the extracted gases were admitted into an all-metal line and purified using Ti-Zr getters.

Abundances of He and Ar isotopes were measured using a static mass spectrometer (MI 1201) with a resolving power of ~ 1000, which allows a complete separation of $^3He^+$ from the hydrogen doublet $^3H^+$ and HD^+. The sensitivity for He was 5×10^{-5} A torr^{-1}, the lowest measurable $^3He/^4He$ ratios being less than 10^{-8}. A mixture of pure ^3He, helium from a high-pressure tank ($^3He/^4He = 2 \times 10^{-8}$) and atmospheric Ne, Ar, Kr and Xe was used as a standard for calibration of the mass-spectrometer. The ratios of $^3He/^4He = 1.6 \times 10^{-6}$ and $^4He/^{20}Ne = 47$ in the mixture were measured using air as the primary standard in the Apatity laboratory; later on these ratios were verified at CRPG (Nancy). Uncertainties in the $^3He/^4He$ ratios of $\sim 10^{-6}$ and $\sim 10^{-8}$ were ±2% and ±20%, respectively, and uncertainties in the $^{40}Ar/^{36}Ar$ ratios of 300 and 50,000 were ±0.3% and ±25%, respectively. The concentrations of ^4He and ^{40}Ar were determined by the peak height method with an uncertainty of ±5 % (1σ). The analytical blanks were within 1×10^{-9}, and 1×10^{-10} cm^3 STP for ^4He, and ^{36}Ar, respectively, for both fusion and crushing experiments.

The contents of U and Th in whole rock samples and mineral separates were measured by X-radiographic techniques (in the Neva Expedition, St. Petersburg, Russia) with a detection limit of about 0.5 ppm. K and Li contents were determined by spectrophotometric techniques after acid digestion and dissolution in distilled water in the Geological Institute, Apatity (Murmansk Region, Russia). The reproducibility of the analyses of these four elements was within ±10 %.

3. Geological and Hydrological Setting of the Permo-Carboniferous Trough (PCT), Northern Switzerland

There are several upper Palaeozoic troughs in the Alp foreland (Figure 1), PCT being one of those. According to Pearson et al. (1991) it is about 2 km deep and 10 to 12 km wide (Figure 2). This location was thoroughly studied as a potential site for radioactive/toxic waste disposal. In particular 7 deep boreholes were drilled within the frame of the Program of the Swiss National Cooperative for the Disposal of Radioactive Waste (Nagra, 1989), samples from one of these, the Weiach borehole, were investigated.

The Weiach borehole is located in northern Switzerland near Rhine river, 25 km north-west of Zürich. Down to a depth of 991 m below the surface, the borehole penetrated the Quaternary overburden of the Molasse Basin, the Tertiary sediments of the Lower Fresh Water Molasse, the Jurassic and Triassic sediment sequences. From 991 m to 2020 m depth Permian and Carboniferous sediments of the Permo-Carboniferous Trough (PCT) were opened and the borehole intruded into the crystalline basement from 2020 m to its final depth, 2482 m (Figure 3).

The crystalline basement underlying the PCT sediments consist of a monotonous series of highly metamorphosed biotite-plagioclase paragneisses and aplitic dykes. The Carboniferous is represented by a ca. 600 m thick sequence of terrigenic sediments of Stephanian age (about 299-305 Ma). The Permian (251-299 Ma) consists of ≈ 450 m of sediments similar to the

Carboniferous and is overlain by the fluvial sediments of the Buntsandstein (242-251 Ma) and the shallow marine and Sabkha-type sediments of the Muschelkalk (227-242 Ma). The Permo-Carboniferous sediments (PCT) consist of alternating layers of shales, clay-, silt- and sandstones and subordinate conglomerates of mainly alluvial and lacustrine origin (Matter et al., 1988). The shale / sandstone mass ratio is approximately 1. Several groups of meter-thick coal seams occur between 1551 m and 1751 m depth.

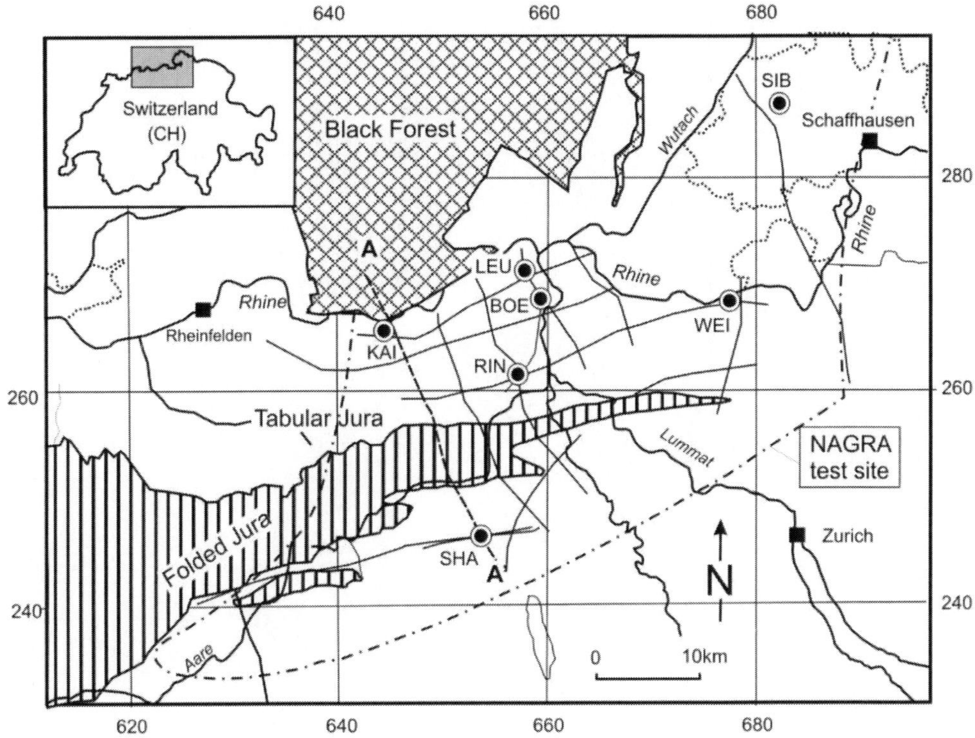

Simplified after Pearson et al., 1991.

Figure 1. Geotectonic map of Northern Switzerland (see inset in the left upper corner). Names of the boreholes: KAI= Kaisten, LEU=Leuggern, BOE=Böttstein, RIN= Riniken, SHA=Schafisheim, WEI=Weiach, SIB= Siblingen. Lines are faults and fissure zones. Line AA' traces the cross section presented in Figure 2.

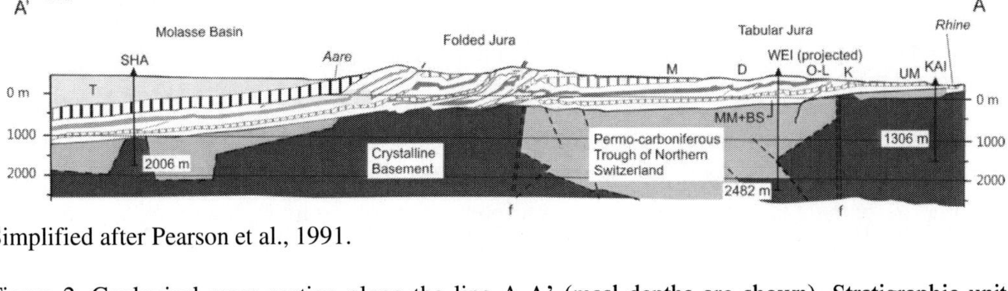

Simplified after Pearson et al., 1991.

Figure 2. Geological cross-section along the line A-A' (masl depths are shown). Stratigraphic units: T=Teriary, M=Malm, D=Dogger, O+L=Opalinus-Clay and Liassic, K=Keuper, UM=Upper Muschelkalk, MM+BS=Middle Muschelkalk to Buntsandstein, f-fault / flexure zones. Names of the boreholes are as in Figure 1.

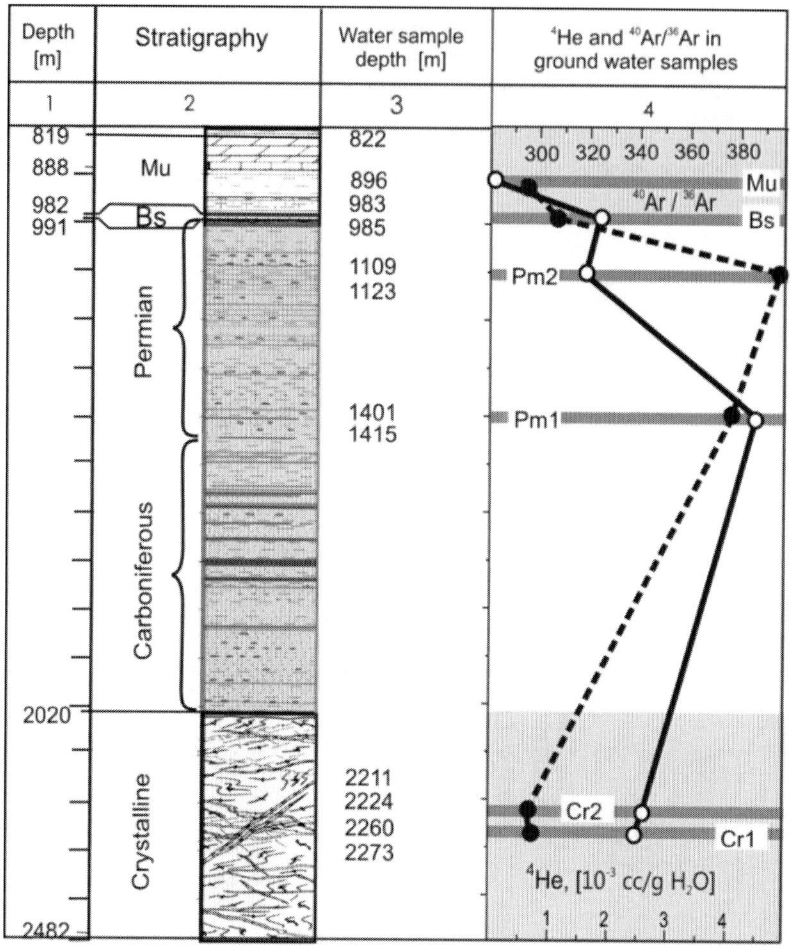

Compiled from Pearson et al. (1991).

Figure 3. The Weiach borehole, Northern Switzerland (coordinates x: 676'750.0, y: 268'620.0, altitude: 369.0 masl, Kanton Zürich, modified from Nagra, 1989). C*olumn 1*- depth, masl. *Column 2* – the principal lithological units and stratigraphy of the test interval. *Column 3* – depth intervals of the aquifers (and the groundwater samples). *Column 4* – measured $^{40}Ar/^{36}Ar$ ratios (black circles, top scale) and He concentrations (open circles, bottom scale) in groundwater samples from the Weiach borehole (data from Pearson et al., 1991). Thick horizontal lines schematically show aquifers, from which water samples were collected (Muschelkalk, Mu, 822-896 m; Buntsandstein, Bs, 981-989 m; Permian, Pm2, 1109-1123 m, and Pm1, 1401-1415 m; crystalline basement, Cr2, 2211-2224 m and Cr1, 2260-2273 m). Note a high He concentration in the Permian groundwater Pm1 and high $^{40}Ar/^{36}Ar$ in both Permian aquifers compared to the almost atmospheric ratios in the groundwater from the overlying Muschelkalk formation and the underlying crystalline basement.

The present day temperature of the PCT ranges between 55°C at the top and 88°C at the bottom. The porosity decreases from values above 15% at ≈ 1060 m depth to below 3% in the crystalline basement. In the Carboniferous sediments the hydraulic conductivity is generally very low ($<10^{-13}$ m s^{-1}) and enhanced values ($\sim 10^{-10}$ m s^{-1}) are restricted to a few occurrences in the upper part of the cross section (Nagra, 1989).

The Permian ground waters were collected from two intervals with elevated hydraulic conductivity (average depths of 1117 m and 1411 m), whereas the Carboniferous sediment sequence contained only stagnant ground waters (Nagra, 1989).

According to Pearson et al. (1991) the degree of the mineralization, enrichment in ^{18}O and ^{2}H, absence of ^{14}C and high contents of radiogenic noble gases indicate a very long residence time for the ground water within the PCT; this is especially valid for the samples from the low permeability section (below depth 1400 m) suggesting stagnant conditions. The two Permian ground water samples show $^{40}Ar/^{36}Ar > 375$, whereas the ground water in the Buntsandstein aquifer just on top of the Permian sediments has a ratio of only 306 and $^{40}Ar/^{36}Ar$ ratios in ground waters from the overlying Muschelkalk and the underlying crystalline basement aquifers are similar the atmospheric value of 296 (Figure 3). The deeper Permian ground water (Pm1) also displays the highest ^{4}He concentration (4.48×10^{-3} cc STP per g H_2O, Figure 3). Slightly lower concentrations are observed in the shallower Permian ground waters (Pm2, 1.59×10^{-3} cc STP per g H_2O), in those near the Muschelkalk formation (1.8×10^{-3} cc STP per g H_2O), and in the deep crystalline aquifers (with an average ^{4}He concentration 2.35×10^{-3} cc STP per g H_2O). The data discussed above suggest local, production for ^{3}He, ^{4}He and $^{40}Ar^*$ in the PCT rock-water system because fluxes from external sources could only dilute the highly mineralised, He- and $^{40}Ar^*$-rich PCT ground waters.

The rock samples (Table 1) were collected from the Permo-Carboniferous Formation opened by the Weiach borehole.

Table 1. Whole rock samples and mineral separates from these rocks, investigated for noble gas isotope abundances (Weiach borehole, Permian sediments, the Permo-Carboniferous Trough of Northern Switzerland)

Depth meters	Sample code	Sample description
1099	WEI-1099-SH	Shale
1116	WEI-1116-GW	groundwater (sampling interval 1109-1123)
1118	WEI-1118-SS	Sandstone
1121	WEI-1121-SS	Sandstone
1126	WEI-1126-SH	Shale
1408	WEI-1408-GW	groundwater (sampling interval 1401-1415)
1415.1	WEI-1415-SS	Sandstone
1415.1	WEI-1415-kf	Microcline
1415.1	WEI-1415-pl	Plagioclase
1415.1	WEI-1415-qz	Quartz
1417.68	WEI-1417-SH	Shale
1417.68	WEI-1417-cm	carbonate cement
1417.68	WEI-1417-mi	mica
1417.68	WEI-1417-fp	feldspar (mixed)
1417.68	WEI-1417-qz	quartz
1417.7	WEI-1418-SH	Shale
1417.7	WEI-1418-cm	clay minerals ~ 100% [20% of total rock]
1417.7	WEI-1418-mu	muskovite ~95-100%, organic + chlorite + biotite <5% [3% of total rock]
1417.7	WEI-1418-org	organic material ~60%, muskovite+chlorite+illite ~40% [20% of total rock]
1417.7	WEI-1418-bi	biotite ~ 70%, chlorite ~30% [7% of total rock]
1417.7	WEI-1418-pl	plagioclase ~ 80-85%, quartz ~ 15-20%, microcline <5% [17% of total rock]
1417.7	WEI-1418-qz	quartz ~ 70-75% plagioclase ~ 20-25% microcline<10% [15% of total rock]
1417.7	WEI-1418-kf	microcline ~75-80%, quartz ~10-15%, plagioclase ~5% [18% of total rock]

The rocks are mainly composed of quartz (~50 %), K-feldspar (~20 %), plagioclase (~10%), and matrix mineral cement (~20 %). The latter includes very fine-grained quartz, sericite, chlorite, biotite and clay minerals. Quartz grains (generally with ~ 10 % contribution of feldspars) were separated from arkosic sandstones, the average grain size is 0.22 ± 0.15 mm. Quartz separates are identified by the depth (m), from which the parental cores were recovered.

4. Origin and Re-Distribution of Helium (Argon) Isotopes in the PCT Rocks, Minerals and Waters

Radiogenic ^{4}He and $^{40}Ar^{*}$ in terrestrial rocks are mainly produced via ^{238}U, ^{235}U, ^{232}Th and ^{40}K decay, whereas the principal source of ^{3}He is reaction between ^{6}Li nuclei and thermal neutrons, $^{6}Li(n,\alpha)^{3}H \rightarrow \beta- \rightarrow {}^{3}He$. Important characteristic of the PCT rocks is the enhanced contents of the parent elements, U, Th and especially Li in shales compared to sandstones (Table 2, Figure 4).

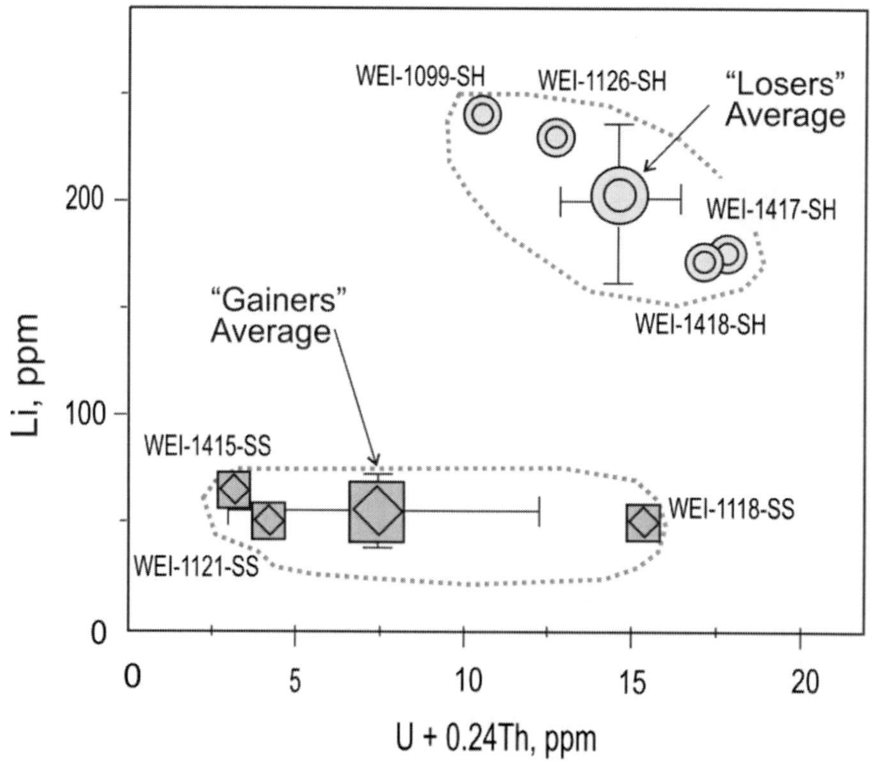

After Tolstikhin et al. (2011).

Figure 4. Parent element contents in whole-rock samples. The sum U + 0.24Th is proportional to radiogenic He production in rocks younger than ≈ 300 Ma. Labels are sample names in the Tables and text. "Losers" are shales, which almost completely lost their *in situ* produced radiogenic helium; "gainers" are sandstones containing excess helium. Note that shales contain more U, Th and much more Li than sandstones: shales are major producers of radiogenic helium in PCT shale / sandstone alternation.

Table 2. Measured concentrations of He and Ar isotopes, parent elements and calculated radiogenic Ar ($^{40}Ar^*$) in rocks and mineral separates from Weiach borehole, Permian sediments, the Permo-Carboniferous Trough of Northern Switzerland

Sample code	U ppm	Th ppm	K wt.%	Li ppm	^4He cc/g $\times10^{-6}$	^{40}Ar cc/g $\times10^{-6}$	$^3He/^4He$ $\times10^{-8}$	$^{40}Ar/^{36}Ar$	$^{40}Ar^*$ cc/g $\times10^{-6}$	$^4He/^{40}Ar^*$
1	2	3	4	5	6	7	8	9	10	11
WEI-1099-SH	3.60	29	3.50	240	92	44	2.70	3400	40	2.3
WEI-1118-SS	8	31	3.40	50	120	40	6.80	3200	36	3.3
WEI-1121-SS	0.92	10	3.30	52	130	34	8.13	2900	30	4.3
WEI-1126-SH	5.20	31	3.50	230	57	38	2.10	2700	34	1.7
WEI-1415-SS	2	8.6	3.22	55	180	52	8.00	3800	48	3.8
WEI-1415-kf	3.8	5.2	10.9	6.5	14	150	7.35	3630	140	0.10
WEI-1415-pl	<1	1.5	0.08	15	200	4.5	7.30	430	1.38	143
WEI-1415-qz	<1	3.8	0.12	15	160	5.1	9.09	470	1.89	84
WEI-1417-SH	11.7	26	3.24	170	57	53	3.50	3800	49	1.2
WEI-1417-cm	12	30	3	130	15	41	5.49	1140	30	0.50
WEI-1417-mi	12.6[1]	16[1]	2.9	360	22	37	6.58	1840	31	0.71
WEI-1417-fp	2.9[1]	2[1]	6.6	21	22	120	8.00	3560	114	0.20
WEI-1417-qz	4.7[1]	3[1]	0.32	27	110	8.1	7.52	1020	5.75	19
WEI-1418-SH	10.8	26	3.5	170	25	47	3.91	4450	43	0.58
WEI-1418-cm	10.5	18	3.2	180	13	29	4.81	1930	25	0.51
WEI-1418-mu	7.7	17	4.8	170	54	88	4.29	1910	75	0.72
WEI-1418-org	13.4	31	3.5	130	13	28	6.49	2730	25	0.52
WEI-1418-bi	12.6	16	3	375	24	43	6.49	2210	38	0.65
WEI-1418-pl	5.8	2.5	4.1	50	27	59	5.41	3360	54	0.49
WEI-1418-qz	4.7	3	1.2	50	71	23	7.87	1950	19	3.7
WEI-1418-kf	2.9	2	7	50	13	87	9.62	2850	78	0.16

[1] Concentrations are assumed to be the same as in sample WEI-1418.

Therefore shales produce the major portion of ^4He and almost all ^3He in the alternating shale - sandstone sequence. A similar distribution of U and Th between shales and sandstones has generally been observed in terrigenic sediments (e.g., Smyslov, 1974; Andrews, 1985; Agarwal et al., 2006; Pekala et al., 2009). Potassium contents in the two rock types are similar, with an average value for PCT whole rock samples of 3.44 ± 0.09 wt. % and similar values were reported by Matter et al. (1988). However, potassium varies substantially in the rock-forming minerals, being especially low in the quartz separates (0.32 wt. %) and high in the potassium bearing minerals such as K-feldspar (up to 10.9 wt.% K, Table 2).

To understand behaviour of radiogenic helium and argon isotopes in the rocks / minerals / waters of PCT, the actually measured concentrations (Table 2) were compared with the calculated ones (Table 3). The total ^4He and $^{40}Ar^*$ concentrations, produced since the time of rock formation ($^4He_{(C)}$ and $^{40}Ar^*_{(C)}$ in Table 3) are calculated from the measured concentrations of parent isotopes (Table 2) and the rock ages (Figure 3). $^3He_{(C)}$ is calculated as a product of the integrated neutron flux, the cross section of the nuclear reaction (960 barn), the ^6Li abundance (0.075 × [Li] in Table 2), and the sedimentation age. The neutron flux in a rock originates mainly from (α,n) reactions and can be calculated from the whole-rock U and Th contents (Table 2), the chemical composition of the rock (Matter et al., 1988), and the contents of neutron-interactive trace elements, i.e., Be, B, and Gd. The method for calculation of $^3He/^4He$ ratio in radiogenic helium was originally proposed by Morrison and Pine (1955) and then discussed in contributions of Gorshkov et. al (1966), Mamyrin and Tolstikhin (1984), Andrews (1985), and Lehmann et al. (1992). According to comparisons of measured and calculated neutron fluxes performed by Gorshkov et al. (1966), the accuracy of these calculations is from 30 to 50 %.

Shales, enriched in the parent elements, are major generators of ^4He in the PCT sediments; assuming no gain / loss of species of interest since the mean sedimentation age, 285 Ma ("closed system" assumption) a calculated average $^4He_{SH,CAL}$ concentration approaches 500×10^{-6} cc STP g^{-1}, exceeding the sandstone value by a factor of ≈ 2, $^4He_{SS,CAL} = 260 \times 10^{-6}$ cc STP g^{-1}. The difference in the productivity between the sandstones and the shales is even higher in case of ^3He. According to abundances of the parent elements, shales produced almost 10 times more ^3He than sandstones; especially productive are Li-rich minerals, clay and carbonate cements as well as micas (WEI-1417-cm, -mi and WEI-1418-cm, -org, -bi, Tables 2, 3, Figure 5).

Table 3. Calculated neutron fluxes and he bulk concentrations of *in-situ* produced He and Ar isotopes, assuming the closed system evolution of the Permian sediments and their average age of 285 Ma. Also given are the retention coefficients, *L*, of measured (m) versus calculated (c) values for ^{3}He, ^{4}He and ^{40}Ar*. The "apparent K-Ar age" is only semi- quantitative, but its average value (295 Ma) is consistent with that assumed for the calculations of the in-situ production (see text)

Sample code	n-flux cm^{-2}/s $\times10^{-5}$	^{3}He(c) cc/g $\times10^{-12}$	^{4}He(c) cc/g $\times10^{-6}$	^{3}He/^{4}He (c) $\times10^{-8}$	^{4}He(c)/^{40}Ar*(c) Ratio	K/Ar "age" Ma	L(^{3}He) (m)/(c) Ratio	L(^{4}He) (m)/(c) Ratio	L(^{40}Ar*) (m)/(c) Ratio
1	2	3	4	5	6	7	8	9	10
WEI-1099-SH	6.3	31	365	8.49	8.7	267	0.08	0.25	0.96
WEI-1118-SS	8.2	8.4	537	1.56	13.2	250	0.97	0.22	0.89
WEI-1121-SS	2.1	2.2	115	1.91	2.9	218	4.81	1.13	0.77
WEI-1126-SH	7.2	34	438	7.76	10.5	227	0.04	0.13	0.81
WEI-1415-SS	1.7	1.96	141	1.39	3.7	338	7.35	1.27	1.24
WEI-1415-kf	1.7	0.23	177	0.13	1.4	297	4.57	0.08	1.08
WEI-1415-pl	1.7	0.53	48	1.10	51.7	401	27.34	4.20	1.49
WEI-1415-qz	1.7	0.53	67	0.79	46.3	357	26.75	2.36	1.31
WEI-1417-SH	12.8	46	623	7.38	16.1	344	0.04	0.09	1.26
WEI-1417-cm	12.8	33	666	4.95	18.5	235	0.03	0.02	0.84
WEI-1417-mi	12.8	94	578	16.26	16.8	255	0.02	0.04	0.91
WEI-1417-fp	12.8	5.50	119	4.62	1.5	388	0.32	0.19	1.44
WEI-1417-qz	12.8	7.1	191	3.72	50.6	408	1.20	0.59	1.52
WEI-1418-SH	12.2	42	599	7.01	14.3	288	0.02	0.04	1.04
WEI-1418-cm	12.2	45	519	8.67	13.5	185	0.01	0.02	0.65
WEI-1418-mu	12.2	42	412	10.19	7.2	354	0.05	0.13	1.30
WEI-1418-org	12.2	32	732	4.37	17.5	171	0.03	0.02	0.60
WEI-1418-bi	12.2	94	578	16.26	16.1	290	0.02	0.04	1.05
WEI-1418-pl	12.2	12	225	5.33	4.6	303	0.12	0.12	1.10
WEI-1418-qz	12.2	12	191	6.28	13.3	362	0.45	0.37	1.33
WEI-1418-kf	12.2	12	119	10.08	1.4	260	0.10	0.11	0.93

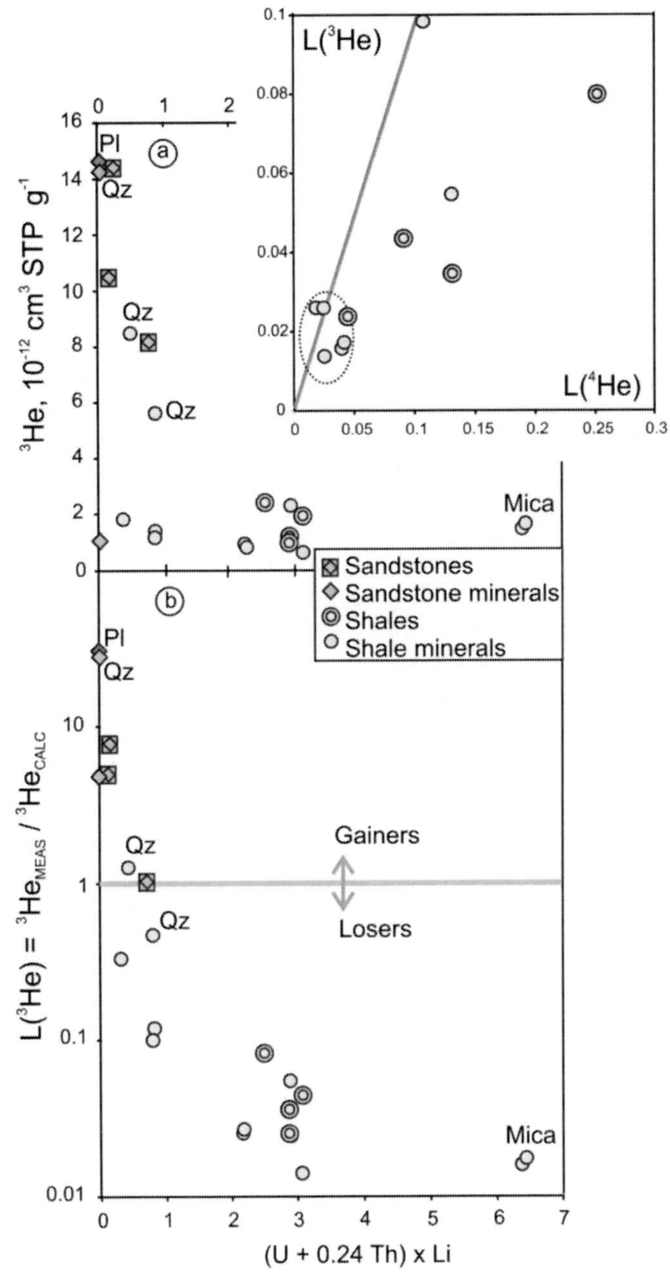

After Tolstikhin et al. (2011).

Figure 5. Measured ^{3}He concentrations (a) and the ^{3}He retention coefficient (b) versus the parent element content combination (proportional to ^{3}He production rate, arbitrary units). The retention coefficient (L) in plot (b) is the measured over the calculated (maximum in-situ produced, Table 3) concentration ratio, L(^{3}He) ≡ $^3He_{MEAS}$ / $^3He_{CALC}$. The closed system evolution (no gain, no loss of species of interest) envisages a ratio 1; a ratio <1 indicates a loss of ^{3}He and vice versa. Shales have lost most of the helium produced from U, Th and Li, whereas sandstones (especially quartz and plagioclase in these rocks) generally contain excess (trapped) helium. The inset shows ^{3}He and ^{4}He loss from shale whole rock samples and mineral separates. The thick grey line shows equal retention of L(^{3}He) ≡ L(^{4}He); note the preferential release of ^{3}He from whole-rock shale samples. Dotted oval bounds samples with enhanced production of helium isotopes (see text for discussion).

The ratio of the measured concentration over the calculated one is defined as the retention coefficient (*L*), which allows better understanding of the noble gas evolution in a rock-water system. The retention of ^{4}He and ^{3}He varies widely (Table 3), by factors of up to ~ 200 for *L*(^{4}He) and ~ 1000 for *L*(^{3}He).

The retention of both helium isotopes is rather low in the shale whole rock samples, with the average *L*(^{4}He) = 0.13, and *L*(^{3}He) = 0.05 (Table 3, Figure 5b). Especially low retention coefficients (≤ 0.03 for both isotopes) are typical of mineral separates enriched in U and Th (WEI-1417-cm and -mi, WEI-1418-cm, -org and -bi, Tables 2 and 3): almost all helium produced in these rocks has been lost. Besides, ^{3}He has been lost preferentially, compared to ^{4}He, from the shale whole rock samples (see inset in Figure 5a). This reflects the different occurrence of the parent elements in different minerals. Lithium, the parent element of ^{3}He, is mainly concentrated in tiny, only a few micrometres thick, mica laths. The radiogenic damage tracks produced by the exothermic (^{6}Li,n) reaction are long enough to cut through the crystalline structure of micas allowing ^{3}He-atoms to be liberated (Figure 5): ^{3}H decays with a half-life of 12 years, and the He-atom residence time in an opened track (under room temperature) is ~ 1 year (Tolstikhin et al., 1999). In contrast, U and Th, the parent elements for ^{4}He, are mainly concentrated in accessory minerals, such as zircon and monazite; they were also identified as U- and Th-bearing phases in shales and claystones from the Opalinus Clay formation in North Switzerland (e.g., Pekala et al., 2009). These minerals retain radiogenic helium isotopes better than micas (see discussion of this issue in Tolstikhin et al., 1996). A similar behaviour of radiogenic He isotopes has also been observed in magmatic and metamorphic rocks, where nucleogenic ^{3}He has been lost preferentially, compared to radiogenic ^{4}He (Mamyrin and Tolstikhin, 1984).

The comparison of measured and calculated values further highlights the difference between shales and sandstones: in the sandstone whole rock samples helium concentrations exceed the in-situ production values (*L*(^{4}He) > 1), except for one sample with shale-type parent element contents (Tables 2 and 3). Moreover, in quartz and plagioclase separated from WEI-1415-SS measured ^{4}He concentrations exceed the *in-situ* production values up to a factor of ≈4 in plagioclase and ≈ 2 in quartz (Table 3). ^{3}He retention coefficients are even higher than those of ^{4}He, especially in quartz and plagioclase separates, where *L*(^{3}He) approaches 27 (Table 3, Figure 5b). These observations indicate that both helium isotopes in these samples are excessive (trapped).

Comparison of measured and calculated ^{3}He/^{4}He ratios (Tables 2 and 3) sheds more light on helium isotope evolution in PCT rocks. The average calculated ^{3}He / ^{4}He ratio for shales, $^3He\,/\,^4He_{SH,CAL} = 7.7 \times 10^{-8}$, exceeds substantially the average measured one, 3×10^{-8}, thus reflecting a preferential loss of ^{3}He from these rocks (inset in Figure 5a). Taking into account the almost complete loss of helium isotopes from shales into the ground (pore) waters and assuming a good preservation of helium in this complementary reservoir, the ^{3}He/^{4}He ratio in PCT waters is expected to be similar to the shale production ratio, or could be slightly higher due to enhanced ^{3}He loss.

This assumption is confirmed by the rather high concentration of ^{4}He in the Pm1 aquifer, 4.48×10^{-3} cc STP per g H_2O and, in accord with the above expectation, the ^{3}He / ^{4}He ratio in Pm1 waters, 9.1×10^{-8} (Pearson et al., 1991), is indeed very similar to the average shale production value presented above. Moreover, a similar ratio is also observed in PCT sandstones, the average value being 7.7×10^{-8} (Table 2), in contrast to the much lower production ratio in these rocks, 1.7×10^{-8} (Table 3). Note that the difference between these measured and calculated ratios is much larger than the estimated uncertainty of the calculated ^{3}He production.

Hence, comparison of He isotope abundances in the shales and sandstones suggests that in the PCT interlayering both helium isotopes were mainly produced in (and almost completely lost by) U, Th, Li rich shale minerals (micas, cements), whereas some sandstone minerals (quartz, plagioclase) contain substantial amounts of trapped helium; important is that ^{3}He / ^{4}He ratios observed in the sandstones and in the PCT ground waters are indistinguishable from the shale production ratio. These ratios are well above the generally accepted crustal ratio of 2×10^{-8} (Mamyrin and Tolstikhin, 1984).

In principal, the excess ^{3}He in the sediments could originate from three different sources: (1) interplanetary dust particles (IDP), (2) spallation reactions during surface exposure of a rock and (3) transfer by volatiles emanated from mantle magmas or/and young mantle rocks.

(1) Regarding the first source, Patterson et al. (1998) interpreted measured ^{3}He concentrations in 480 Ma year old marine limestone as extra-terrestrial He preserved in IDPs. For pelagic clays of 0.5-90 Ma, Mukhopadhyay and Farley (2006) showed (using mineral separation and sequential leaching) that ^{3}He is mainly associated with (extraterrestrial) silicates, Fe-Ni sulfides and possibly magnetite with ^{3}He / ^{4}He ratios and ^{3}He concentrations of $\sim 2 \times 10^{-4}$ and $\sim 10^{-10}$ cc STP g^{-1}, respectively. In contrast, the high ^{3}He concentrations in PCT rocks are observed in quartz grains. Moreover, Ballentine and Burnard (2002) estimated that even if all IDPs falling on continents were preserved over the last 70 Ma, the amount of ^{3}He contained in IDPs would only be about 1% of the ^{3}He produced in the continental crust by the (^{6}Li + n) reaction.

(2) Spallation reactions during the surface exposure of rocks also produce ^{3}He in minerals (e.g., Trull et al., 1991). The ^{3}He /^{4}He ratio in spallation helium is estimated at ~ 0.25 (Eugster, 1988) and the maximum concentration of $^3He_{SPAL}$ in terrestrial samples is about 10^{-10} cc STP g^{-1} (measured in diamond grains selected from sedimentary rocks, Kamensky and Tolstikhin, 1992). The extent to which spallation-produced $^3He_{SPAL}$ is retained in quartz grains is, however, a matter of debate. Niedermann (2002) concluded that $^3He_{SPAL}$ may be retained in quartz more or less completely under favourable conditions, but is more commonly almost completely lost. As it is shown below, a very fast equilibration of He concentrations inside and outside quartz grains discards the spallation hypothesis.

Indeed neither of these two hypotheses is supported by the distribution of He isotopes in the PCT sandstones. Quartz and plagioclase separates from these rocks are the principal carriers of excess He (Tables 2, 3 and Figure 5) and the amounts of helium extracted from quartz separates by crushing and by fusion are similar (Lehmann et al., 2003b; Tolstikhin et al., 2005). This means that vesicles (directly observable in thin sections of PCT sandstones) are the major carriers of excess helium-3, in contrast to what would be expected from spallation production. Moreover, the excess of ^{3}He is always coupled with an excess in ^{4}He in almost constant proportions as indicated by the very small deviation of the average measured ratio of ^{3}He/^{4}He = $(7.7 \pm 0.7) \times 10^{-8}$ (see samples containing excess helium, Tables 2 and 3). Mixing of terrestrial and interplanetary and/or spallogenic helium is unlikely to generate an almost constant ^{3}He/^{4}He ratio in different samples.

(3) Excess ^3He combined with excess ^4He in rocks, minerals and ground waters are commonly interpreted to reflect He transfer from deep mantle or mixed crust-mantle sources. Available data (Pearson et al., 1991; Figure 3) show that helium concentrations and contribution of radiogenic argon in both underlying and overlaying aquifers are substantially lower than in waters of a central core of a sedimentary basin, convincingly indicating an intra-basin origin of noble gas isotopes in the PCT rock-water system.

For the K – Ar* system, radiogenic $^{40}Ar^*$ contents are derived from measured ^{40}Ar and $^{40}Ar / ^{36}Ar$ ratios (Table 2) with $^{40}Ar_{(M)}^* = {}^{40}Ar \times (1 - (296 / (^{40}Ar / ^{36}Ar)))$. $^{40}Ar^*$ contents are similar for shale and sandstone whole rock samples, the average values being $41.6 \pm 6.3 \times 10^{-6}$ cc g^{-1} and $38.2 \pm 8.7 \times 10^{-6}$ cc g^{-1}, respectively. The apparent K/Ar ages of PCT rocks and mineral separates vary within a factor of 2 (Table 3) with an average value of 295 ± 70 Ma. The large standard deviation of this age is not surprising considering (1) the detrital origin of most of the minerals, (2) the occurrence of trapped excess $^{40}Ar^*$ in some minerals, and (3) the re-crystallisation / weathering / diagenetic processes to which some of these minerals were subjected after sedimentation. Whereas in the first and second cases the mineral age would be older than the sedimentation age, it would (generally) be younger in the third case. The K-Ar apparent age of the PCT samples overlaps with the time period of the sedimentation (299-251 Ma; Gradstein et al., 2004). This indicates that the sediment pile never experienced an event that might have caused substantial loss of radiogenic argon, and that an average age of sedimentation, 285 Ma, can be adopted for this study.

Using this age and the closed system approximation (supported by the above comparison) allows the retention coefficient for radiogenic argon, $L(^{40}Ar^*) = {}^{40}Ar_{(M)}^* / {}^{40}Ar_{(C)}^*$, to be calculated. $L(^{40}Ar^*)$ varies from 0.6 (WEI-1418-org) to 1.52 (WEI-1417-qz), indicating that some minerals lost and others gained radiogenic argon (Table 3) and also reflecting the low mobility of $^{40}Ar^*$ compared with helium isotopes.

Concluding this Section, it worth to highlight that the source of radiogenic (nuclegenic) isotopes of helium and argon in the rocks, minerals, and waters of PCT are the parent elements, containing in the rocks (predominantly in shales). Helium, being by far more mobile than argon, almost completely released from the host rocks into the ground (pore) water and accumulated in this (the only complementary to the rocks) reservoir. Then He isotopes penetrated into some minerals (quartz, plagioclase) thus ensuring excess helium in there as well as in the whole rock sandstone samples. This scenario is convincingly supported by helium isotopic compositions: $^3He / ^4He$ ratios measured in these minerals are very similar to that calculated for shales and to those observed in the ground water samples. In the next Section we describe the respective processes of helium isotope redistribution between minerals and related pore waters in a quantitative way.

5. Sites and Mobility of Noble Gases in the Natural Minerals

Subsurface migration of noble gases through pore water media and their removal by ground water flow are popular topics of terrestrial hydrology (e.g., Mazor, 1991; Cook and

Herczeg, 2000; Rübel et al., 2002; Plummer, 2005). However, the concept of mineral grains being sinks for noble gas atoms is relatively new (Lehmann et al., 2003b; Baxter, 2003; Tolstikhin et al., 2005) and thus deserves some additional discussion.

To describe noble gas migration and accumulation in the quartz grains (separated from PCT sandstones), three structural components constituting of the grains were considered: (1) crystalline lattice domains characterized by low defect density, low helium solubility and penetrability; (2) vesicles including (observable) gas and liquid phases inside the grains; (3) helium-penetrable domains or channels connecting vesicles with the grain surfaces and thus with surrounding environment. Concerning the component (1), available estimates of the noble gas solubility in the quartz crystalline lattice vary over many orders of magnitude, from $\sim 3 \times 10^{-3}$ cc STP g^{-1} atm^{-1} (Watson and Cherniak, 2003) to $<10^{-6}$ cc STP g^{-1} atm^{-1} (Roselieb et al., 1997; Argunova et al., 2003). Watson and Cherniak (2003) impregnated crystalline quartz with Ar (at temperatures from ~ 500 to ~ 1200 ^{O}C and partial Ar pressures from 1 to 185 MPa) and derived from these experiments an Ar partition coefficient of $\sim 3 \times 10^{-3}$ cm^3 STP g^{-1} atm^{-1} for < 100 nm thick surface rims (the coefficient being practically independent of temperature). Their samples were natural and synthetic quartz crystals grown under hydrothermal conditions, optically clear but containing localized patches and fluid inclusions of variable size. The authors extrapolated this high solubility to the entire quartz grains without further investigation of their subsurface domains. It is unclear how argon atoms with a Van der Waals diameter of 3.60 Å can intrude into the quartz lattice, since the largest diameter of the [0001] channel is ≈ 2.5 Å. Roselieb et al. (1997) also saturated quartz grains (separated from high-quality quartz crystals) of different sizes (from 11 to 80 micron) under high temperatures (1300 ^{O}C) and high Ar pressures (up to 8000 bar) and, in contrast to the previous result, obtained the upper limit of Ar partitioning to be $< 2 \times 10^{-6}$ cc STP g^{-1} atm^{-1}.

Moreover, even much smaller He atoms (2.8 Å) can hardly penetrate into an "ideal" quartz crystal. Argunova et al. (2003) studied He concentrations and migration rates in different segments of a quartz crystal, grown in an autoclave at a temperature of 330 ^{O}C and He partial pressure of ~20 atm. According to their results, the amount of helium dissolved in the quartz lattice decreases along with the defect density; for segments of the crystal with the minimal density $<10^2$ cm^{-2}, the helium partitioning into the crystal was $< 5 \times 10^{-7}$ cc STP g^{-1} atm^{-1}. This value should be considered as an upper limit for helium solubility in a defect-free quartz lattice. Kalashnikov et al. (2003) argued that only (metastable) helium atoms in the triplet state are able to move through defect-free quartz crystalline lattice. These results are in conflict with the Funk's et al. (1971) estimate of He solubility in crystalline quartz of $\approx 3 \times 10^{-4}$ cc STP g^{-1} atm^{-1}. These authors did not characterize the sample studied.

The simplest explanation of these discrepancies appears to be the different quality of the quartz crystals used in the various experimental studies: the noble gas solubility and diffusivity vary from (almost) zero in an "ideal" quartz crystalline lattice to the high values of these parameters observed in fused quartz, vitreous silica, obsidians, etc. Thus, Carroll and Stolper (1991) impregnated Ar into SiO_2 glass and interpreted their results in terms of a solubility model involving a fixed population of available sites or "holes." They estimated the density of sites available for Ar occupancy in SiO_2 glass at $\approx 10^{21}$ cm^{-3} (in some sense this is a defect density). A similar approach applied by Watson and Cherniak (2003) gave ~ 10^{20} cm^{-3}, indicating that the structure of a substantial portion of surface rims of their samples was similar to SiO_2 glass (or includes other defects enhancing noble gas solubility).

The second component (fluid inclusions) are observed microscopically in the PCT quartz separates, and the total local sink capacity (in the terminology of Baxter, 2003) or the He-accessible volume (in the terminology of Lehmann et al., 2003b) varies from 10^{-4} to 10^{-3} parts of the grain volume, as derived from the helium saturation experiments (Tolstikhin et al., 2005, Gannibal, 2012).

The third structural component of the natural quartz, e.g., the helium-penetrable domains or channels (network of defects), was introduced to allow helium migration inside the grains. Comparison of Tables 2 and 3 for sample WEI-1415-qz shows, that the ^4He / $^{40}Ar^*$ ratio in the trapped gases (i.e., the ratio of measured minus calculated concentrations of ^4He and $^{40}Ar^*$) approaches 200. Lehmann et al. (2003b) directly measured ^4He / $^{40}Ar^*$ = 330 in gases released by crushing of a neighbour sample WEI-1409-qz. These ratios substantially exceed the average production ratios in shales (^4He / $^{40}Ar^*$ = 14) and in sandstones (7). Even though some fractionation between radiogenic ^4He and $^{40}Ar^*$ occurs during their migration from the minerals (production sites) to the pore waters, this is not expected to be that large (Tolstikhin et al., 1996; Mamyrin and Tolstikhin, 1984).

Thus, argon gives another characterization of the noble gas penetrable domains between pore water and vesicles inside quartz grains: this gas also penetrates through, but at a much slower rate compared with helium.

According to the above model the total volume of a quartz sample V_Σ (eqn 1) includes volumes of defectless lattice, V_{CR}, channels, V_{CH}, and fluid inclusions, V_{VES}:

$$V_\Sigma = V_{CR} + V_{CH} + V_{VES}. \qquad (1)$$

Fluid inclusions, however, may include gas, V_G, and liquid, V_L,

$$V_{VES} = V_G + V_L. \qquad (2)$$

In eqns (1,2) each volume (except V_{CR}) contains helium atoms N; $n_\Sigma = N_\Sigma/V_\Sigma$ is the bulk atomic concentration; $n_{CH} = N_{CH}/v_{CH}$ is the concentration in channels, and so on.

The equilibrium helium concentrations in fluids and channels relate to the concentration in the gas phase via the dimensionless Henry constants (eqn 3), H_L (liquid) and H_{CH} (channels):

$$n_L = H_L\, n_G,\ n_{CH} = H_{CH}\, n_G, \qquad (3)$$

Then the total amount of helium atoms and the bulk concentration in a sample are

$$N_\Sigma = n_{CH}\, V_{CH} + n_L\, V_L + n_G\, V_G = (H_{CH}\, v_{CH} + H_L\, v_L + v_G)\, n_G \equiv V_{HAV}\, n_G \qquad (4)$$

and

$$n_\Sigma = N_\Sigma / V_\Sigma = (V_{HAV}/V_\Sigma)\, n_G \equiv \varepsilon_{HAV}\, n_G. \qquad (5)$$

In eqns (4) and (5) V_{HAV} is the apparent internal helium accessible volume [cc per cc of a sample] and ε_{HAV} is the effective porosity [dimensionless].

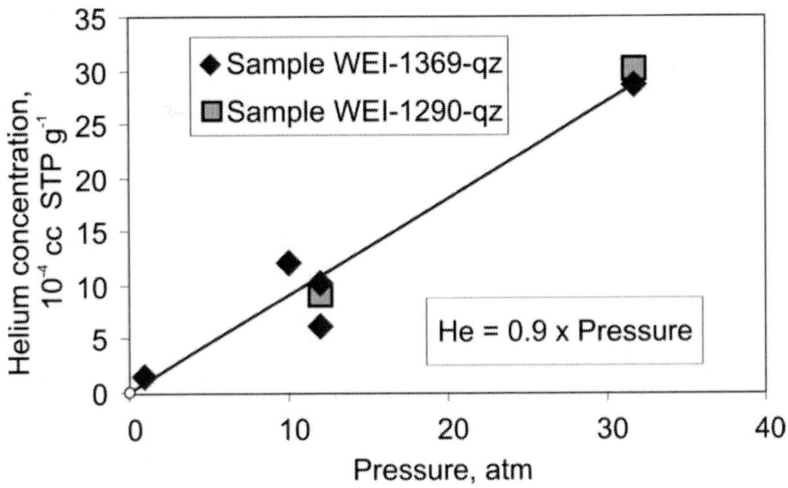

After Tolstikhin et al. (2005).

Figure 6. Pressure – concentration relationship. The concentration increases linearly with the pressure indicating "ideal" behaviour of helium in the internal helium-accessible volumes (HAVs). Slope of the regression line (forced to pass through X = Y = 0) is the effective Henry constant the quartz samples studied, $H_\Sigma = 0.9 \times 10^{-4}$ cc STP g^{-1} atm^{-1}.

Eqn (4) highlights different contributions of the phases constituting HAV: the gas phase enters with weight 1, whereas the weight of liquid and channels (which might at best be represented by amorphous silica) equals to the respective Henry constants. Helium solubility in fresh water is about 0.01 (e.g., Top et al., 1987). In brines and silicate glasses it decreases down to ≈ 0.001. This diminishes the effective contributions of V_L and V_{CH} into V_{HAV} by 2-3 orders of magnitude, the respective phases therefore could be ignored in most cases; correspondingly, from Eqns (4, 5) it follows that $V_{HAV} \approx v_G$ and $\varepsilon_{HAV} \approx \varepsilon_G$.

Thus the principal site of helium in quartz grains separated from PCT sandstones is the gaseous phase of fluid inclusions. Direct observations of the thin sections, extraction of almost all helium by crushing (see Table 4) and measurements of sizes of the post-crushing powder indicate large sizes of inclusions, > 10 μm; correspondingly, helium should behave in vesicles as an ideal gas. To check this several saturation experiments with different helium pressure were carried out (Figure 6). Even though results for two different samples are plotted, the quality of the correlation is reasonably good. The deviations of the data-points from the regression are generally less than the total analytical error plus the error related to heterogeneity of the samples. Thus, ideal gas equations may be applied to describe concentration – pressure relationships in the samples.

6. Helium Concentration Equilibrium in Quartz Fluid Inclusions and Pore Waters: A New Method in Hydrology

6.1. Sample Preservation During the Experiments

Using quartz crystals as indicators of helium concentrations in surrounding pore waters requires sample preservation in the course of the respective experiments: the internal helium accessible volumes (HAV, see Eqn 4), quantified via quartz saturation with helium, must correspond to the volumes available in natural unprocessed samples.

To meet this requirement, a moderate processing temperature was selected. Stepwise heating of sample 1409.83 showed that intense helium release from the quartz grains starts at temperatures above 500 ^{O}C, indicating that helium-accessible volumes would be preserved below this temperature (Lehmann et al., 2003b). Also an investigation of synthetic quartz crystals using X-ray diffraction tomography with a resolution of about 5 micron did not reveal any new vesicles, or cracks, or other defects developed in the course of heating up to 500 ^{O}C (Argunova et al., 2003). Correspondingly, a temperature of about 350 ^{O}C appeared to be reasonable for He saturation. Furthermore, this temperature also ensures reasonably short, within ~1000 hours, duration of helium saturation (and helium isothermal extraction) experiments.

Our experiments show that the helium accessible volumes in quartz grains separated from PCT sandstones did not develop in the course of repeated extraction / saturation procedures. Thus, helium release patterns, obtained for isothermal heating of unsaturated samples, are indistinguishable from or very similar to those of the respective saturated ones (Figure 7). Moreover, the repeated saturation / extraction performed with the very same sample-aliquot resulted in fully identical release patterns (see 3 release patterns for sample 1389, T_{SAT} = 350 ^{O}C, Figure 7). Hence, the saturation does not affect the rate of helium migration to / from the samples (sample 1389, Figure 7), nor the amount of He impregnated under given *PT* conditions (same sample, Table 4).

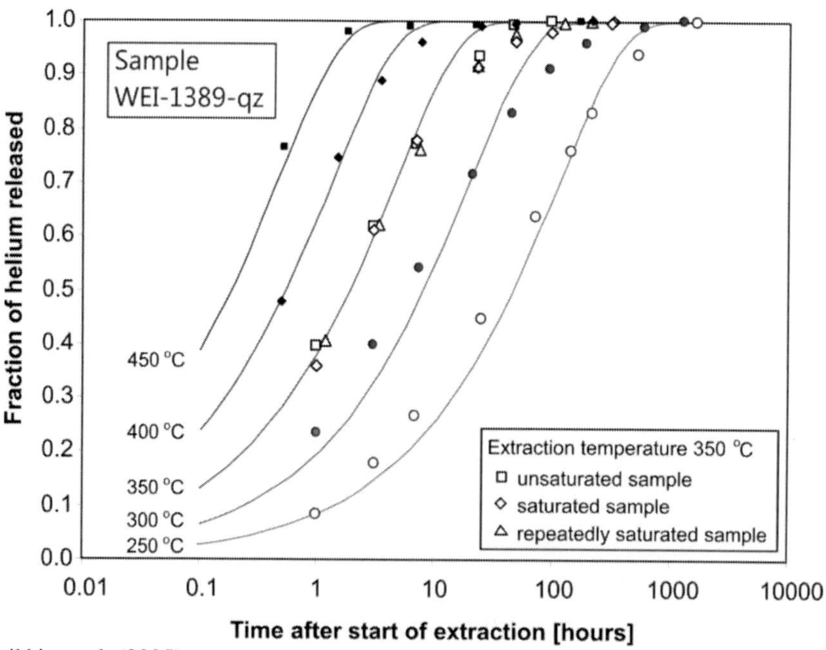

After Tolstikhin et al. (2005).

Figure 7. Isothermal helium extraction from sample WEI-1389-qz. Note a very similar release pattern for unsaturated and saturated aliquots of this sample obtained for the temperature 350 ^{O}C: the saturation / extraction procedures do not affect the quartz grains and the same sample can be used for measurements of the initial (non-saturated) He concentration and those in the (repeatedly) saturated sample. These release patterns were transferred into the Arrhenius plot (Figure 9).

Finally, comparison of an apparent pressure derived from He-solubility and measured He-concentrations in PCT ground waters (0.2 to 0.6 atmospheres) and the pressure derived from the extraction / saturation experiments (see below Table 4) definitely shows that HAVs have not changed substantially due to isothermal extraction / saturation experiments: both values are very similar. Also, the sample weight was preserved in our experiments within 0.07 %, as repeated weightings of sample 1389 have shown.

Summarizing, the available data indicate preservation of quartz separates in the course of our experiments and show, that one sample-aliquot can be used for both measurements, i.e., the initial helium concentration and the helium accessible volume.

6.2. Equilibration Time Scale

A numerical modelling allows the release patterns of Figure 7 to be converted into a temperature – time relationships. The model envisages the diffusive gas transfer characterized by the diffusion coefficient D and diffusion path L, which is a mean length of “channels” in a mineral. The two parameters, D and L, are both unknown; however they enter the diffusion equation as a ratio of $L^2/D \equiv \tau$, and τ is the diffusion time available from the measurements (Figure 7).

Solution of the 3D diffusion equation for sphere with zero boundary conditions at a given temperature links the time interval τ and the fraction of helium released during this interval (Figure 8ab). The curves, derived from the diffusion model, have been calculated under assumption of a constant τ-value for given temperature T. The lower-temperature curves (250 and 300 ^{O}C, Figure 7) in some cases slightly underestimate the actual He loss for shorter time intervals and overestimate for longer ones. This apparent discrepancy most probably results from non-homogeneity of natural crystals in regards to the diffusion path and/or diffusivity of different parts of a crystal. Assuming that the diffusion path is responsible and that path lengths vary according to the power low (favouring number of short lengths, Figure 8a), then a much better agreement is achieved between the measured and calculated release curves (Figure 8b).

The constant-τ variable-diffusion-lengths model gives the best fit τ value for each extraction temperature, which value is then converted into an Arrhenius plot, ln τ [yrs] versus $1/T$ [K^{-1}]. The data points are regressed by a straight line (Figure 9):

$$\ln \tau = a + b\,(1/T) = a + (E/R)\,(1/T) = -(4.4 \pm 0.8) + (10450 \pm 500)\cdot(1/T) \qquad (6)$$

The constant b in Eqn (6) relates to an activation energy E as $b = E/R$, where R is the universal gas constant (R = 8.31 J mol^{-1} K^{-1}). From our fit a value of E = 86.8 kJ mol^{-1} results. Important is that 95%-saturation is achieved after a time $t = \tau/4$, and the calculated 95%-saturation values (τ/4) vary within 1.6 h to 385 h for the temperature variation from 450 °C to 250 °C, respectively.

Extrapolation to the temperature of the Permian-Carboniferous formation, the “initial” temperature of our samples ($T_{INI} = 70^oC = 343\ K$), gives an equilibration time τ = 6500 years (95% saturation after $t = \tau/4$ = 1600 years). It should be noted that such an extrapolation in a logarithmic plot (Figure 9) to a much lower temperature results in a rather large uncertainty in spite of the quite good quality of the experimental data in the temperature range of 250-450°C. Nevertheless the equilibration time scales derived from our saturation / extraction experiments are orders of magnitude less than the He residence times in PCT rock – water system (see Table 5 and Figure 11 and 12 below); these comparisons prove that the equilibration of He concentrations inside the quartz grains and in hosted pore waters has indeed been achieved.

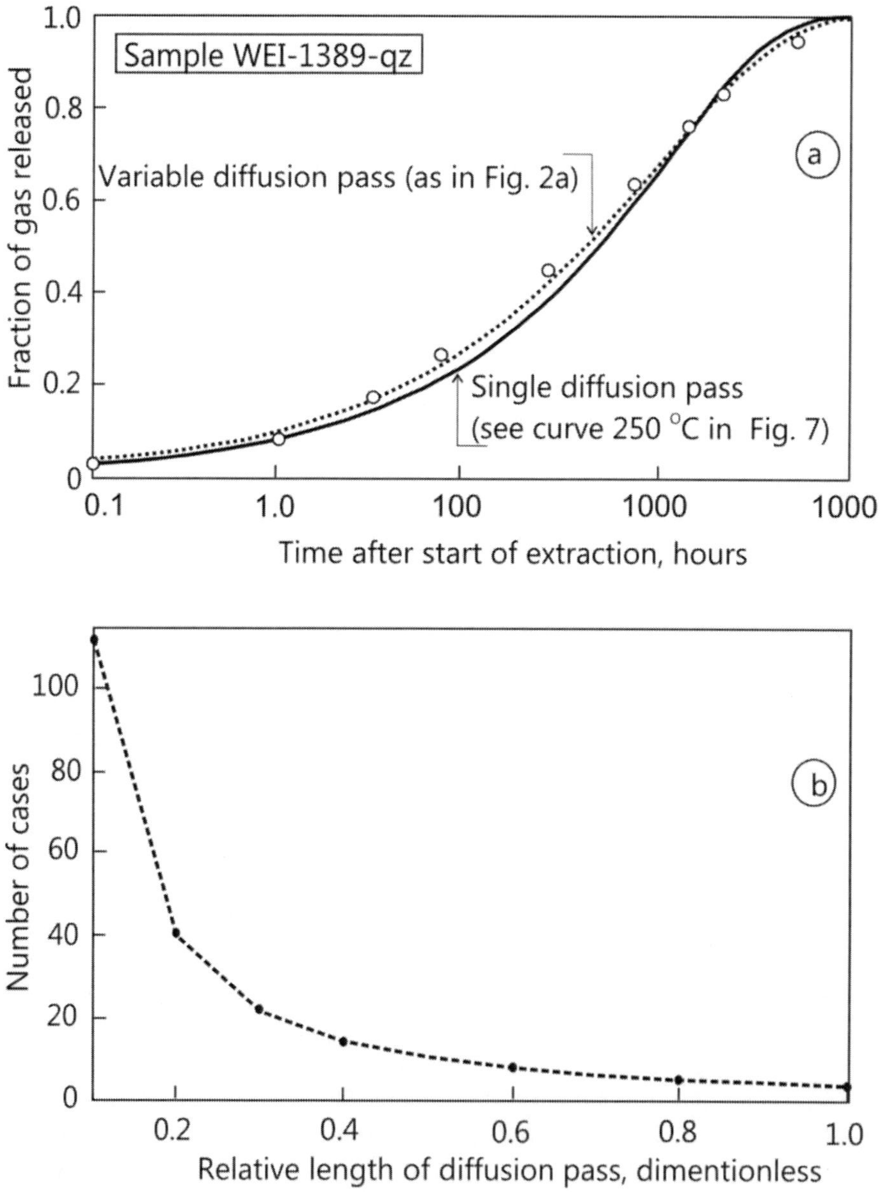

After Tolstikhin et al. (2005).

Figure 8. Relationship between observed and model-generated release patterns. The model-derived release pattern (plot a, dotted line) obtained for the variable diffusion path lengths (favouring smaller lengths, plot (b)) fits better to the observed pattern (open circles in (a)), than the result from a single-diffusion-path model (solid line in (a), see also Figure 7).

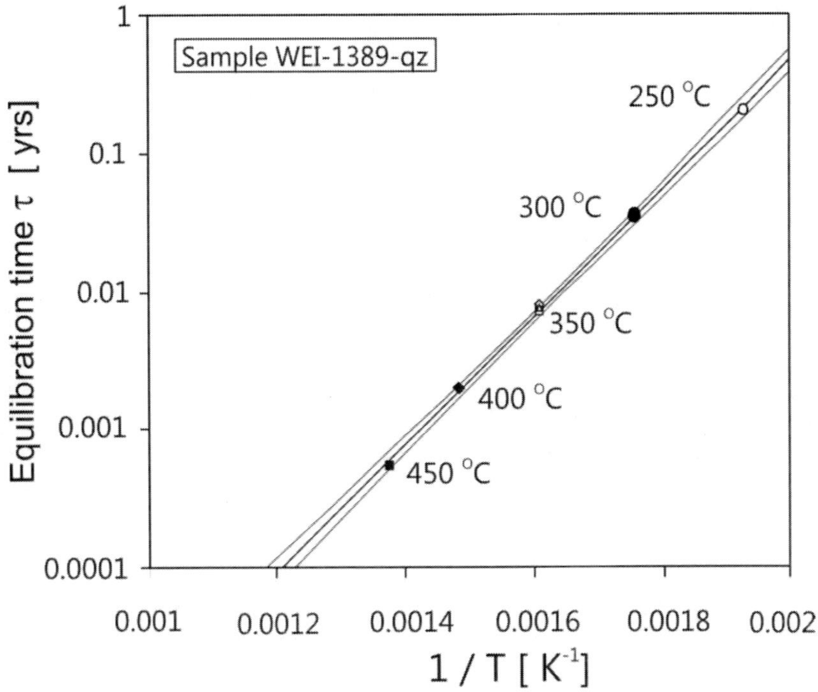

After Tolstikhin et al. (2005).

Figure 9. Arrhenius plot for sample WEI-1389-qz. Instead of poorly known diffusion radii L and diffusion coefficients D, their ratio, which is the time parameter, is shown (left axis) versus the reverse temperature. Regression of this correlation to the average PCT temperature, 70 °C (corresponding to 0.0029 K^{-1}), gives $\tau \approx 6{,}000$ years: helium concentrations inside the quartz grains and in the respective pore waters achieve equilibrium during this time.

Comparison of mineral-derived and measured ground water concentrations of helium (see Figure 10 below) also indicates the concentration equilibrium between the quartz crystals and the porewaters. It is important that the above conclusion relies also on *a priori* data, i.e., results of measurements of He concentrations in ground waters (see Table 5 below). These data could have been absent for another hydrological structure. An internal criterion of the helium concentration equilibrium should be worked out in the course of future studies.

7. Results of the Reconstruction: Helium Concentrations and Residence Times in the Porewater

From the above considerations (see Eqns 4,5 and related Text) it follows, that $V_{HAV} \approx v_G$ and $\varepsilon_{HAV} \approx \varepsilon_G$ and that helium behaves as an ideal gas inside quartz samples, hence

$$N_\Sigma = V_{HAV}\, p_G / (k\,T) \approx v_G\, p_G /(k\,T) \tag{7}$$

and

$$n_\Sigma = \varepsilon_{HAV}\, p_G / (k\,T) \approx \varepsilon_G\, p_G /(k\,T). \tag{8}$$

In Eqs. (7, 8) V_{HAV} is the apparent internal helium accessible volume [cc per cc of a sample] and ε_{HAV} is the effective porosity [dimensionless], p_G is the helium partial pressure in gas phase [atm], k is the Boltzmann constant [1.36×10^{-22} cc atm K^{-1}] and T is the temperature [K]. Then measurements of initial (subscript INI) and saturated (SAT) concentrations allow ε_{HAV} (see Eqn 5) and the helium partial pressure p_{INI} (under PCF-conditions) to be derived:

$$\varepsilon_{HAV} = k\, T_{SAT}\, n_{\Sigma}(sat)/p_{SAT} \quad (9)$$

$$p_{INI} = p_{SAT}\, N_{INI}\, T_{INI} / N_{SAT}\, T_{SAT} \quad (10)$$

Table 4 presents the experimental data obtained for quartz separates from the PCT and p_{INI} and ε_{HAV} values calculated from these data using Eqns (9, 10).

The equilibrium between helium pressure in HAVs and in external pore waters allows the pore water concentration (C_{PW}) to be quantified using the Henry law:

$$C_{PW} = H(T,M)_{H2O} \times p_{INI} \quad (11)$$

For salinity M varying from 40 to 100 g l^{-1} and temperature T = 343 K as measured in PCF aquifers (Pearson et al., 1991), the helium solubility $H(T,M)_{H2O}$ derived from Weiss (1971) and Top et al., (1987) varies in a narrow range from 8×10^{-3} to 6.7×10^{-3} cc STP atm^{-1} per cc H_2O and the mean value 7.35×10^{-3} cc STP per cc H_2O is used hereafter; a similar value was reported in by Yakutseny (1968) for given T, M parameters. Product of this solubility (assumed to be constant for the depth interval studied) and the internal helium pressure (with an average value of $p_{INI} = 0.49 \pm 0.21$, calculated from Table 4) gives the mineral-derived helium concentration in respective pore water (Eqn 11). All these concentrations (with one exception) vary within a narrow range between the values actually measured in adjacent aquifers (Figure 10) and the average concentration for all 36 samples, $C_{PW} = 0.00361 \pm 0.00157$ cc STP per cc H_2O (Table 4), is very similar to those observed in the overlaying and underlaying aquifers (same units), 0.0016 and 0.00448, respectively. The excellent agreement seen in Figure 10 validates the proposed approach.

The concentrations of He isotopes in the interstitial pore water of the PCT sediments cannot be measured directly. However, assuming closed-system conditions and complete loss of He from the minerals into the pore water one can assess a maximum pore water concentration (Figure 11). For example, substituting the maximum value of in-situ produced ^{4}He in the shale sample WEI-1417-SH, $^4He_{SH,CALC} = 623 \times 10^{-6}$ cc STP g_{ROCK}^{-1} (Table 3), an average shale porosity $\Phi_{SH} = 0.12$, an average shale density $\rho_{SH} = 2.5$ (Matter et al., 1988), and a pore water density $\rho_{PW} = 1$ in the balance equation (12) gives

$$^4He_{PW}(\text{WEI-1417-SH}) = {}^4He_{SH,CALC} \times [(1-\Phi_{SH})/p_{SH}] \times (\rho_{SH}/\rho_{PW}) = 11 \times 10^{-3} \text{ cc STP } g_{H2O}^{-1} \quad (12)$$

Using average U and Th contents in shales (Table 2) gives $^4He = 9.3 \times 10^{-3}$ cc STP g_{H2O}^{-1} in the respective pore water, other things being equal, which is only a factor of ~ 2 higher than the value actually measured in the ground water Pm1 (Figure 3 and 11).

Table 4. Initial ($^4He_{INI}$) and saturated ($^4He_{SAT}$) helium concentrations, initial partial helium pressure ($p(He_{INI})$) in helium accessible volumes (ε_{HAV}), and the helium concentrations in pore waters, calculated from $p(He_{INI})$ and He solubility in water (all samples are quartz separates from PCT sandstones opened by Weiach borehole, so only depth of each sample is shown)

Depth	Extraction	temperature of extraction	$^4He_{INI}$	conditions of saturation			$^4He_{SAT}$	p (He)$_{INI}$	ε_{HAV}	^{4}He in pore water
				time	P (He)	T				
m		°C	× 10^{-4} cc STP /cc	hours	atm.	°C	× 10^{-2} cc STP /cc	atm.	× 10^{-4} cc/cc	× 10^{-3} cc STP /cc H_2O
1	2	3	4	5	6	7	8	9	10	11
1130.24	crushing	20 °C	0.74	1003	12	285	0.15	0.369	2.53	2.72
1130.24	fusion	1700	0.64	1003	12	285	0.10	0.461	1.75	3.39
1150.71	crushing	20 °C	1.12	504	32	290	0.55	0.396	3.55	2.91
1150.71	fusion	1700	0.94	504	32	290	0.28	0.643	1.83	4.73
1150.71	crushing	20 °C	1.12	1412	32	290	0.49	0.442	3.18	3.25
1150.71	fusion	1700	0.94	1412	32	290	0.31	0.599	1.97	4.40
1189.71	isotherm.	350	2.89	1008	30.7	290	2.39	0.226	16.05	1.66
1250.34	isotherm.	350	3.16	504	32	290	1.58	0.389	10.21	2.86
1250.34	isotherm.	350	3.16	1412	32	290	2.01	0.306	12.97	2.25
1290.08	crushing	20 °C	1.38	1003	12	285	0.24	0.417	4.15	3.07
1290.08	fusion	1700	1.22	1003	12	285	0.14	0.626	2.46	4.60
1290.08	crushing	20 °C	1.38	1003	31.7	298	0.80	0.328	5.27	2.41
1369.38	fusion	1700	0.83	1104	31.7	298	0.41	0.388	2.70	2.86
1369.38	fusion	1700	0.83	1020	10	298	0.15	0.335	3.13	2.47
1369.38	fusion	1700	0.83	1028	1	285	0.02	0.277	3.79	2.04
1389.67	isotherm.	350	2.83	1003	12	285	0.65	0.324	11.00	2.38
1389.67	isotherm.	250	2.83	1003	12	285	0.56	0.373	9.53	2.74
1389.67	isotherm.	300	2.83	1003	12	285	0.64	0.324	10.97	2.38
1389.67	isotherm.	400	2.83	1003	12	285	0.64	0.326	10.92	2.40
1389.67	isotherm.	450	2.83	1003	12	285	0.51	0.409	8.70	3.01
1409.83	crushing	20 °C	2.36	504	32	290	0.93	0.495	5.99	3.64
1409.83	crushing	20 °C	2.36	1412	32	290	0.34	1.352	2.19	9.94
1409.83	fusion	1700	1.12	1412	32	290	0.39	0.558	2.51	4.10
1409.83	stepwise heating	200 - 1700	3.40	504	32	290	0.73	0.908	4.71	6.68
1407.5	isotherm.	350	4.30	504	37.5	290	1.51	0.650	8.31	4.78
1407.5	isotherm.	350	4.27	504	37.5	290	1.62	0.602	8.91	4.43
1526.91	isotherm.	350	4.54	504	37.5	290	1.82	0.570	10.01	4.19
1647.18	isotherm.	350	3.91	504	37.5	290	1.34	0.667	7.37	4.90
1737.56	isotherm.	350	7.07	504	37.5	290	2.08	0.777	11.45	5.71
1737(1)	isotherm.	350	2.61	500	38	290	1.16	0.520	6.31	3.82
1737(2)	isotherm.	350	2.61	500	38	290	1.27	0.474	6.92	3.48
1856.97	isotherm.	350	4.08	504	37.5	290	1.49	0.626	8.20	4.60
1950.74	isotherm.	350	2.76	504	37.5	290	1.15	0.548	6.33	4.03
2012(1)	isotherm.	350	1.96	500	38	290	1.67	0.272	9.05	2.00
2012(2)	isotherm.	350	1.96	500	38	290	1.44	0.315	7.83	2.31
2012.09	isotherm.	350	2.13	504	37.5	290	1.34	0.363	7.37	2.67
AVERAGE								0.49	6.67	3.61
STDEV								0.21	3.58	1.57

This close agreement implies good preservation of He in the Permian rock-water system at Weiach, a long residence time of He atoms in this system, and highly stagnant conditions for the Permian ground water Pm1.

The high helium concentration in the shale pore water and the ground water and the long residence time of He minimise the effect of helium isotope fractionation during He loss from the shales, as indicated by the almost indistinguishable $^3He/^4He$ values of the Pm1 ground water and the shale whole rock production ratio.

Applying Eqn (12) to the sandstone rock - water system and using the same parameters with the average $^4He_{SS,CALC} = 264 \times 10^{-6}$ (Table 3) gives $^4He_{PW,CALC} = 4.8 \times 10^{-3}$. Thus a helium concentration gradient is apparent between shale pore waters (with the maximum concentration of ~10^{-2} cc STP ^4He per g_{H2O}) and sandstone pore waters (4.8×10^{-3}) and much steeper gradient is expected for ^3He.

Once released to the pore water, the He atoms migrate along concentration gradients to domains of lower He concentrations. On a small scale (several metres to tens of metres) this migration occurs from the highly productive shale layers to the less productive sandstone layers, where He is incorporated inside some rock-forming minerals, i.e., quartz and plagioclase. The time required for equilibration between He isotope concentrations in the pore water and in the He-accessible volume inside a mineral (the "transmissive timescale" according to the definition of Baxter, 2003) appears to be short, ~10 Ka as discussed above.

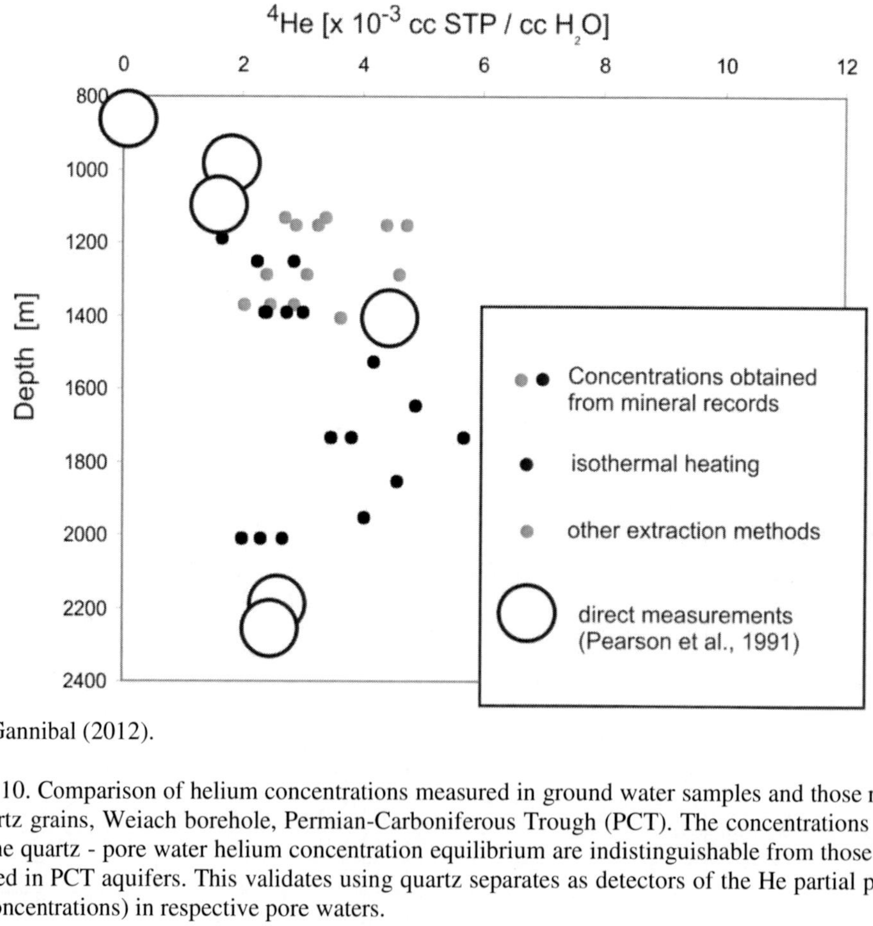

After Gannibal (2012).

Figure 10. Comparison of helium concentrations measured in ground water samples and those recorded by quartz grains, Weiach borehole, Permian-Carboniferous Trough (PCT). The concentrations inferred from the quartz - pore water helium concentration equilibrium are indistinguishable from those directly observed in PCT aquifers. This validates using quartz separates as detectors of the He partial pressures (and concentrations) in respective pore waters.

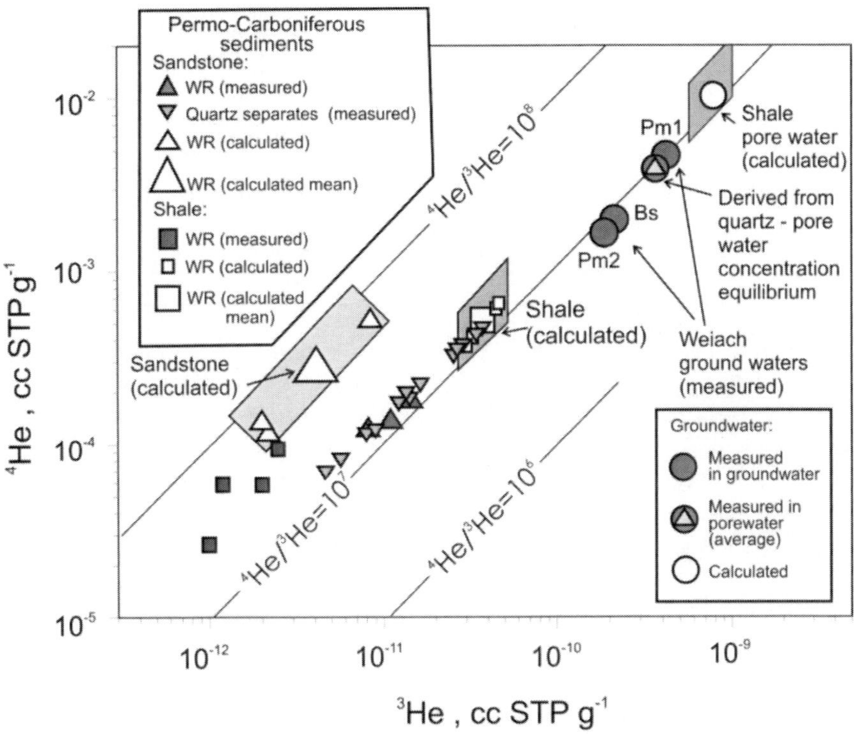

Figure 11. Measured and calculated helium isotope abundances in rocks, mineral separates, and ground (pore) water samples, Weiach borehole, Permian-Carboniferous Trough (PCT). The $^3He/^4He$ ratios measured in sandstones (whole-rocks and mineral separates) and in groundwater samples are indistinguishable from the production ratio in the shales and the observed ratios in ground water samples. The calculated helium concentration for stagnant pore water (the closed system approximation, see text) is only slightly above the measured concentrations, indicating a very long residence time for radiogenic He atoms in PCT.

This time is orders of magnitude shorter than the mean residence time of He atoms in Permian sediments of the PCT at Weiach. This validates migration of He atoms into quartz and plagioclase grains (see data for WEI-1415-pl in Tables 2 and 3). Note that both (i) He atom incorporation into quartz and plagioclase grains and (ii) He removal by water flow in sandstone-related aquifers would increase the He concentration gradient between shale and sandstone layers and thus the rate of migration of shale-produced He.

On a large scale (i.e., several hundred meters), He and $^{40}Ar^*$ atoms migrate to the nearest aquifer with the lower concentrations of these species and then the water flow carries them out of the PCT rock-water system. Thus He migrates from the pore waters to those of the upper Permian Pm2 and Buntsandstein Bs aquifers, which have lower He concentration, and to the Muschelkalk (Mu) ground waters with extremely low He concentration. $^{40}Ar^*$ atoms move to Muschelkalk and crystalline aquifers with almost atmospheric ^{40}Ar / ^{36}Ar ratios (Figure 3). Comparison of helium inventory in the water - rock system with the rate of helium generation inferred from contents of the parent elements allows the apparent helium residence time scales to be constrained (Table 5). In the last line of Table 5 the minimal helium inventory is combined with the maximal rate of radiogenic production (calculated from the maximum concentrations of the parent elements). It is obvious that even this (the shortest) time indicates that PCT ground waters are stagnant.

Table 5. Estimates of the residence times , $<\tau>$, of radiogenic helium isotopes in the Permian sediments of the Permo-Carboniferous Trough (Northern Switzerland)

Sample	U meas	Th meas	$^4He_{WR}$ meas.	He_{PW} [1] calc.	$^4He_{TOT}$ calc.	Production	$<\tau>$
	ppm	ppm	10^{-3} cm^3 STP He /cm^3 rock	10^{-3} cm^3 STP He /cm^3 water	10^{-3} cm^3 STP He / cm^3 sample	10^{-12} cm^3 STP He /cm^3 rock /year	Ma
Shale 1418-SH	10.8	26.5	0.064	5.6	0.74	5.3	140
Shale mean [2]	7.83	28.0	0.145	3.6	0.57	4.4	130
Permian rock minimal [3]	11.7	31	0.025	1.7	0.24	5.3	46

[1] $^4He_{PW}$ is derived assuming equilibrium between [^{4}He] in quartz grains and the pore water. For shale WEI-1418-SH, He_{PW} is the average value obtained for the pore waters of neighbour sandstone samples WEI-1407-SS and WEI-1409-SS, see Table 4.

[2] Calculated using the mean concentrations of U, Th and ^{4}He for shales (Table 2) and the mean calculated pore water concentration (Table 4).

[3] Hypothetical rock composition with maximum concentrations of U and Th (from samples WEI-1417-SH and WEI-1126-SH, respectively, see Table 2), minimum concentrations of ^{4}He (WEI-1418-SH) and minimum $^4He_{PW}$ (sample 1189.71, Table 4), using a mean porosity of 12 %.

It should be noted however that the residence times presented in Table 5 relay on the helium balance in individual samples and do not take into account helium migration between shale and sandstone layers and through the PCT as a whole. For instance, the balance estimates for sandstones approach (or even exceed) the rock age due to presence of excess (shale-produced) helium.

1D diffusion-removal model (to be discussed in details elsewhere) allows more reliable constraints on the helium residence times in PCT.

The model envisages the following data, processes and conditions.

- Measured concentrations of the parent elements, U, Th, K and Li, which allow radiogenic helium and argon production to be quantified (see Tables 2,3 and also Weber et al., 1986, Tolstikhin et al., 1996).
- Measured He and Ar isotope abundances in PCT rocks and rock-forming minerals (Table 2).
- Stratigraphic and hydrological cross-section of the Weiach borehole (along with respective chronologies, Pearson et al., 1991), which determines specific layers, included in the model, and times of their formation.
- Measured He and Ar isotope abundances in water samples collected from PCT and adjacent aquifers (Figure 12, Pearson et al., 1991).
- He concentrations in pore waters recorded by quartz grains separated from PCT sandstones (Table 4).

The initial and boundary conditions of the model are as follows.

(1) Consequent formation of the Crystalline basement (315 Ma ago), the Permian (285 Ma), and the Muschelkalk (240 Ma) layers; each of them includes relevant aquifers considered as thin separate layers (Figure 3, Table 1).

(2) Duration of formation of each layer "*i*", Δt_i, and the initial He concentration are zero, $\Delta t_i \equiv 0$, $C_i(t = t_{i0}) \equiv 0$. The running time, *t*, varies from 315 Ma ago till present.
(3) A lower boundary of the crystalline basement is assumed to be at a depth $z \to \infty$.
(4) The helium production rate, P_i, is derived for each layer from its average U and Th concentrations.
(5) The rock - ground water system is able to lose helium by two removal mechanisms.
5.1. He diffusion along the concentration gradients towards a "surface" layer, where He concentration is assumed to be zero, and to the aquifers seen in Figure 3.
5.2. He removal by the ground water flow. The removal rate is defined as θ_i [year^{-1}] and it is assumed to be constant since the formation time t_i. $\theta_i \geq 0$ is applied only to the aquifers, *i* = Cr1, Cr2, Pm1... and $\theta_i \equiv 0$ elsewhere. The product $\theta_i \times [He]_i$ gives the helium flux from the PCT rock – water system via aquifer *i*. The removal rate θ_i and the mean residence time of He atom τ_i relate as $\tau_i \equiv (2 \ln 2)/\theta_i = 1.4 / \theta_i$.
(6) Helium diffusion coefficient D_i is assumed to be constant within each stratigraphic layer since the formation time t_i. D_i value depends on the depth of the layer, i.e., $D_{CRYST} \leq D_{PCT} \leq D_{MUSH}$. Helium concentration on the open surface of the topper strata C_{SURF} is assumed to be zero, $C_{SURF} \equiv 0$. For example, when the crystalline basement was formed, He dissipated from its surface during age interval from 315 to 285 Ma, $C_{SURF,CRYST}$(315-285 Ma) $\equiv 0$. Then this surface was covered by Permian-Carboniferous sediments, and He dissipated from the very top of the PCT rocks, $C_{SURF,PERM}$(285-240) $\equiv 0$. After formation of the Muschelkalk sediments 240 Ma ago, He concentration in the aquifer Mu1 at depth z = 859 m is assumed to be negligible, $C_{Mu1} \equiv 0$ (the measured concentration is by a factor ~1000 below that in the neighbour aquifer Pm3, see Figure 3).

Taking into account the above initial and boundary conditions, ^{4}He concentration inside each layer i is expressed by eqn (13):

$$\partial_t C_i = \partial_z \left(D_i \partial_z C_i\right) + P_i - \theta_i C_i \tag{13}$$

At the boundaries between adjacent layers an equality of helium concentrations and helium fluxes from both sides of boundary is assumed:

$$C_i\big|_- = C_{i+1}\big|_+, \qquad - D_i \partial_z C\big|_- = - D_{i+1} \partial_z C_{i+1}\big|_+ \tag{14}$$

where signs "minus/plus" in eqn (14) correspond to lower/upper surface of the respective boundary.

Applying the simplest scenario, i.e., the constant diffusion coefficient for the Permian-Carboniferous sediments, $D_{PC} \equiv$ Const, while these coefficients in the underlying and overlaying strata are by a factor of 3 lower and higher than D_{PC}, respectively, then a reasonable agreement between the calculated concentrations and those observed in PCT pore and ground waters holds for $1 \times 10^{-4} \leq D_{PC} \leq 1 \times 10^{-2}$ m^2 yr^{-1}, with the best fit value 1×10^{-3}

m^2 yr^{-1} (Figure 12). This value is exactly the same as derived by Rubel et al. (2002) for the Opalinus Clay formation at Mont Terri, Switzerland. Substituting this value, 1×10^{-2} m^2 yr^{-1}, into relationship for the diffusion pass, $R = (D\ t)^{1/2}$, gives R ~100 m during t ~10 Ma, or t ~100 Ma if the whole ~ 1000 m thickness of PCT rocks is considered.

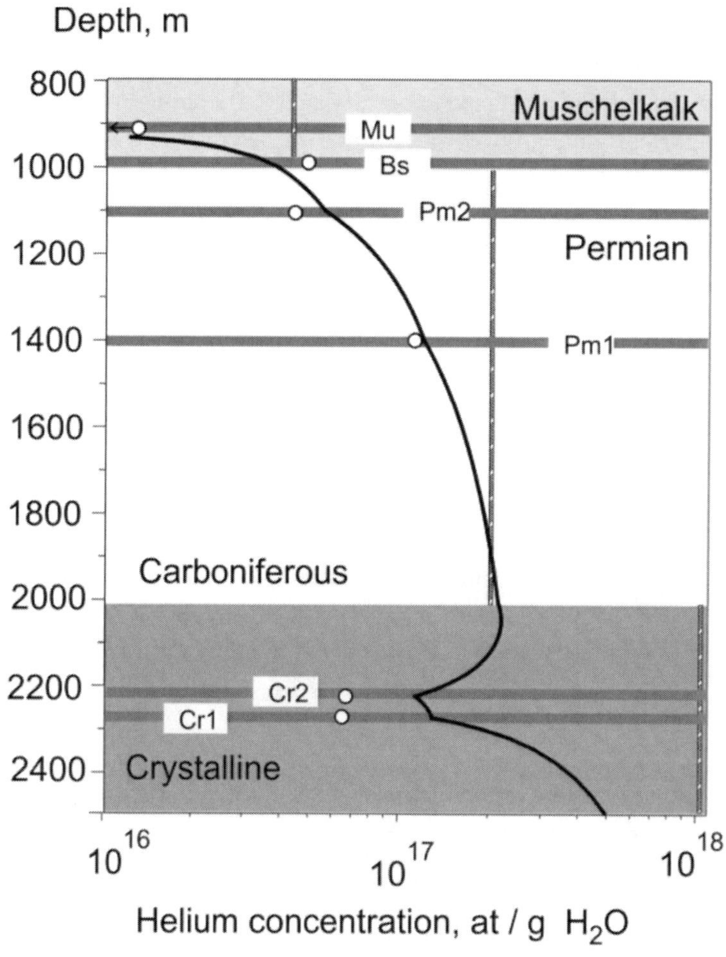

Figure 12. Helium concentration profiles derived from the modelling of helium production (by PCT rocks), migration into pore space, removal by diffusion (through porous media) and by groundwater flow in aquifers. Permo-Carboniferous and adjacent strata, Weiach borehole. Aquifers are shown as horizontal bands (see caption to Figure 3); circles are ^{4}He-concentrations in the aquifers; vertical bands are the maximal calculated ^{4}He-concetrations in pore waters of the given cross section assuming: (i) no gain/loss of the parent elements and helium within the PCT rock – water system; (ii) all radiogenic helium produced by PCT rocks is concentrated in the pore water. Solid curve presents the concentration profile derived from He modelling; the parameters derived from modelling are: the diffusion coefficients ($\times 10^{-3}$ m^2 $year^{-1}$): $D_{CR} = 0.33$, $D_{PCT} = D_{MU} = 1$; the water flow removal rates ($\times 10^{-5}$ $year^{-1}$): $\theta_{Cr1} = 36$, $\theta_{Cr2} = 30$, $\theta_{Pm1} = 4.3$, $\theta_{Pm2} = 17$, $\theta_{Bs} = 0.77$.

Mechanism of He removal by water flow is included in order to reconcile the observed He concentrations in the Cr1 and Cr2 aquifers with those calculated, which otherwise differ by an order of magnitude. The residence time of helium atom in the Permian aquifers is τ_{Pm1}

≈ 30000 years thus indicating the long residence times even considering helium removal by water flow.

CONCLUSION

Analysis of the parent element contents (U, Th, K, Li) and He and Ar isotope abundances in rocks, mineral separates, pore- and ground waters of shale-sandstone interlayering of the Permo-Carboniferous Trough (PCT), Northern Switzerland, shows that intra-basin production and loss of radiogenic He and Ar isotopes are responsible for the observed present-day inventories.

The shales contain enhanced concentrations of U, Th, Li and produce larger amounts of radiogenic ^{4}He and nucleogenic ^{3}He compared with the sandstones, by factors of ~2 and ~10, respectively.

The measured concentrations of both He isotopes in the shales are much lower than the calculated production values for whole-rock samples and mineral separates. This indicates a major loss of He from the shales into its complementary reservoir, the pore water. For the shale, pore water a ^{3}He/^{4}He production ratio approaching $\approx 1 \times 10^{-7}$ and a high concentration of ^{4}He $\approx 10 \times 10^{-3}$ cc STP $g_{H2O}{}^{-1}$ were calculated assuming (i) a closed system evolution of the PCT shale - porewater system since the time of sedimentation (285 Ma ago) and (ii) that all helium produced resides in the water. Comparing these values and those observed in the ground water from the Permian aquifer Pm1 shows that: (i) the observed and calculated ^{3}He/^{4}He ratios are almost indistinguishable; (ii) the observed concentration is only ≈ 2 times lower than the value calculated for the pore water, which implies a very long residence time of He atoms in the PCT rock-water system.

After being released from the shales, He atoms migrate through pore waters along the concentration gradient towards the sandstone and then (partially) penetrate into suitable minerals, i.e., quartz and plagioclase. This is indicated both by of the helium excess in the sandstones and especially in quartz and plagioclase from these rocks, and by ^{3}He/^{4}He ratios in the sandstones, indistinguishable from the shale production ratio and the ratio observed in PCT ground waters but different from that generated within the sandstones.

Quartz grains, separated from PCT sandstones, have recorded helium concentrations of surrounding pore waters. These mineral-recorded concentrations are in excellent agreement with those actually observed in aquifers of the PCF. The agreement validates a promising new tool: recovery of rock – pore waters helium inventory without extraction and investigations of the pore waters themselves. The inferred pore water concentrations and measured whole-rock showed a large contribution of water-hosted helium in the inventory of PCF rocks.

The abundances of the parent elements and He isotopes in the whole rocks, mineral separates and ground and pore waters allow the He residence times in PCT be estimated. Respective modelling have shown that the diffusion / removal rate of helium from the PCT rock - ground waters system is very slow, the best-fit diffusion coefficient being $D = 1 \times 10^{-3}$ $m^2\ yr^{-1}$, which is very close to the value obtained by Rubel et al. (2002) for the Opalinus Clay formation. The residence time of helium in PCT rock – pore water system is of the order of 10 Ma.

Acknowledgments

This contribution was supported by the Russian Foundation for Fundamental Investigations (RFFI), grants 10-05-00306-a and 13-05-00647-A. The authors thank Dr. N. Waber for his permanent help with this project.

References

Agarwal M., Gupta S. K., Deshpande R. D., and Yadava M. G., 2006. Helium, radon and radiocarbon studies on a regional aquifer system of the North Gujarat–Cambay region, India. *Chem. Geol.,* 228, 209– 232.

Andrews J.N., 1985. The $^{3}He/^{4}He$ ratios of radogenic helium in crustal rocks and its application in growndwater circulation studies. *Chem. Geol.* 49, 339-351.

Andrews J.N. and Fontes J.-C., 1991. Importance of the in situ production of ^{36}Cl, ^{36}Ar and ^{14}C in hydrology and hydrogeochemistry., in: *Isotope Techniques in Water Resources Development,* IAEA, Vienna, 1991, Vienna, pp. 245-269.

Argunova T. S., Sorokin L. M., Pevzner B. Z., Je J. H., Hwu Y., and Tsai W.-L., 2003. X-ray imaging study of lattice defects related to diffusion of helium in quartz. *J. Phys. D: Appl. Phys.* 36, 1-5.

Ballentine, C.J., Burnard, P.G., 2002. Production, release and transport of noble gases in the continental crust. In: Porcelli, D., Ballentine, C.J., Wieler, R. (Eds), *Noble Gases in Geochemistry and Cosmochemistry.* Reviews in Mineralogy and Geochemistry, 47, Geol.Soc. & Min Soc. of America, Washington D.C., USA, 481-538.

Ballentine, C.J., Burgess, R., Marty, B., 2002. Tracing fluid origin, transport and interaction in the crust. In: Porcelli, D., Ballentine, C.J., Wieler, R. (Eds), *Noble Gases in Geochemistry and Cosmochemistry.* Reviews in Mineralogy and Geochemistry, 47, Geol.Soc. & Min Soc. of America, Washington D.C., USA, 539-614.

Baxter E. F., 2003. Quantification of the factors controlling the presence of excess ^{40}Ar or ^{4}He. *Earth Planet. Sci. Lett.* 216, 619-634.

Broadhurst C. L., Drake M. J, Hagee B.E., Bernatowicz T.J., 1992. Solubility and partitioning of Ne, Ar, Kr, and Xe in minerals and synthetic basaltic melts. *Geochim. Cosmochim. Acta* 56, 709-723.

Carroll M. R. and Stolper E. M., 1991. Argon solubility and diffusion in silica glass: Implications for the solution behavior of molecular gases. *Geochim. Cosmochim. Acta* 55, 211-225.

Carroll M. R. and Stolper E. M., 1993. Noble gas solubilities in silicate melts and glasses: New experimental results for argon and the relationship between solubility and ionic porosity. *Geochim. Cosmochim. Acta* 57: 5039-4051.

Castro M.C., Stute M., Schlosser P., 2000. Comparison of ^{4}He ages and ^{14}C ages in simple aquifer systems: implications for ground water flow and chronologies. *Appl. Geochem.* 15, 1137-1167.

Cook P. G. and Herczeg A. L., 2000. *Environmental Tracers in Subsurface Hydrology.* Kluwer Academic Publishers, pp. 529.

Dunai T. J. and Roselieb K., 1996. Sorption and diffusion of helium in garnet: implications for volatile tracing and dating. *Earth Planet. Sci. Lett*. 139, 411-421.

Eugster, O., 1988. Cosmic-ray production rates for ^{3}He, ^{21}Ne, ^{38}Ar, ^{83}Kr, and ^{126}Xe in chondrites based on ^{81}Kr-Kr exposure ages. *Geochim. Cosmochim. Acta* 52, 1649-1662.

Funk H., Baur H., Frick U., Schultz L., and Signer P., 1971. On the diffusion of He in quartz. *Zeitschrift fur Kristallographie* 133, 225–233.

Gerling E.K., 1940. About helium solubility in melts. *Dokl. Akad. Nauk SSSR.* 27, 23 – 24.

Gerling E.K., Pushkarev Yu.D., and Kotov N.V., 1965. Behaviour of some minerals under heating and increased Ar pressure. *Izv. Akad. Nauk SSSR.* 11, 3 – 13.

Gorshkov G. V., Zjabkin V. A., Ljatkovskaya N. M., and Tsvetkov O. S., 1966. *Natural neutron background of atmosphere and Earth's crust.* Atomizdat, Moscow, pp. 410 (in Russian).

Gradstein F. M., Ogg J. G., Smith A. G., Bleeker W., and Lourens L. J., 2004. A new geologic time scale, with special reference to Precambrian and Neogene. *Episodes* 27, 83-100.

Gannibal M. A., 2012. Equilibrium partitioning of helium between rock and pore water: new approach for dating of old ground waters. *Geokhimiya* 1, 1-12 (in Russian).

Ikorsky S. V. and Kusch V. D. (1992) Nitrogen in fluid inclusions of alkaline rocks of Khibina massif and method of its chromatographic determination. *Geochimiya* 7, 962-970 (in Russian).

Ivanov, V.V., Medovy, V.I., Dobrovolskaya, V. I., 1978. Fields of helium concentrations in sedimentary sequences. *Geol. Rev*. 21, 891-903.

Jambon A. and Shelby J., 1980. Helium diffusion and solubility in obsidians and basaltic glass in the range 200-300°C. *Earth Planet. Sci. Lett.* 51, 206-214.

Kalashnikov E., Tolstikhin I. N., Lehmann B. E., and Pevzner B. Z., 2003. Helium transport along lattice channels in crystalline quartz. *J. Phys. Chem. Solids* 64, 2293-2300.

Kamensky I. L., Tolstikhin I. N., 1992. High He-3/He-4 ratios in diamonds - Constraints on alluvium age. *Geochimiya* 4, 561-569.

Kamensky I.L., Tolstikhin I.N., Sharkov I.V., Pushkarev Y.D., 1984. The first results of measuring helium isotopic composition on single-cascade mass-spectrometer MI-1201. *Geochimiya* 3, 439-443.

Karpinskaya T.B, Shanin L.L., and Borisevich and Borisevich I.V., 1965. Artificial impregnation of Ar into mica, olivine and pyroxene. *Izv. Akad. Nauk SSSR* 11, 14 –16.

Kipfer R., W. Aeschbach-Hertig Peeters F., Stute M. 2002. Noble gases in lakes and ground waters. *Noble gases in geochemistry and cosmochemistry*. D. Porcelli, Ballentine, C.J., Wieler, R. (editors) Washington, Mineral. Soc. Amer. 47: 615-700.

Lehmann B.E., Davis S.N., Fabryka-Martin J.T., 1992. Atmospheric and subsurface sources of stable and radioactive nuclides used for groudwater dating. *Water Resource Res.* 29, 2027-2040.

Lehmann B.E., Love A., Purtschert R., Collon P., Loosli H.H., Kutschera W., Beyerle U., Aeschbach-Hertig W., Kipfer R., Frape S.K., Herczeg A., Moran J., Tolstikhin I.N., Groning M., 2003a. A comparison of ground water dating with Kr-81, Cl-36 and He-4 in four wells of the Great Artesian Basin, Australia. *Earth Planet. Sci. Lett.* 211, 237-250.

Lehmann B. E., Waber H. N., Tolstikhin I. N., Kamensky I. L., Gannibal M., Kalashnikov E., Pevzner B., 2003b. Helium in solubility equilibrium with quartz and porefluids in rocks - A new approach in hydrology. *Geophys. Res. Lett.* 30, 1128-1132.

Mamyrin B.A., Tolstikhin I.N., 1984. *Helium isotopes in nature*. In: Developments in Geochemistry, Vol. 3. Elsevier Science Publisher, Amsterdam, pp. 273.

Marty, B., Criaud, A., Fouillac, C., 1988. Low enthalpy geothermal fluids from the Paris sedimentary basin -1. Characteristic and origin of gases. *Geothermics* 17, 619-633.

Marty B., Torgersen T., Meynier V., O'Nions R.K. and DeMarsily G., 1993. Helium isotope fluxes and ground water ages in the Dogger aquifer, Paris Basin. *Water Resources Research* 29, 1025-1035.

Matter A., Peters T.J., Blaesi H.-R., Meyer J., and Ischi H., 1988. Sondierbohrung Weiach, Geologie; Nagra Technical Report NTB 86-01.

Mazor E., 1972. Paleotemperatures and other hydrological parameters deducted from noble gases dissolved in ground waters; Jordan Rift Valley, Israel. *Geochim. Cosmochim. Acta* 36, 1321-1336.

Mazor E., 1991. Applied chemical and isotopic ground water hydrology. *Open Univrsity Press,* Milton Keynes, pp. 274.

Morrison P. and Pine J., 1955. Radoigenic origin of the helium isotopes in rock. *Annals New York Acad. Sci.* 62, 69-92.

Mukhopadhyay S., Farley K.A., 2006. New insights into the carrier phase(s) of extraterrestrial ^{3}He in geologically old sediments. *Geochim. Cosmochim. Acta* 70, 5061–5073.

Nagra, 1989. Sondierbohrung Weiach, Untersuchungsbericht. *Nagra Technischer Bericht* NTB 88-08, Nagra, Wettingen, Schweiz.

Niedermann, S., 2002. Cosmic-Ray-Produced Noble Gases in Terrestrial Rocks: Dating Tools for Surface Processes. In: Porcelli, D., Ballentine, C.J., Wieler, R. (Eds), *Noble Gases in Geochemistry and Cosmochemistry.* Reviews in Mineralogy and Geochemistry, 47, Geol.Soc. & Min Soc. of America, Washington D.C., USA, pp. 731-784.

Osenbrück K., Lippmann J. and Sonntag C., 1998. Dating very old pore waters in impermeable rocks by noble gas isotopes. *Geochim. Cosmochim. Acta* 62, 3041-3045.

Patterson D.B., Farley K.A. and Schmitz B., 1998. Preservation of extraterrestrial ^{3}He in 480-Ma-old marine limestones. *Earth Planet. Sci. Lett.* 163, 315-325.

Pearson, F.J., Balderer, W., Loosli, H.H., Lehmann, B.E., Matter, A., Peters, T., Schmassmann, H., Gautschi, A., 1991. *Applied isotope hydrogeology- A case study in Northern Switzerland.* Elsevier Science Publisher, Amsterdam, pp. 439

Pekala M., Kramers J. D., Waber H. N., Gimmi T., and Alt-Epping P., 2009. Transport of U-234 in the Opalinus Clay on centimetre to decimetre scales. *Applied Geochemistry* 24, 138-152.

*Plummer L. N., 2005. Dating of young ground water. In *Isotopes in the Water Cycle: Past, Present and Future of a Developing Science* (ed. P. K. Aggarwal, J. R. Gat, and K. F. O. Froehlich), Springer. pp. 193-220.

Roselieb K., Blanc P., Buttner H., Jambon A., Rammensee W., Rosenhaljer M., Vielzeuf D. and Walter H., 1997. Experimental study of argon sorption in quartz: Evidence for argon incompatibility. *Geochim. Cosmochim. Acta* 61, 533-542.

Rübel A.P., Sonntag C., Lippmann J., Pearson F.J. and Gautchi A., 2002. Solute transport in formations of very low permeability: Profiles of stable isotope and dissolved noble gas contents of porewater in the Opalinus Clay, Mont Terri, Switzerland. *Geochim. Cosmochim. Acta* 66, 1311-1321.

Savchenko V.P., 1935. The problems of geochemistry of helium. *Natural gases* 9, 53-197 (in Russian).

Shelby J. E., 1971. Diffusion of helium isotopes in vitreous silica. *Phys. Rev* 4, 2681 -2686.

Shelby J. E., Keeton S.C. and Ianucci J.J., 1976. Effect of gas composition on the pressure dependence of helium solubility in vitreous silica. *Journ. Appl. Phys.* 47, 3952 – 3955.

Sheldon A. L., Solomon D. K., Poreda R. J. and Hunt A., 2003. Radiogenic helium in shallow ground water within a clay till, southwestern Ontario. *Water Resour. Res.* 39, 1331, doi:10.1029/2002WR001797.

Smyslov A. A., 1974. *Uranium and Thorium in the Earth crust*. Leningrad, Nedra, pp. 231.

Solomon D. K., 1996. Source of radiogenic helium-4 in shallow aquifers: Implications for dating young ground water. *Water Res. Res.* 32, 1805-1813.

Stute M., Sonntag C., Deak J. and Schlosser P., 1992. Helium in deep circulating ground water in the Great Hungarian Plain: Flow dynamics and crustal and mantle helium fluxes. *Geochim. Cosmochim. Acta* 56, 2051-2067.

Tolstikhin, I.N., Lehmann, B.E., Loosli, H.H., Gautschi, A., 1996. Helium and argon isotopes in rocks, minerals and related ground waters: A case study in Northern Switzerland. *Geochim. Cosmochim. Acta* 60, 1497-1514.

Tolstikhin I. N., Lehmann B. E., Loosli H. H., Kamensky I. L., Nivin V. A., Orlov S. P., Ploschansky L. M., Tokarev I. V. and Gannibal M. A., 1999. Radiogenic helium isotope fractionation: The role of tritium as ^{3}He precursor: geochemical applications. *Geochim. Cosmochim. Acta* 63, 1605-1611.

Tolstikhin I.N., Gannibal M., Tarakanov S., Pevzner B., Lehmann B.E., Ihly B. and Waber H.N., 2005. Helium transfer from water into quartz crystals: A new approach for porewater dating. *Earth Planet. Sci. Lett.* 238, 31-41.

Tolstikhin I., Waber H.N., Kamensky I., Loosli H.H., Skiba V. and Gannibal M., 2011. Production, redistribution and loss of helium and argon isotopes in a thick sedimentary aquitard-aquifer system (Molasse Basin, Switzerland). *Chemical Geology*, 286, 48-58.

Top Z., Eismont W. C. and Clarke, W.B., 1987. Helium isotope effect and solubility of helium and neon in distilled water and seawater. *Deep-Sea Res.* 34, 1139-1148.

Torgersen T., 1989. Terrestrial helium degassing fluxes and the atmospheric helium budget: Implications with respect to the degassing process of continental crust. *Chem. Geol. (Isot. Geosci. Sec.)* 79, 1-14.

Torgersen T. and Clarke W.B., 1985. Helium accumulation in ground water. I: An evaluation of sources and the continental flux of ^{4}He in the Great Artesian Basin, Australia. *Geochim. Cosmochim. Acta* 49, 1211-1218.

Torgersen T. and Clarke W.B., 1987. Helium accumulation in ground water. III: Limits on helium transfer across the mantle-crust boundary beneath Australia and the magnitude of mantle degassing. *Earth Planet. Sci. Lett.* 84, 345-355.

Torgersen T., Habermehl M.A. and Clarke W.B., 1992. Crustal helium fluxes and heat flow in the Great Artesian Basin, Australia. Chem. Geol. (Isot. Geosci. Sec.) 102, 139-152.

Trull T.W., Kurz M.D. and Jenkins W.J., 1991. Diffusion of cosmogenic ^{3}He in olivine and quartz: Implications for surface exposure dating. *Earth Planet. Sci. Lett.* 101, 241-256.

Watson E. B. and Cherniak D. J., 2003. Lattice diffusion of Ar in quartz, with constraints on Ar solubility and evidence of nanopores. *Geochim. Cosmochim. Acta* 67, 2043–2062.

Weber H. P., Sattel, G. and Sprecher C., 1986. Sondierbohrungen Weiach, Riniken, Schafisheim, Kaisten, Leuggern - geophysikalishe daten. *Technical Report NTB-85-50.* Wettingen, NAGRA.

Weiss R.F., 1971. Solubility of helium and neon in water and seawater, *J. of Chem. and Engin. Data* 16, 235-241.

Yakutseny V.P., 1968. *Geology of helium*, Nedra, Leningrad, pp 232. (in Russian).

In: Sandstone
Editor: Afsoon Moatari Kazerouni

ISBN: 978-1-62948-382-5

Chapter 2

WATER IMBIBITION INTO SANDSTONES: INFLUENCE OF FLOW RATE ON WATER DISTRIBUTION AND ACOUSTIC RESPONSE

***Sofia Lopes*[1,2*]*, Maxim Lebedev*[1,3] *and Tobias M. Müller*[2]

[1]Curtin University, Department of Exploration Geophysics, Australia
[2]CSIRO, Earth Science and Resource Engineering, Australia
[3]CO2CRC, Curtin University, Australia

ABSTRACT

Fluid injection into a porous medium in order to displace another fluid is a common procedure while dealing with the recovery of subsurface fluids (e.g., secondary waterflooding, oil recovery). The seismic method has been widely used to image underground structures and, consequently, map reservoirs. More specifically, the time-lapse method consists in acquiring successive 3D seismic data of the same area over a period of time using acoustic sensors arrays. The aim of our experimental work is to observe changes in seismic signals resulting from changes in the acoustic impedance. Since the acoustic impedance is the product of velocity and density, time-lapse signals are affected by the compressibilities of the reservoir rock and of pore fluids. This acoustic monitoring is also the basis to track saturation front displacements. The interaction of seismic waves with the fluid-rock system is complex and complicates the interpretation of time-lapse signals. In order to better understand this interaction, experiments at laboratory scale were performed. Distilled water is injected into a dry sandstone sample (from the Otway Basin, South Australia, Australia), porosity: 19.5%; permeability: 25.0 mD. Ultrasonic P-wave velocities (V_P) across the sample are calculated using the first-break pick of the output signal recorded by an oscilloscope. Simultaneously, the water displacement is visualized using X-ray Computed Tomography (CT). The CT scans are maps of the fluid distribution from which the water saturation (S_W) is estimated. The evolution of V_P and S_W with the injection of water at low and high injection rates (LIR and HIR, respectively) compared to the natural imbibition (NI) rate is studied. It is also

* Corresponding author: s.correialopes@postgrad.curtin.edu.au; GPO Box U1987, Perth WA 6845, Australia; Tel.: +61 8 9266 3519; Fax.: +61 8 9266 3407.

investigated how flow rate influences the displacement and geometry of the saturation front. A clear dependence of the evolution of V_P and S_W with the flow rate is experimentally observed. The saturation front passing the ultrasonically monitored position is associated with a characteristic change in V_P. There is a sudden increase in V_P, sharper for the HIR but still noticeable for the LIR. The HIR promotes a flat, compact saturation front and a fast increase in V_P and S_W while the LIR promotes a diffusive saturation front and both V_P and S_W increase slowly as water is injected. NI reveals a pattern on the evolution of V_P and S_W similar to the HIR, sustaining a flat saturation front. For the NI there is a continuous decrease of the imbibition rate revealing that at a certain point gravity forces overcome capillary forces. A set of experiments is also performed where the injection rate is increased and decreased within a single injection. Once more, a strong influence of the injection rate on the evolution of V_P and S_W is spotted. Decreasing (increasing) the injection rate directly decreases (increases) V_P and S_W. Another important result is that both experiments (at constant and variable injection rates) show that the same S_W is associated to very different V_P. Since the same S_W can relate to different fluid distributions, this result reinforces that acoustic waves are not only sensitive to the amount of fluid present in the pore space, but also how it is distributed within the sample.

Keywords: Sandstone, imbibition, acoustic monitoring, x-ray computed tomography, rock physics

1. Introduction

The injection of a fluid into a porous medium in order to displace another fluid is a common procedure when dealing with the recovery of subsurface fluids (for example, secondary oil recovery, waterflooding, CO_2 sequestration). The seismic method has been widely used to image underground structures and, consequently, map reservoirs. A more recent technique is the time-lapse or 4D seismic method providing information on changes of fluid contents over time (Calvert 2005). The time-lapse seismic method comprises of acquiring successive 3D seismic data surveys of the same area over a period of time using acoustic sensors arrays. The recorded seismic data is the basis to characterize reservoirs in terms of their shape (thickness, width and dip) and properties such as porosity, pore fluid and degree of saturation (Bjorlykke 2010).

The aim of the time-lapse seismic method is to observe changes in seismic signals resulting from changes in the acoustic impedance. It is a method that remotely analyzes the movement of reservoir fluids and might be used to track fluid fronts. The acoustic impedance is the product of velocity and density. Therefore, time-lapse signals are affected by the compressibilities of the reservoir rock and of the pore fluids. For example, soft rocks, such as sandstones, will generate higher impedance contrasts than stiff rocks such as carbonates. Fluids with strong density contrasts like gas-oil or gas-water will have distinguishable signals while heavy oils flooded by water may be very challenging to monitor seismically (Lumley 2001).

Imbibition, the displacement of a non-wetting fluid by a wetting fluid, is driven by surface energy through the action of capillary pressure (Akin et al., 2000; Morrow & Mason 2001) and particularly in a vertical-directed injection is the result of the balance between capillary forces (upward) and gravity forces (downward, inhibiting the imbition). Natural or spontaneous imbibition is the process by which the wetting fluid is drawn into the pore space by capillary action (suction). In a dynamic forced imbibition, the rate is imposed by means of an injection pump and the non-wetting fluid is displaced into the porous medium under a pressure gradient.

When dealing with the injection of fluids to displace another fluid in a porous medium the main goal is to maximize displacement efficiency, i.e., maximize the amount of residual fluid that can be expelled. It is a challenging problem as the effectiveness of the imbibition depends significantly on the nature of the porous medium (porosity, permeability and pore structure) and of the fluid system (viscosity of each fluid and the viscosity ratio between fluids) (Sudaryanto & Yortsos 2000; Li & Horne 2010) and on the rate of the injection (Rapoport & Leas 1953; Babadagli & Ershaghi 1992). The flow rate has a strong impact on immiscible-phase displacement: at high flow rates the saturation front is sharp with a straight piston-like displacement while at low flow rates capillarity controls the distribution of the fluid in the micropore-space causing a diffusive broadening of the saturation front (Blunt & Scher 1995; Meleán, Broseta & Blossey 2003). Thus, understanding the impact of flow rate on fluid distribution and its associated acoustic response may have significant implications in reservoir surveillance.

There are a considerable number of variables that influence the propagation of acoustic waves: 1) the rock type (mainly its mineralogical composition); 2) the type of pores (predominance of micro- or macro-porosity, closed or open pores, etc.); 3) the type of fluids present in the pore space (for example, gas-water or oil-water systems) (Lumley 2001); 4) fluid distribution (Cadoret, Marion & Zinszner 1995); and 5) pressure (Guéguen & Palciauskas 1994). There is also a significant dependence of the acoustic response with the injection rate (Meleán, Broseta & Blossey 2003; Karimaie & Torsaeter 2007).

The most influential forces governing fluid flow are viscous, gravity and capillary forces as well as the applied external pressure (Leverett 1940; Karimaie & Torsaeter 2007; Riaz et al., 2007). The relative importance of capillary, viscous and gravitational forces will determine the overall flow characteristic. Capillary and gravity forces are the governing forces for water imbition. When injecting water in an air-saturated system, viscous forces are negligible as the residual air offers virtually no resistance to the water flow. Viscous forces are only relevant if the viscosity of the injecting fluid is comparable to the viscosity of the displaced fluid. For example, when water is injected in an oil-saturated system, viscous forces govern the fluid flow (Akin et al., 2000). Gravity forces are caused by differences in the density of the fluids. The influence of gravity increases with the increase in density contrast between the fluids (for example, water/air). More particularly, gravity forces tend to stabilize the interface between fluids and inhibit fluid displacement (Birovlej et al., 1991). Capillary forces are present in multiphase flow and they are always active at the interface between two immiscible fluid phases (Skjæveland & Kleppe 1992). In water imbibition experiments, capillary forces control the microscopic fluid distribution of water-driven displacements. Changes in fluid distribution and pore fluid pressure are translated into changes in acoustic wave propagation properties reflecting solid/fluid interactions at the macroscopic level. Cadoret, Marion & Zinszner (1995), Rangel-German & Kovscek (2002), Meleán, Broseta & Blossey (2003) and Karimaie & Torsaeter (2007) performed fluid injection experiments where specific velocity-saturation relationships were established within a variety of injection rates. All these relationships relate to single, constant-injection-rate experiments. However, the procedure of changing injection rates during a continuous fluid injection is a common procedure to control pore pressure and avoid reservoir depletion (Teufel, Rhett & Farrell 1991; Yousef et al., 2006). Lopes & Lebedev (2012) performed laboratory experiments where the injection rate was decreased and increased in a single, continuous experiment. It was experimentally observed that changing the injection rate has a major impact on the acoustic response and evolution of fluid saturation. For an extensive review on the influence of saturation and fluid distribution on seismic wave velocities refer to Müller, Gurevich & Lebedev (2010).

X-ray Computed Tomography (CT) has been widely used to investigate fluid flow patterns and characterize pore-space (quantification and interconnectivity of pores). It adds a new dimension to the understanding of solid/fluid interactions through the direct visualization of fluid distribution. X-ray CT is a non-destructive method that allows 2D (and 3D) visualization by creating images that map the variations in X-ray attenuation derived from the different densities of the scanned object (Ketcham & Carlson 2001). The raw data provided by the X-ray CT scanner is converted into "CT numbers" (or "CT values", given in Hounsfield units, HU) which threshold is set by the computer system. The CT numbers are mapped and the resulting images are presented in a gray-scale: shaded gray (or black) for low CT numbers and light gray (or white) for high CT numbers. As the density changes with the intrusion of water, the CT number of each pixel changes. So, X-ray CT scanning provides us with a visual display of fluid distribution in the rock and the possibility of tracking its displacement with time. Garg, Zwahlen & Patzek (1996) performed water imbibition into a dry sandstone and the saturation front was imaged through X-ray CT. A positive correlation between water imbibition and weight gain of the sample was established. Meleán, Broseta & Blossey (2003) used X-ray CT to analyse the dispersion of saturation fronts in oil saturated samples. Distilled water mixed with a dopant was injected at various injection rates and at different initial wetting fluid saturations. At low velocities, the saturation front was distinctively dispersive. Schembre & Kovscek (2003) performed spontaneous water imbibition in sandstones. From the CT scans, saturation profiles were acquired and used to obtain relative permeabilities for both phases. Tang & Kovscek (2011) injected water into an oil-saturated sandstone at various injection rates. X-ray CT scans imaged the fluid displacement and the growth of fingering in the water/oil interface.

The aforementioned works provide insight into solid/fluid interactions based on X-ray CT scanning but no acoustic waveforms were acquired during imbibition. However, matching the acoustic response with the evolution of fluid saturation provides additional means to analyse solid/fluid interactions occurring during imbibition. X-ray CT scanning can complement the acoustic data with an image of the saturation front and its geometry. Though the complementary information given by this type of laboratory experiment has the potential to provide new insights, not many experiments have been reported. Monsen & Johnstad (2005) developed an acoustic sample holder that was transparent to X-rays. 3D images were reconstructed from the scans and velocity-saturation relationships were established thus linking qualitative and quantitative measurements. Lebedev et al., (2009) performed forced water imbibition into a sandstone at different injection rates. Compressional (P) and shear (S) wave velocities were recorded while the sample was scanned. Velocity-saturation relationships were found and complemented with fluid distribution images. Though these experimental works combine acoustic response with evolution of water saturation, the data set is quite minute. Moreover, the X-ray CT scans were radial and the geometry of the saturation front could not be visualized. More recently, Lopes & Lebedev (2012) acquired data of P-wave velocities and water saturation with axial CT scanning such that the evolution of the saturation front could be imaged. A characteristic evolution of acoustic velocities and water saturation as well as distinct saturation front geometries were observed for different injection rates. Thus, it was experimentally confirmed that fluid flow regimes and associated acoustic responses depend on the injection rate.

This chapter continues and extends the experimental work of Lopes & Lebedev (2012). The ultrasonic P-wave velocity, water saturation and saturation front displacement are monitored at low and high injection rates (forced imbibition by the means of an injection pump). It is specifically intended to distinguish the evolution of all three measured quantities for each injection rate and comprare it to when natural imbibition is performed on the same sample. The comparison with the natural imbibition rate is important as it provides essential information for evaluating the displacement efficiency of a reservoir (Sudaryanto & Yortsos 2000).

This chapter is organized as follows: first, the setup and the methodology (simultaneous ultrasound monitoring and X-ray CT scanning) for the forced water imbibition and natural imbibition experiments are explained. The "Results" section presents: (1) the temporal evolution of the acoustic wave velocities and water saturation at different injection rates, (2) the CT scans depicting the displacement and geometry of the saturation front for each injection rate and (3) the results of the variable injection rate experiment. In the discussion, the evolution of P-wave velocities and water saturation and the displacement and geometry of the saturation front are related to changes in the pore-scale. The final conclusion pinpoints the most important implications of our experimental results on the flow rate dependence of velocity-saturation relationships.

2. Methods

Forced and natural imbibition of distilled water is performed into a reservoir sandstone (Otway Basin, South Australia, Australia) with the intention of spotting differences in fluid distribution (evolution of water saturation and displacement and geometry of the saturation front) and acoustic response (evolution of ultrasonic P-wave velocities) at different injection rates. All experiments are performed using the same sample whose main petrophysical properties are shown in Table 1.

The porosity and permeability of the sample are measured using helium gas (Automated Porosimeter Permeameter, *Coretest Inc. model AP-608*). The sample was cored from its original block and left to dry on open-air; no other treatment (thermal or pressure) was done before the experiments. The outer surface of the cylindrical sample is covered by epoxy, thus enforcing a unidirectional flow. Two plastic rings are glued to the top and bottom of the sample. Each ring has a hole in the centre. The hole at the top ensures direct contact with atmospheric pressure, avoiding pressure build up as the water is injected. The bottom hole is connected to the injection pump. The experimental setup consists of an injection pump (*Shimadzu LC-20AT*), where the injection rate is set; a pulser/receiver (*Olympus 5077-PR*), where the acoustic impulse is triggered manually; a pair of P-wave piezoelectric transducers (*Olympus D7207*); and an oscilloscope (*Agilent DSO-3202-A*), where the acoustic waveform is visualized (Figure 1).

Table 1. Main petrophysical characteristics of the Otway Sandstone

OTWAY SANDSTONE	
Length (cm)	6.33
Diameter (cm)	3.80
Porosity (%)	19.5
Pore Volume (mL)	14.0
Permeability (mD)	25.0

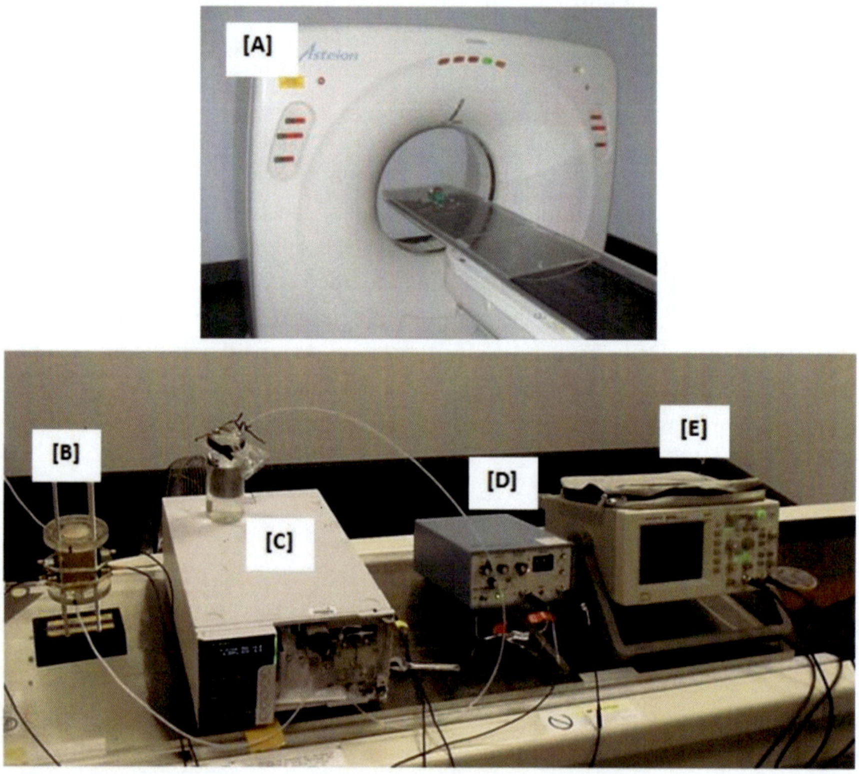

Figure 1. Experimental setup: [A] medical X-ray CT scanner; [B] sample set for vertical injection; [C] injection pump; [D] pulser/receiver and [E] oscilloscope.

During each imbibition, P-waves and CT scans are acquired simultaneously (Figures 2 and 3). Through this procedure a P-wave velocity is matched with a water saturation level and related to the position of the saturation front.

P-waveforms are acquired using P-wave transducers with 1.5 cm of diameter and acquisition frequency of 1 MHz. These transducers are always located in the same position (Figure 2). P-wave velocities are calculated using the first-break picks of the output signal recorded by the oscilloscope. The distance travelled by the P-wave is the distance between transducers which is always the same as the transducers are fixed to the walls of the sample with glue. The delay in arrival time is calibrated using an aluminium rod, which P-wave velocity is well known. The experimental error in calculating the P-wave velocity, V_P, $\sigma(V_P)$ is given by Equation 1 (Taylor 1997).

$$\sigma(V_P) = V_P.\sqrt{\left(\frac{\sigma(D)}{D}\right)^2 + \left(\frac{\sigma(t_P)}{t_P}\right)^2}. \tag{1}$$

D and $\sigma(D)$ are the distance travelled by the P-wave (the distance between transducers) and the error in measuring D, respectively. $\sigma(D)$ is constant for all measurements as the distance, D, between the transducers is always the same. t_P is the P-wave arrival time for each P-wave recorded and $\sigma(t_P)$ is the error in measuring t_P. When applying Equation 1 to our data, the average error in V_P results in 5 m/s, reflecting the high precision of our measurements.

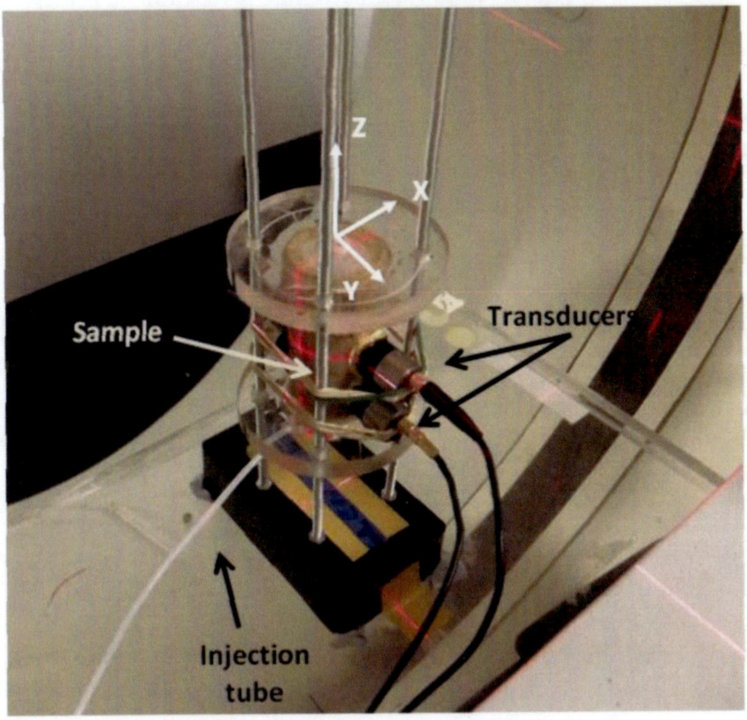

Figure 2. Sample with two pairs of transducers glued to the walls of the sample, set for vertical injection. Only the pair closest to the point of injection was actively monitored. The metallic frame holds the sample in a fixed position.

The CT scans provide maps of the fluid distribution from which the water saturation is estimated. The sample is scanned along the same axial section (Figure 3) so the fluid distribution and geometry of the fluid front can be compared over the same scanning plane.

Our study of solid/fluid interactions is optimized by injecting solely water in a dry sample. In this way, a higher contrast in the air- and water-filled areas is obtained and dry and wet zones are distinguished by direct observation of the CT scans. Another reason for choosing water injection in dry samples relates to the strength of the acoustic signatures. P-wave arrival times between dry and water-filled areas are much more discernible than the ones between water- and oil-saturated areas, for example. Moreover, no chemicals are added to the injected water to increase X-ray absorption. With this procedure, several imbibitions are performed on the same sample assuring that there is no structural damage (or chemical erosions) between consecutive experiments.

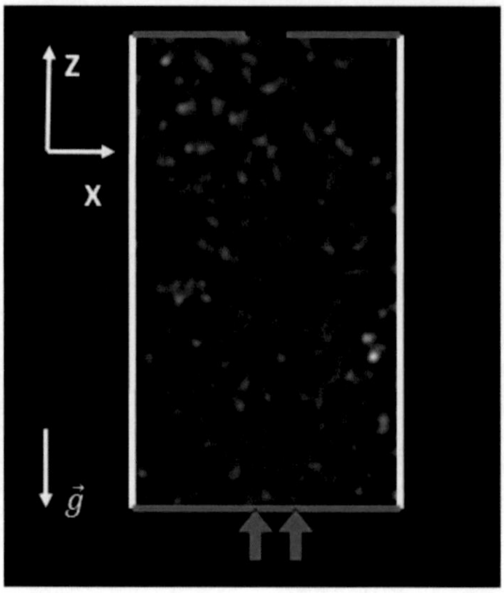

Figure 3. CT scan of the dry sample. Dashed red circle represent the area covered by the transducers in the scanning XZ plane. The sample is covered laterally by epoxy (white lines). Blue lines represent the two plastic rings: one glued to top with an open end and the other glued to the bottom with a connection to the injection pump. Blue arrows indicate the direction of injection opposite to gravity, $\vec{g}$.

Three sets of experiments are performed at room temperature and with no axial or confining pressure applied to the sample. Distilled water imbibes in a dry sample in the direction against gravity (Z-direction, Figure 3):

1. At constant injection rate.

The rates of 2.0 and 0.5 mL/h are chosen. Based on our experience dealing with fluid flow at laboratory scale, 2.0 mL/h is a high injection rate for Otway sandstones, and 0.5 mL/h is a low injection rate (fast and slow imbibition of water, respectively). Between each experiment the sample is dried in an oven, on vacuum at 50°C for two days and then left on vacuum (without heating) for a week. One day before the experiment, the sample is left outside the oven to equilibrate with room conditions.

2. Through natural imbibition.

After the forced imbibitions at constant injection rate, the sample dried following the same drying procedure in (1.) and was prepared for natural imbibition. The sample was placed on top of a water surface allowing a constant natural uptake of water through the bottom hole.

3. At variable injection rate.

The forced imbibition starts at 2.0 mL/h and is decreased to 0.1 mL/h after approximately 3 hours. The low injection rate is kept for almost 24 hours and then increased back to 2.0 mL/h until the end of the experiment.

2.1. CT Image Processing

By performing X- Computed Tomography (CT), the results of the ultrasonic acoustic monitoring is complemented with a dimensional image representing the evolution of the fluid distribution in the rock. X-ray CT is a non-destructive method that allows 2D (and 3D) visualization by creating images that map the variations in X-ray attenuation derived from the different densities of the scanned object. Most X-ray CT scanners consist of a single X-ray source and a ring of detectors that rotate around the scanned object. The X-ray source emits a beam of photons and the set of detectors measure the intensity of the received X-ray. The sent signal is attenuated by scattering, while passing the scanned object, and absorption. The level of attenuation for a monoenergetic beam and a homogenous material is given by Beer's Law (Equation 2) (Wellington & Vinegar 1987).

$$I = I_0 \exp(\mu x). \tag{2}$$

I_0 is the initial X-ray intensity, I is the X-ray intensity measured by the detectors, μ is the linear attenuation coefficient for the scanned material and x is the length of the X-ray path through the scanned material. For a heterogeneous material, for each material i, with attenuation coefficient μ_i and over a length of x_i, the X-ray intensity is given by Equation 3.

$$I = I_0 \exp[\textstyle\sum_i(-\mu_i x_i)]. \tag{3}$$

The raw data provided by the X-ray CT scanner is converted into "CT numbers" (in Hounsfield units, HU) which threshold is set by the computer system. These CT numbers are mapped and the resulting scans are presented in a gray-scale: shaded gray (or black) for low CT numbers and light gray (or white) for high CT numbers. The CT number of each pixel of the scans is approximately equal to the average of the CT numbers related to the mineral grains, pore-spaces and pore-fluids (Toms-Stewart et al., 2009). In this case, each CT scan contains 512 x 512 pixels (one voxel = 0.2 mm x 0.2 mm). Each scanning covers 1 mm of the sample's thickness over 3 seconds.

For a solid homogeneous matrix, the CT number of the pixel is equal to the CT number of the grains (see Equation 4).

$$CT = CT_{matrix} = CT_{grain}. \tag{4}$$

In the presence of pores (heterogeneous material with porosity ϕ), CT of the matrix, CT_{matrix}, is given by Equation 5.

$$CT_{matrix} = (1 - \phi)CT_{grain}. \tag{5}$$

If the pores are filled with a fluid, Equation 5 becomes Equation 6.

$$CT_{matrix} = (1 - \phi)CT_{grain} + \phi CT_{fluid\,1}. \tag{6}$$

If the pores are filled with two different fluids ($S_{fluid\,1}$ is the saturation of fluid 1), CT_{matrix} is given by Equation 7.

$$CT_{matrix} = (1-\phi)CT_{grain} + S_{fluid\,1}\phi CT_{fluid\,1} + \left(1-S_{fluid\,1}\right)\phi CT_{fluid\,2} \tag{7}$$

For the specific case of a porous rock, for example, *fluid 1* is water and *fluid 2* is air, ($S_{fluid\,1} = 0, CT_{fluid\,2} = CT_{air}$), the CT number for the sample when it is dry is given by Equation 8.

$$CT_{dry} = (1-\phi)CT_{grain} + \phi CT_{air}. \tag{8}$$

If this porous rock is saturated with water ($S_{fluid\,1} = S_w$), the CT_{S_w}number is given by Equation 9.

$$CT_{S_w} = (1-\phi)CT_{grain} + S_w\phi CT_w + (1-S_w)\phi CT_{air}. \tag{9}$$

Therefore, the variation in the CT number due to the presence of water is given by subtracting Equation 8 from Equation 9. The result is Equation 10.

$$CT_{S_w} - CT_{dry} = S_w\phi(CT_w - CT_{air}). \tag{10}$$

For most medical CT scanners, the CT number of air is "-1000" and the CT number of water is "0". Substituting in Equation 10 the variation of CT numbers between saturated sample and dry sample, ΔCT, is given by Equation 11.

$$\Delta CT = CT_{S_w} - CT_{dry} = S_w\phi 1000. \tag{11}$$

The water saturation, S_w, calculated by knowing the difference in CT numbers (ΔCT), is then given by Equation 12.

$$S_w = \frac{\Delta CT}{\phi 1000}. \tag{12}$$

Therefore, directly through the CT scans, solid/fluid interactions can be qualitatively evaluated (with the variation of the gray scale and by identifying areas of high or low CT numbers) and quantitatively (by calculating S_w). It is possible to plot the CT numbers with length of the sample at several moments of the experiment (Figure 4) and calculate ΔCT.

Digital manipulation of the raw CT data is made to enhance contrast between the dry areas and wet areas and delimitate their interface (i. e., the saturation front). In order to do so, the CT scan of the dry sample is subtracted from the consecutive CT scans of the partially water-saturated sample. The resulting image is a map that features the pixels which CT numbers have changed due to the presence of water (Figure 5).

The error in calculating Sw, $\sigma(S_W)$ is given by the sum of the relative errors in measuring the CT numbers (see Equation 13). Both the porosity, ϕ, and the factor "1000" are constants, so

$$\sigma(S_W) = \sigma\left(CT_{dry}\right) + \sigma(CT_w) = 2.\sigma(CT) \tag{13}$$

because $\sigma\left(CT_{dry}\right) = \sigma(CT_w)$. The relative error in measuring CT, $\sigma(CT)$, is given by the average deviations in the CT numbers, ΔCT, relative to the average CT value, $\overline{CT}$ (Equation 14). For our medical X-ray CT scanner, the settings are $\Delta CT{\sim}15$ and $\overline{CT}{\sim}1500$ so that

$$\sigma(CT) = \frac{\Delta CT}{\overline{CT}} = \frac{15}{1500} = 1\%. \tag{14}$$

Therefore, the error in estimating S_W is $\sigma(S_W) = 2 \times 1\% = 2\%$.

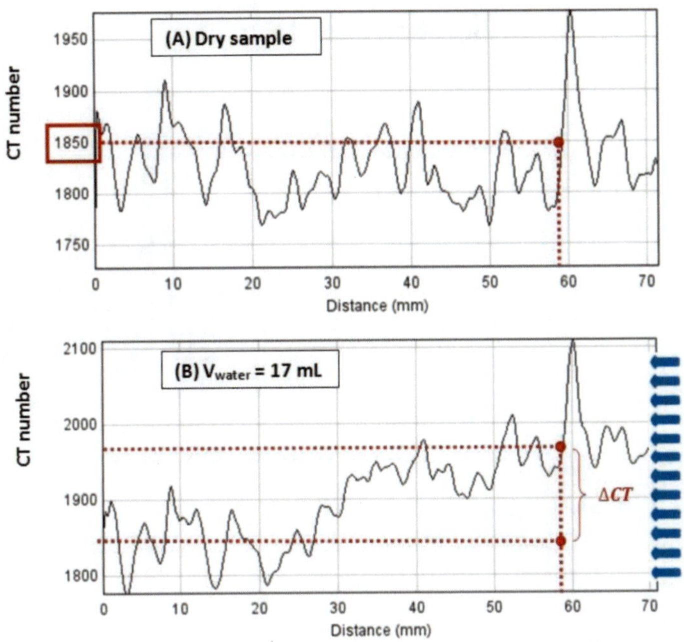

Figure 4. CT number profile for (A) dry sample and (B) after injecting 17 mL of water on a sandstone sample. The dots close to 60 mm mark the position of the P-wave transducers. Note the progressive increase of CT numbers with volume of water injected from right to left of the sample (ΔCT = 130 corresponding to S_W =76%). Blue arrows indicate direction of injection (against gravity).

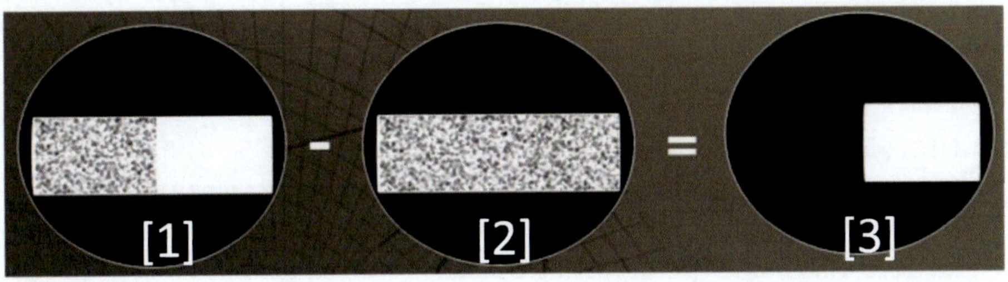

Figure 5. Digital manipulation of the raw CT scans: the CT scan of the dry sample [2] is subtracted from the CT scan of the partially water-saturated sample [1] resulting in a scan where the background is subtracted and only the areas where the CT number has changed with the presence of water are featured [3].

3. Results

3.1. Constant Injection Rate

3.1.1. High and Low Injection Rates

Figure 6 shows the waveforms of the P-waves recorded when the sample is dry and when 1 and 4 mL of water is injected at the injection rate of 2.0 mL/h. Within two hours of injection, the wave amplitude decreases considerably. Changes in the P-wave arrival time are also recorded. At 4 mL of injected water, the P-wave arrival time decreases significantly resulting in an increase of P-wave velocities of almost 170 m/s from when the sample is dry.

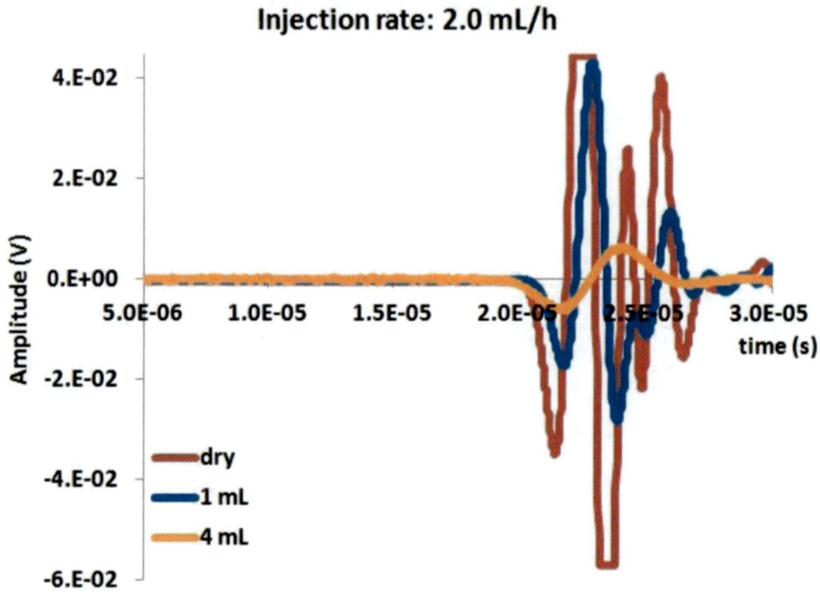

Figure 6. P-waveforms for dry sample, 1 and 4 mL of water injected at the injection rate of 2.0 mL/h. Note the change in wave amplitude and P-wave arrival times as water is injected.

Figure 7 shows the evolution of P-wave velocities, V_P, and water saturation, S_W, with the duration of injection for the high injection rate of 2.0 mL/h (HIR) (Figure 7-A), and for the low injection rate of 0.5 mL/h (LIR) (Figure 7-B). Note how both V_P and S_W evolve in the same manner for each fluid flow regime with the approaching of the saturation front: a sharp increase in the HIR and a much slower increase in LIR. In more detail:

1. HIR: after an initial decrease of V_P during the first hour, V_P increases 200 m/s within 2 hours (at a rate of 100 m/s/h) reaching the maximum value of 2190 m/s. This is followed by an almost constant evolution. S_W follows the same pattern. A sharp increase 1 hour after the beginning of the experiment is observed. S_W increases 60% within 3 hours. This sharp increase is then followed by a slow but steady increase, up to the maximum value of 66%.
2. LIR: the evolution of V_P and S_W is much more gradual if compared to the HIR case. V_P only starts to increase 4 hours after the beginning of the imbibition, at the slow

rate of 40 m/s/h. S_W increases to less than 50% within 8 hours. No CT scans were acquired during the last 2 hours of the imbibition due to technical issues but the imbibition was not stopped. S_W is expected to keep increasing continuously and slowly as seen for V_P.

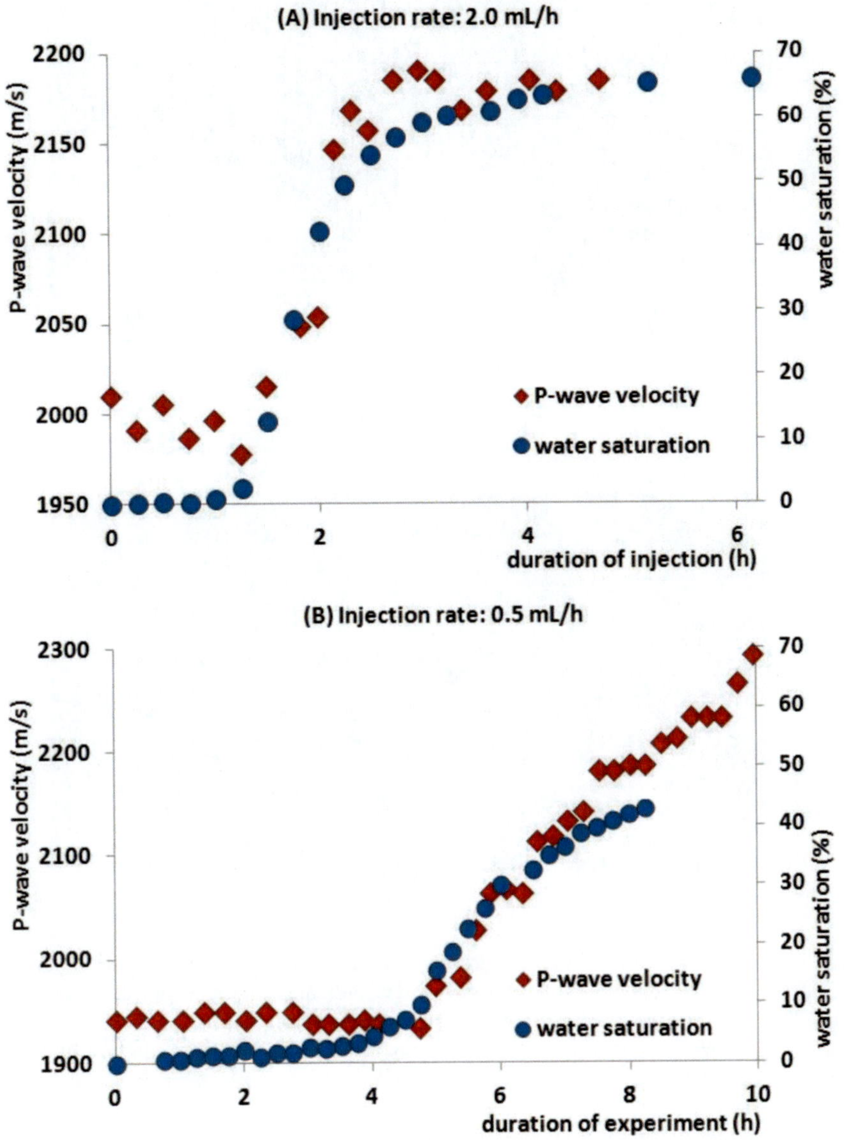

Figure 7. Evolution of P-wave velocities and water saturation with duration of injection for (A) high injection rate (HIR): 2.0 mL/h and (B) low injection rate (LIR): 0.5 mL/h. No CT scans were acquired after 8 hours in (B).

The evolution of V_P and S_W is intrinsically related to the advancing saturation front and its geometry. This can be verified by direct inspection of the CT scans. Figure 8 shows the advancement of the saturation front for 1, 2 and 4 mL of water injected during the HIR (Figure 8-A) and the LIR (Figure 8-B).

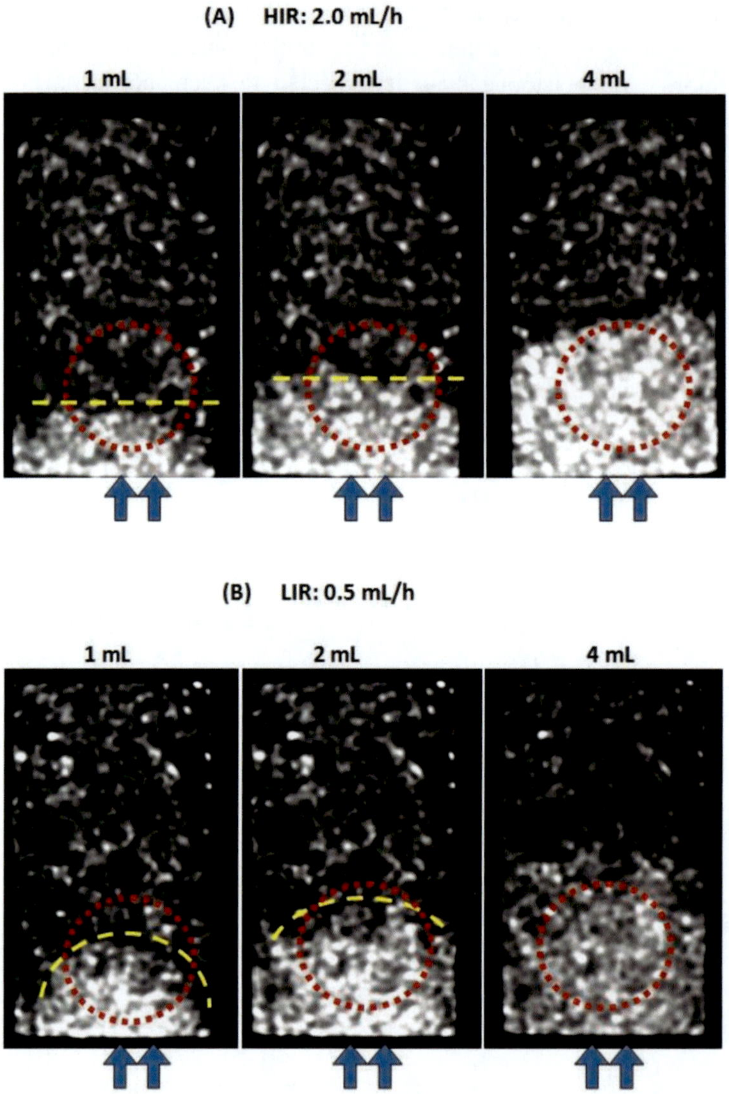

Figure 8. CT scans for 1, 2 and 4 mL of water injected for (A) high injection rate (HIR): 2 mL/h and (B) low injection rate (LIR): 0.5 mL/h. More "white" indicates higher water saturation. Blue arrows indicate direction of injection. Note in (A) a flat saturation front in all cases while in (B) there is a progressive increase in of curvature as more water is injected (long yellow dashed lines). Dashed red circles represent the area covered by the transducers in the scanning plane.

The HIR sustains an almost flat, compact saturation front throughout the imbibition (indicated by the yellow dashed line in Figure 8-A). Though there is an initial curvature of the saturation front promoted by the point injection (not shown), it soon becomes an almost flat saturation front. This initial curvature persists in the LIR for at least 4 hours. However, at 4 mL, the saturation front is considerably flatter (Figure 8-B).

The changes between the CT number at a certain amount of water injected and the CT number when the sample is dry (ΔCT) with the sample's height (Z-direction) is shown in Figure 9. In this way, specific patterns of fluid displacement are matched to specific injection rates.

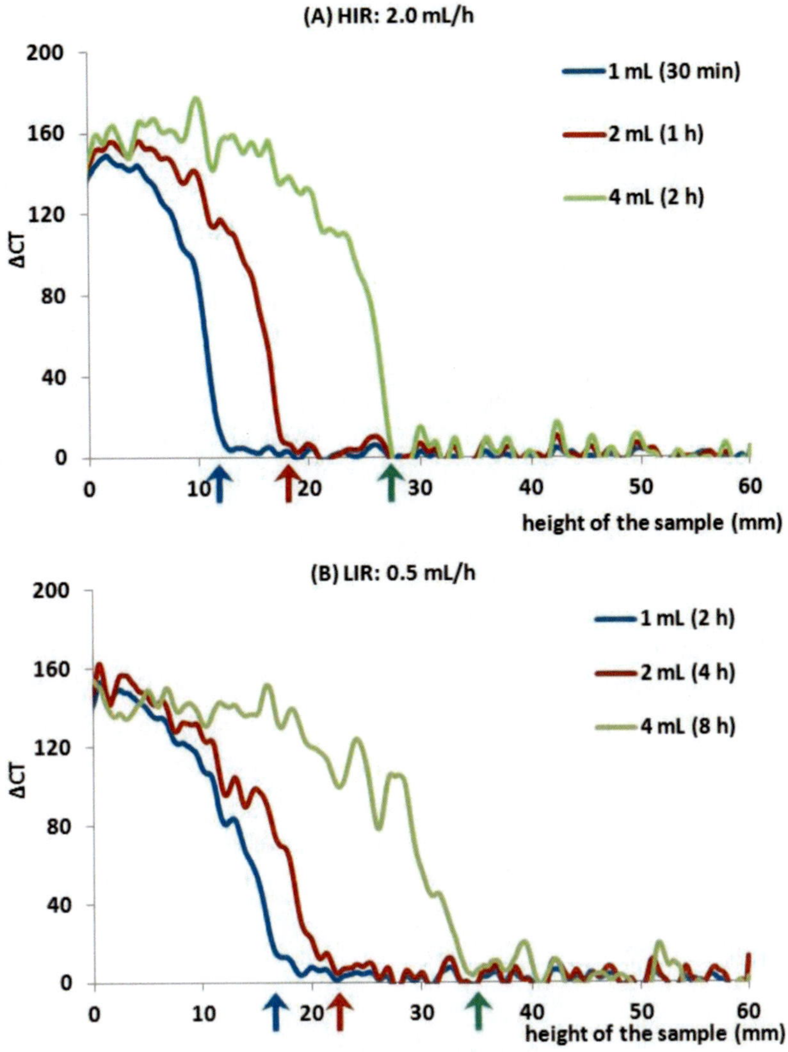

Figure 9. Variation of CT numbers, $\Delta CT = CT(w) - CT(dry)$, with height of the sample, for (A) high injection rate (HIR): 2 mL/h and (B) low injection rate (LIR): 0.5 mL/h. Note that for the same amount of volume of water injected, the saturation front is farther up for LIR (compare position of coloured arrows).

The following observations can be made:

1. The sharp decrease of ΔCT (down to zero) after 1, 2 and 4 mL of injected water at the HIR characterizes a steep, piston-like saturation front. Conversely, the slow decrease at LIR characterizes a more diffusive saturation front. Compare: for 1 mL of water injected, ΔCT = 160 is at 12 mm for HIR (Figure 9-A) while for LIR is at 28 mm (Figure 9-B);
2. The height of the saturation front tends to be farther up for the LIR as the imbibition progresses. For example, after injecting 4 mL of water, the saturation front is at 28 and 35 mm at HIR and LIR, respectively. This means that the same volume of water injected occupies a larger pore volume in the LIR case and consequently,

corresponds to a more diffusive distribution of water. This is visible from the intense white in HIR at 4 mL when compared to LIR (less bright) for the same volume.

It is important to note that in Figure 9-B, the plots for 1 and 2 mL of water injected are relatively close together when compared to Figure 9-A. This is due to the fact that between 1 and 2 mL the water spreads laterally before it starts advancing in the height (Z-direction) (see the increase in radius of the yellow dashed line between 1 and 2 mL in Figure 8-B).

3.1.2. Natural Imbibition

It is intended to compare the evolution of V_P and S_W of HIR and LIR with the natural imbition (NI) rate and classify it as a low or high injection rate.

The amount of air expelled during the NI is strongly related to the imbibition rate which can be estimated through the volume of water imbibed. The volume of water imbibed is usually estimated by weighing the sample with a scale at several stages of the imbibition (Li & Horne 2010).

This method cannot be applied to this experiment because of the continuous X-ray CT scanning. All the experimental setup sits inside the CT scanner ring and must be metal-free as X-ray/metal interactions blur the CT scans.

The total volume of imbibed water is estimated by comparing the CT scans of the dry sample with the consecutive scans of the water-saturated sample. $\overline{\mathrm{CT}_{\mathrm{dry}}}$ is the average CT number of the dry sample and $\overline{\mathrm{CT}_2}$ is the average CT number at the following moment. $\overline{\mathrm{CT}_2}$ is a superposition of the CT number for water, $\mathrm{CT_w}$, air, $\mathrm{CT_{air}}$ and $\overline{\mathrm{CT}_{\mathrm{dry}}}$. Since estimating the volume of imbibed water, $\mathrm{V_{H_2O}}$, is intended, the contribution of each phase (water, air and dry matrix) is weighed over the sample's volume, V. It is assumed that the pores previously dry are totally filled with water. Thus $\overline{\mathrm{CT}_2}$ is given by Equation 15.

$$\overline{\mathrm{CT}_2} = \frac{\mathrm{V_{H_2O}CT_{H_2O}} + \mathrm{V}\overline{\mathrm{CT}_{\mathrm{dry}}} - \mathrm{V_{H_2O}CT_{air}}}{\mathrm{V}}. \tag{15}$$

Rearranging Equation 15 and noting that for a medical CT scanner $\mathrm{CT_{H_2O}} = 0$ and $\mathrm{CT_{air}} = -1000$, we obtain Equation 16.

$$\mathrm{V_{H_2O}} = \frac{\overline{\mathrm{CT}_2} - \overline{\mathrm{CT}_{\mathrm{dry}}}}{1000}\mathrm{V}. \tag{16}$$

The NI rate is estimated by dividing the imbibed water volume by the time interval. The height used to estimate the water-filled volume is obtained from in the CT scans. No CT scans were acquired between 1 and 3 hours due to technical issues but the injection was not stopped.

Figure 10 shows the evolution of the estimated imbibition rate and volume of water imbibed with the duration of the natural imbibition (NI). The initial fast imbibition rate slows down considerably (by almost 70%), comparable to an exponential decay. Note that the calculated NI rate is an overestimation. For example, in phase (b) of Figure 10, 3 mL of water is imbibed over 3 hours, resulting in a rate of 1 mL/h. The estimated NI rate falls between 1.5 and 2.0 mL/h.

This discrepancy is due to the fact that all the pore volume is assumed to be filled with water which is not the case (S_W = 100% is never reached during the natural imbibition). Therefore, the volume of water imbibed is overestimated and, consequently, the imbibition rate. However, the evolution reflects an expected trend: slowing down of the natural imbibition as the saturation front advances in height.

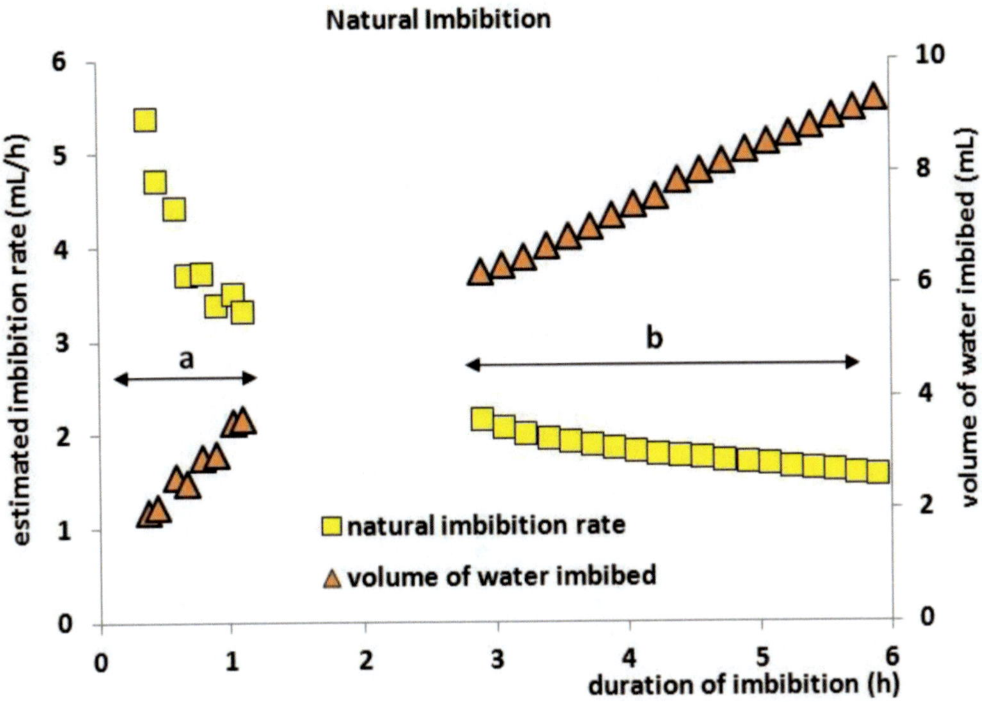

Figure 10. Estimated imbibition rate and evolution of volume of water imbibed during natural imbibition. Note the decrease of the imbibition rate similar to an exponential decay, initially dominated by capillary effects (a) followed by a phase dominated by gravity (b). No CT scans acquired between 1 and 3 hours.

Figure 11 shows the evolution of V_P and S_W for the natural imbibition experiment. A similar trend to the one obtained for the HIR (Figure 7-A) is observed: a sharp increase of V_P and S_W and fast stabilization within 1 hour reaching the final values of 2136 m/s and 66%, respectively. No CT scans were acquired between 1 and 3 hours due to technical issues but the injection was not stopped and P-waves were recorded.

The similarities in the time-evolution of V_P and fluid displacement between the injection rate of 2.0 mL/h (HIR, Figure 7-A) and the natural imbibition (Figure 11) means that the rate of 2.0 mL/h is actually not as high as initially thought.

To accurately establish the evolution of V_P and S_W at a high injection rate, an additional forced imbibition experiment is performed at the injection rate of 5.0 mL/h (Figure 12). The highest V_P value reached in this experiment is 2203 m/s. A flat saturation front is sustained throughout the forced imbibition as seen by the CT scan in Figure 12 but we only reach a maximum S_W of 43%.

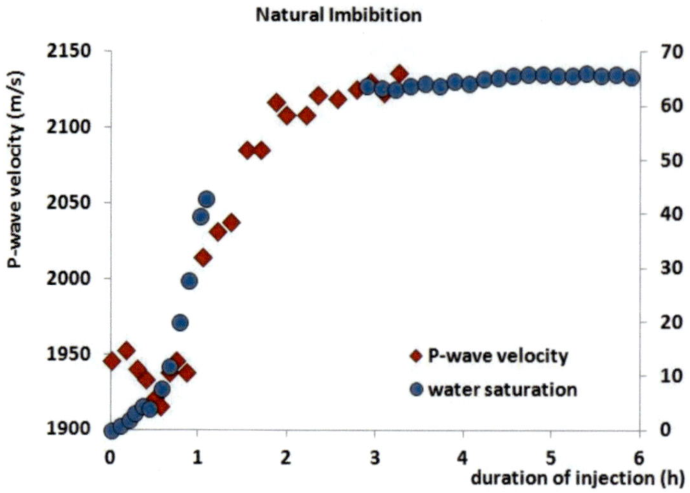

Figure 11. Evolution of P-wave velocities and water saturation with duration of injection for the natural imbibition. No CT scans acquired between 1 and 3 hours.

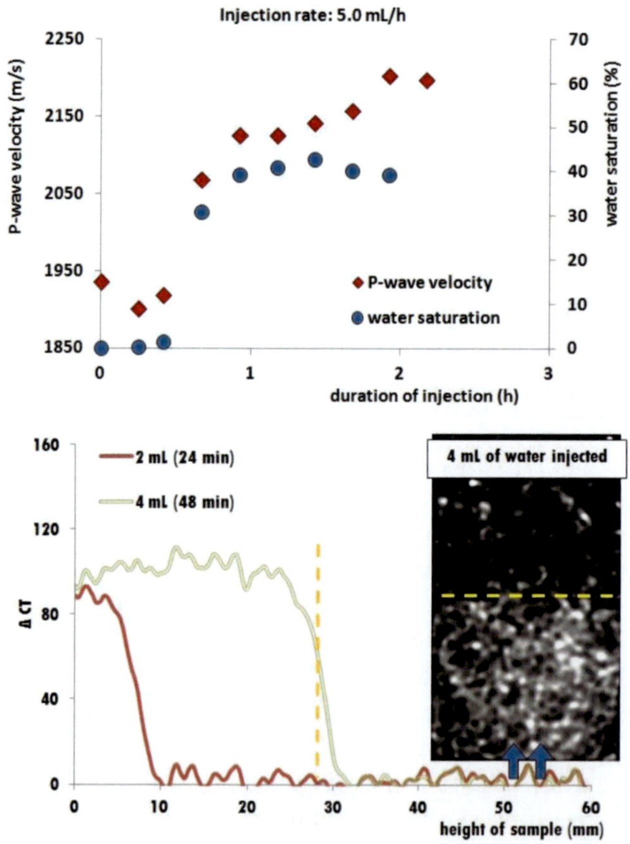

Figure 12. (Top) Evolution of P-wave velocities and water saturation with duration of injection at the constant rate of 5.0 mL/h. (Bottom) Variation of CT numbers with height of the sample for 2 and 4 mL of water injected. Note the steep slope of ΔCT, characteristic of a flat saturation front as confirmed by the CT scan (yellow dashed line), at ~30 mm. Blue arrows indicate direction of injection.

3.2. Variable Injection Rate

A forced imbibition is performed and the injection rate is changed within the continuous imbibition. Distilled water is injected at the injection rate of 2.0 mL/h for approximately 3 hours. Then, the injection rate is decreased to 0.1 mL/h and kept constant for approximately 24 hours. Afterwards, the injection rate is increased back to 2.0 mL/h until the end of the experiment.

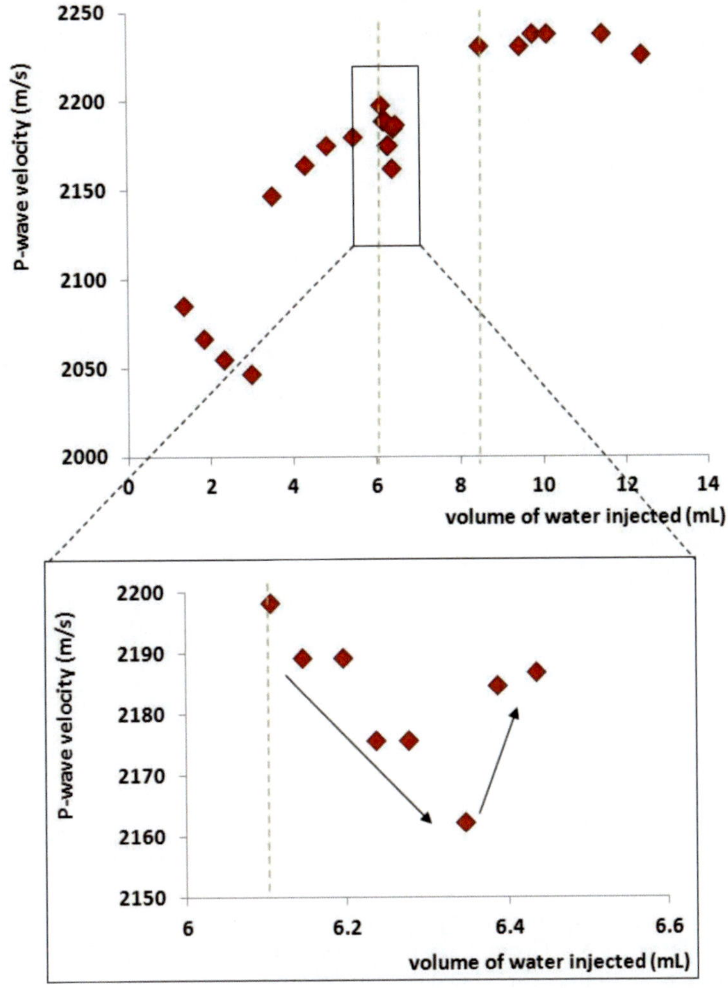

Figure 13. Evolution of P-wave velocities (V_P) with volume of water injected. Note the initial decrease of V_P due to the approaching of the saturation front. In the detail box: decreasing the injection rate decreases V_P for 2h30min followed by a sudden increase. Green dashed lines represent the moments when the injection rate was decreased and increased, from 2.0 to 0.1 and back to 2.0 mL/h.

Figures 13 and 14 show the evolution of V_P and S_W with volume of water injected, respectively. Figure 15 shows a sequence of CT scans from dry sample to 6.1 mL of water injected at the injection rate of 2.0 mL/h. Here we can spot the position of the saturation front relative to the ultrasonically monitored position (red dashed circles in Figure 15) during the first 3 hours. As the saturation front approaches the ultrasonically monitored position, V_P decreases. After 2.9 mL of water injected, V_P quickly increases while S_W increases steadily. At this moment, the saturation front starts crossing the ultrasonically monitored position and at 6.1 mL, the saturation front has fully passed it.

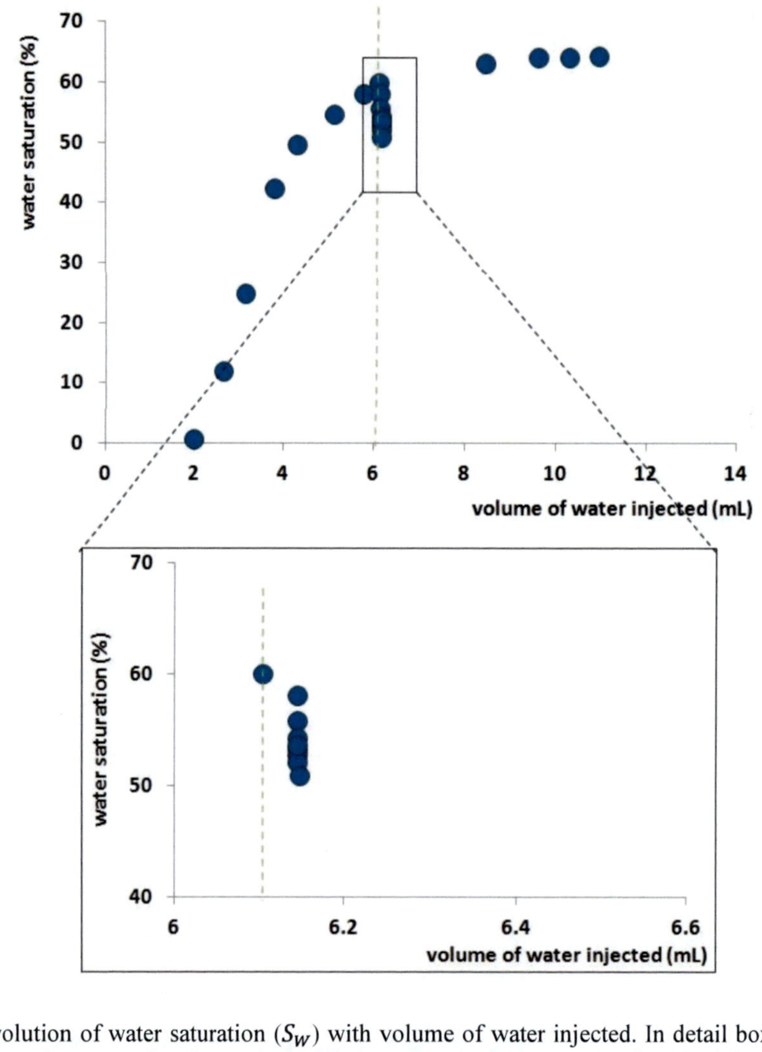

Figure 14. Evolution of water saturation (S_W) with volume of water injected. In detail box: decreasing the injection rate decreases S_W. Green dashed lines represent the moments when the injection rate was decreased and increased, from 2.0 to 0.1 and back to 2.0 mL/h.

We can see the influence of changing injection rates on the evolution of V_P and S_W:

1. Decreasing the injection rate, decreases both V_P and S_W. The decrease in V_P is immediate, from 2199 m/s to 2163 m/s. It lasts for 2h30min and is followed by a sudden increase that continues to the end of the injection at the rate of 0.1 mL/h. S_W decreases immediately: 9% in 20 minutes (from 60% to 51%). By the end of the injection at the rate of 0.1 mL/h, S_W has increased up to 63%.
2. Increasing the injection rate, slightly and steadily increases V_P and S_W.

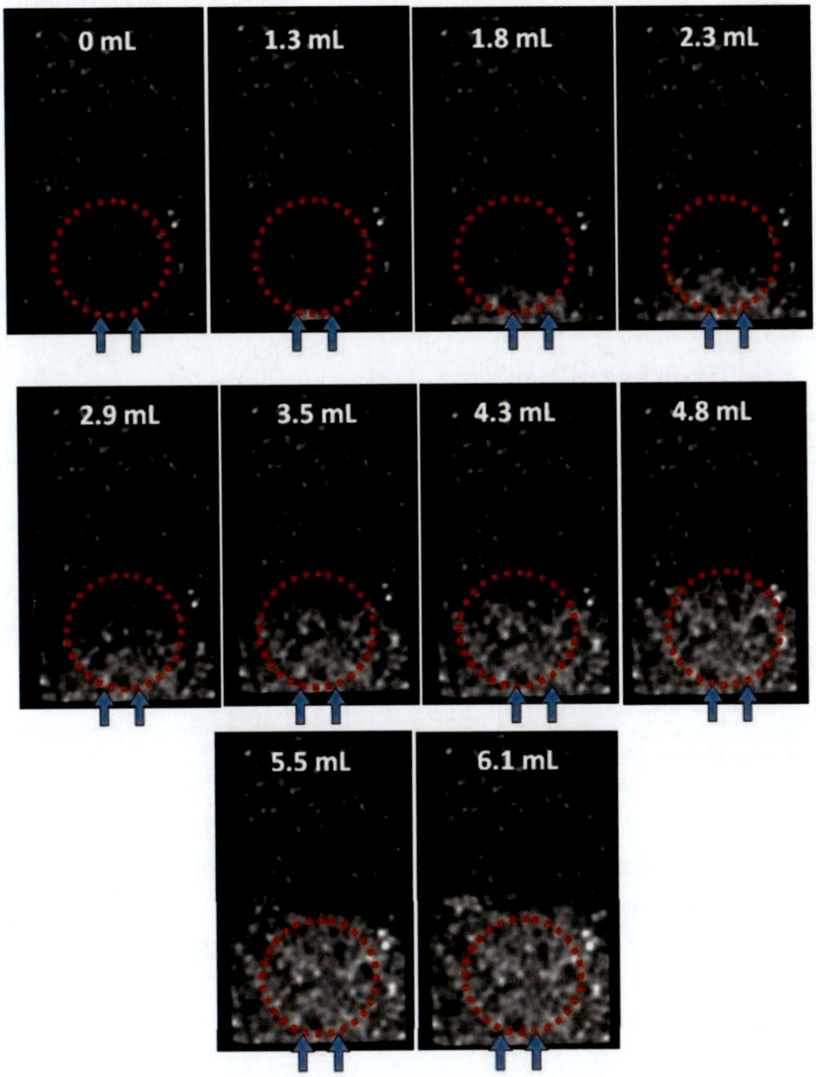

Figure 15. Sequence of CT scans from dry sample to 6.1 mL of water injected at the injection rate of 2.0 mL/h (for the variable injection rate experiment). Dashed red circle represent the area covered by the transducers in the scanning plane. Blue arrows indicate the direction of injection (see Figure 3).

4. Discussion

4.1. Fluid Flow Regime

The capillary number at all chosen injection rates determines which forces, capillary or viscous, are dominant in our fluid flow experiments. Following Riaz et al., (2007), the capillary number, Ca, is given by Equation 17.

$$Ca = \frac{\mu_w . U}{\gamma} \tag{17}$$

where $U = \frac{4.Q}{\pi.D^2}$. μ_w is the viscosity of the injected fluid (water, 10^{-3} Pa.s), γ is the interfacial tension between air and water ($72 \times 10^{-3}\ N/m$ at 25°C). U is the injection velocity, which depends on the injection rate, Q, the dimensions of the fluid piston, D (which is assumed to be the dimensions of the point injection, $0.5\ \times 10^{-2}\ m$). Therefore, for each injection rate (5.0, 2.0, 0.5, 0.2 and 0.1 mL/h):

- $Q = 5.0\ mL/h,\ Ca \sim 5 \times 10^{-9}$
- $Q = 2.0\ mL/h,\ Ca \sim 2 \times 10^{-9}$
- $Q = 0.5\ mL/h,\ Ca \sim 5\ \times 10^{-10}$
- $Q = 0.2\ mL/h,\ Ca \sim 2 \times 10^{-10}$
- $Q = 0.1\ mL/h,\ Ca \sim 1 \times 10^{-10}$

All the injection rates result in $C_a < 10^{-5}$ and therefore all the fluid flows are dominated by capillary forces (Ding & Kantzas 2004).

The viscosity ratio, M, is the ratio between the viscosities of the displaced phase (residual fluid) and displacing phase (injected fluid) (Equation 18). It classifies the stability of the fluid flow. Flow stability is an essential requirement for the development of an efficient displacement of the residual fluid (Yortsos & Huang 1986). In water imbibition,

$$M = \frac{\mu_{displaced\ phase}}{\mu_{displacing\ phase}} = \frac{\mu_{air}}{\mu_{water}}. \tag{18}$$

If M>1, the fluid displacement is unstable and fingering occurs. In all our imbibitions, $M \sim \frac{10^{-5}}{10^{-3}} = 10^{-2} \ll 1$, therefore the fluid displacements are stable (Riaz et al., 2007). This is confirmed by visual inspection of the saturation fronts in the CT scans.

4.2. Constant Injection Rate

The first drop in V_P is associated to the approaching saturation front. This velocity reduction is interpreted as a *moisture weakening effect*. The capillary pressure assumes its highest value at low S_W, i.e., at the beginning of the imbibition (Riaz et al., 2007). And therefore, water is able to propagate easily without the assistance of external forces. Therefore, ahead of the saturation front, there is a small fraction of water (moisture). This moisture can cause a weakening of the matrix. This reduces the bulk modulus and, as a direct consequence, V_P decreases (Gassmann-Wood theory) (Mavko, Mukerji & Dvorkin 1998; Toms et al., 2006). This drop is consistently observed on the imbibitions initiated at high rates (5.0 and 2.0 mL/h) and at natural imbibition rate but not in the low rate (0.5 mL/h). Due to the low injection rates, the displacement of the saturation front is slow enough to create instant equilibrium between neighbouring grains.

The initial decrease of V_P is followed by a continuous increase. The saturation conditions are then set in a preferential way for the bulk modulus to increase and as a direct consequence V_P increases (Gassmann-Wood theory). This increase continues until the maximum S_W is achieved. After 10 hours of imbibition, V_P = 2293m/s, i.e., it is 100 m/s higher than the final V_P for HIR.

Experimental data suggests that the higher injection rate of 5.0 mL/h is not most efficient in displacing the residual fluid (in this case, air). The relatively low value of the final S_W = 43% seems to indicate that many gas bubbles are trapped behind the advancing saturation front. Conversely, the injection rates equal to or lower than the natural imbibition rate lead to a higher S_W = 66%, thus providing a more efficient way to displace the residual fluid. This is because high rates do not allow enough time for the wetting phase (water) to contact with the matrix and pore structure for an efficient imbibition. This means that the water does not completely displace the gas from the pores and part of it remains trapped. This is also supported by microscopic examinations showing that the residual gas is surrounded by water and it is unable to escape by ordinary flow (Katz et al., 1966).

An injection rate much lower than the natural imbibition rate is also production-efficient though more time is required for water pathways to be established. Natural imbibition processes are the result of an ongoing interplay between gravity and capillary forces. These two forces are always present during imbibition processes but one can overcome the other. During vertical imbibition, it is possible to distinguish these forces: initially, the flow is dominated by capillary forces (characterized by a fast imbibition of water), followed by a second stage when the process changes from capillary driven to capillary-gravity driven flow. At this stage, gravity causes a considerable slowdown of the imbibition rate (segregating the flow) but capillary forces are still strong enough to force a slow but continuous flow (Karimaie & Torsaeter 2007).

4.3. Variable Injection Rate

Four distinct stages are observed in the evolution of V_P and S_W (Figures 13 and 14, respectively). The initial evolution of V_P and S_W are intrinsically related to the position of the saturation front (Figure 15).

Stage 1

As the saturation front approaches the ultrasonically monitored position (Figure 15, between 0 and 2.9 mL of water injected), V_P decreases by 53 m/s (Figure 13). As explained in the first paragraph of "Discussion – Constant Injection Rate", this velocity reduction is interpreted as a *moisture weakening effect* that causes a weakening of the matrix, thus reducing the rock skeleton bulk modulus.

Stage 2

When the saturation front penetrates the ultrasonically monitored position, V_P increases with increasing S_W. As explained in the second paragraph of "Discussion – Constant Injection Rate", this is due to the increase of the bulk modulus (as established by the Gassmann-Wood theory).

Stage 3

When the injection rate decreases from 2.0 to 0.5 mL/h, V_P and S_W decrease immediately (stage 3.1). After that, V_P quickly increases while S_W keeps decreasing (stage 3.2) (Figures 13 and 14, between green dashed lines).

Stage 3.1.

Upon decreasing the injection rate, V_P decreases instantly but slowly, 36 m/s in 2h30 (Figure 13). This decrease of V_P with decreasing injection rate seems to be connected to the partially saturated conditions of the sample. At this point S_W is at 60%. This means that 40% of pores are still filled with air. Decreasing the injection rate increases the capillary pressure between the two fluids (Perkins 1957). At this moment, the unconstrained gas is free to move and expand. This is accompanied by a redistribution of the water and air phases leading to a decrease of S_W (Figure 14). This saturation reduction causes a decrease in the effective fluid bulk modulus. Thus, the un-drained bulk modulus decreases and V_P decreases according to the Gassmann-Wood theory.

Stage 3.2

The initial decrease of V_P is followed by a sudden increase while S_W keeps decreasing down to 51% (Figures 13 and 14). After decreasing the injection rate and consequently, the capillary pressure, more empty spaces are being filled with water and V_P increases. The redistribution of fluid caused by the change of injection rate prompts higher values of V_P: 2231 m/s at the end of the injection rate of 0.5 mL/h for 63% of S_W.

Stage 4

V_P and S_W increase slightly and steadily when the injection rate increases from 0.5 to 2.0 mL/h (Figures 13 and 14). As explained in Stage 2, this increase is directly related to the increase of volume of water injected and, therefore, the bulk modulus (Gassmann-Wood theory).

The trend described on the evolution of V_P and S_W with the presence of water and with the change of injection rates has been observed on another sandstone sample (Lopes & Lebedev 2012). Though in the latter case, the variations were higher in absolute value. This might be due to a more compressible solid matrix. Moreover, the same trend was observed on carbonates (Lopes et al., 2013) though in this case, the acoustic signatures were lower in absolute value as characteristic of a stiffer rock matrix.

CONCLUSION

Laboratory-scale experiments were performed in order to quantify the influence of the injection rate on the evolution of ultrasonic P-wave velocities and water saturation and relate these quantities with the geometry and displacement of the saturation front. The X-ray CT scanning allows not only a visualization of the saturation front but also provides data to estimate the water saturation, volume of water imbibed and, consequently, the rate of the natural imbibition.

The saturation front passing the ultrasonically monitored position is detected through a change in the acoustic response. This change is characterized by a sudden increase in P-wave velocities, much sharper for the high injection rate but still noticeable for the low injection rate. The rate at which the P-wave velocities change is intrinsically related to the type of saturation front: sharp and straight for the high injection rate and dispersive for the low injection rate. Interestingly, the evolution's pattern of the P-wave velocities mimics the evolution of the water saturation. This result implies that it is possible to continuously and actively monitor saturation fronts solely through the analysis of the acoustic response and also infer about the geometry of the saturation front.

It is found that that an injection rate higher than the natural imbibition rate is not efficient in the production of the residual fluid whereas an injection rate close to the natural imbibition rate is the most efficient. A low injection rate is also efficient in displacing the resisual fluid but only over a long period of time.

Another important outcome of this study is that the same P-wave velocity can be obtained for different water saturations. Acoustic wave propagation is extremely sensitive to spatial fluid distribution more than to the actual water saturation level (Cadoret, Mavko & Zinszner 1998). Specially, the presence of patchy saturation (fluid clusters at the macroscopic scale) influences how acoustic waves propagate and interact with porous media by creating significant acoustic attenuation and dispersion (Li, Zhong & Pyrak-Nolte 2001; Müller, Gurevich & Lebedev 2010).

These experiments show that acoustic wave velocities are sensitive to the change of the injection rate and that small variations in water saturation can create significant changes in the acoustic response (at ultrasonic frequencies). Decreasing the injection rate decreases the P-wave velocity and water saturation immediately. When the injection rate increases, both the P-wave velocity and water saturation increase. A decrease of the P-wave velocity is also observed in the beginning of the imbibition as the saturation front approaches the ultrasonically monitored position due to capillary effects. All these features have been observed in a previous work on sandstones (Lopes & Lebedev 2012) and on the carbonates (Lopes et al., 2013).

ACKNOWLEDGMENTS

This work was partially funded by Australia-China Natural Gas Technology Fund (Contract Number CTR-DJ-04194-1), by Commonwealth Scientific and Industrial Research Organisation (CSIRO), by the Australian Commonwealth Government through the Cooperative Research Centre for Greenhouse Gas Technologies (CO2CRC) and by the sponsors of the Curtin Reservoir Geophysics Consortium (CRGC).

The authors would like to thank Dr Ben Clennell (CSIRO) for ongoing discussions and interest in this work; Dr Lionel Esteban (CSIRO) and Eva Caspari (Curtin University) for technical support. One of the authors (S. L.) would like to thank CSIRO for financial and technical support.

References

Akin, S., Schembre, J. M., Bhat, S. K., Kovscek, A. R.: Spontaneous imbibition characteristics of diatomite. *Journal of Petroleum Science and Engineering* 25, 149-165 (2000).

Babadagli, T., Ershaghi, I.: Imbibition assisted two-phase flow in natural fractures. *SPE* 24044 (1992).

Birovljev, A., Furuberg, L., Feder, J., Josang, T., Maloy, K. J., Aharony, A.: Gravity invasion percolation in two dimensions: experiment and simulation. *Physical Review Letters* 67-5, 584-588 (1991).

Bjorlykke, K.: Petroleum Geoscience: From Sedimentary Environments to Rock Physics. Springer. ISBN 9783642023316 (2010).

Blunt, M. J., Scher, H.: Pore-level modelling of wetting. *Physical Review E* 52(6), 6387-6403 (1995).

Cadoret, T., Marion, D., Zinszner, B.: Influence of frequency and fluid distribution on elastic wave velocities in partially saturated limestones. *Journal of Geophysical Research* 100 (B6), 9789-9803 (1995).

Cadoret, T., Mavko, G., Zinszner, B.: Fluid distribution effect on sonic attenuation in partially saturated limestones. *Geophysics* 63-1, 154-160 (1998).

Calvert, R.: 4D technology: where are we and where are we going to? *Geophysical Prospecting* S3, 161-171 (2005).

Ding, M., Katnzas, A.: Capillary number correlations for gas-liquid systems, *SEP* 2004-062 (2004).

Fornasini, P. The uncertainty in physical measurements: an introduction to data analysis in the physics laboratory. *Springer*. ISBN-10: 038778649X, 301 pp. (2008).

Garg, A., Zwahlen, E., Patzek, T. W.: "Experimental and numerical studies of one-dimensional imbibition in berea sandstone" in the Proceedings of the Sixteenth Annual American Geophysical Union Hydrology Days, pp. 171-183, Denver, CO, April 15-18 (1996).

Guéguen, Y., Palciauskas, V.: Introduction to the physics of rocks. Princeton University Press. ISBN-10 0691034524, 392 pp. (1994).

Katz, D. L., Legatski M. W., Gorring L., Nielsen, R. L.: How water displaces gas from porous media. *The Oil and Gas Journal* 55-60 (1966).

Karimaie, H., Torsaeter, O.: Effect of injection rate, initial water saturation and gravity on water injection in slightly water-wet fractured porous media. *Journal of Petroleum Science and Engineering* 58, 293-308 (2007).

Ketcham, R. A., Carlson, W. D.: Acquisition, optimization and interpretation of X-ray computed tomography imagery: applications to the geosciences. *Computer & Geosciences* 27, 381-400 (2001).

Lebedev, M., Toms-Stewart, J., Clennell, B., Shulakova, V., Paterson, L., Müller, T., Gurevich, B., Wenzlau, F.: Direct laboratory observation of patchy saturation and its effects on ultrasonic velocities, *The Leading Edge* 28-1, 24-27 (2009).

Leverett, M. C.: *Capillary Behaviour in Porous Solids* (1940).

Li, K., Horne, R. N.: Method to evaluate the potential of water injection in naturally fractured reservoirs, *Transport in Porous Media* 83, 699-709 (2010).

Li, X., Zhong, L., Pyrak-Nolte L. J.: Physics of partially saturated porous media: residual saturation and seismic-wave propagation. *Annual Review of Earth and Planetary Sciences*, 29, 419-60 (2001).

Lopes, S., Lebedev, M.: Research note: Laboratory study of the influence of changing the injection rate on the geometry of the saturation front and on P-wave ultrasonic velocities in sandstone. *Geophysical Prospecting* 60, 572–580 (2012).

Lopes, S., Lebedev, M., Müller, T., Clennell, M. B., Gurevich, B.: Forced imbibition into a limestone: measuring P-wave velocity and water saturation dependence on injection rate. *Accepted for publication. Geophysical Prospecting* (2013).

Lumley, D. E.: Time-lapse seismic reservoir monitoring, *Geophysics* 66-1, 50-53 (2001).

Mavko, G., Mukerji, T., Dvorkin, J.: The rock physics handbook: Tools for seismic analysis in porous media. Cambridge University Press. ISBN-13 9780521861366, 503 pp. (1998).

Meleán, Y., Broseta D., Blossey. R.: Imbibition fronts in porous media: effects of initial wetting fluid saturation and flow rate, *Journal of Petroleum Science and Engineering* 39, 327-336 (2003).

Monsen, K., Johnstad, S. E.: Improved understanding of velocity-saturation relationships using 4D computer-tomography acoustic measurements, *Geophysical prospecting*, 53, 173-181 (2005).

Morrow, N. R., Mason, G.: Recovery of oil by spontaneous imbibition. *Colloid and Interface Science* 6, 321-337 (2001).

Müller, T. M., Gurevich B., Lebedev M.: Seismic wave attenuation and dispersion due to wave-induced flow in porous rocks – A review. *Geophysics*, 75, A147–A164 (2010).

Perkins, F. M.: An Investigation of the role of capillary forces in laboratory water floods. *AIME Technical note*. SPE 840-G (1957).

Rangel-German, E. R., Kovscek, A. R.: Experimental and analytical study of multi-dimensional imbibition in fractured porous media. *Journal of Petroleum Science and Engineering* 36, 45-60 (2002).

Rapoport, L. A., Leas, W. J.: Properties of linear water-floods. *Petroleum Transactions* 18, 139-148 (1953).

Riaz, A., Tang, G. Q., Tchelepi, H. A., Kovscek, A. R.: Forced imbition in natural porous media: comparison between experiments and continuum models. *Physical Reviews E*-75, 036305 (2007).

Schembre, J. M., Kovscek, A. R.: A technique for measuring two-phase relative permeability in porous media via X-ray CT measurements. *Journal of Petroleum Science and Engineering* 39, 159-174 (2003).

Skjæveland, S. M., Kleppe, J.: SPOR monograph: recent advances in improved oil recovery methods for north sea sandstone reservoirs. *Norwegian Petroleum Directorate*, Stavanger, ISBN 82-7257-340-7 (1992).

Sudaryanto, B., Yortsos, Y. C.: Optimization of fluid dynamics in porous media using rate control I, Equal mobility fluids. *Physics of Fluids* 12-7, 1656-1670 (2000).

Tang, G.-Q., Kovscek, A. R.: High resolution imaging of unstable, forced imbibition in Berea sandstone. *Transport of Porous Media* 86, 617-634 (2011).

Taylor, J.: An Introduction to Error Analysis: The Study of Uncertainties in Physical Measurements. University Science Books. ISBN-10 0935702423, 327 pp. (1997).

Teufel, L. W., Rhett, D. W., Farrell, H. E.: Effect of reservoir depletion and pore pressure drawdown on in situ stress and deformation in the Ekofisk Field, North Sea in the

Proceedings of 32nd U. S. symposium for Rock Mechanics, Norman, Oklahoma, 10-12 July (1991).

Toms, J., Müller, T. M., Ciz, R., Gurevich, B.: Comparative review of theoretical models for elastic wave attenuation and dispersion in partially saturated rocks. *Soil Dynamics and Earthquake Engineering* 26, 548–565 (2006).

Toms-Stewart, J., Müller, T. M., Gurevich, B., Peterson, L.: Statistical characterization of gas-patch distributions in partially saturated rocks. *Geophysics* 74-2, WA51 (2009).

Wellington, S. L., Vinegar, H. J.: X-ray computerized tomography. *Journal of Petroleum Technology* 39, 885-898 (1987).

Yortsos, Y. C., Huang, A. B.: Linear-stability analysis of immiscible displacement: Part 1 – Simple basic flow properties, *SEP* 12692 (1986).

Yousef, A. A., Gentil, P., Jensen, J. L., Lake, L. W.: A capacitance model to infer interwell connectivity from production- and injection-rate fluctuations. *Reservoir Evaluation and Engineering*, SPE 95322 (2006).

In: Sandstone
Editor: Afsoon Moatari Kazerouni

ISBN: 978-1-62948-382-5

Chapter 3

EFFECTIVE THERMAL CONDUCTIVITY OF DRY AND FLUID-SATURATED SANDSTONES AT HIGH TEMPERATURES AND HIGH PRESSURES: REVIEW OF THE EXPERIMENTAL METHODS AND MODELING

***Zumrud Z. Abdulagatova*[1,*], *Ilmutdin M. Abdulagatov*[2,3,†] *and Aziz I. Abdulagatov*[1,‡]**

[1]Institute of Physics of the Dagestan Scientific Center of the Russian Academy of Sciences, Makhachkala, Dagestan, Russia

[2]Geothermal Research Institute of the Dagestan Scientific Center of the Russian Academy of Sciences, Makhachkala, Dagestan, Russia

[3]Present address: Thermophysical Properties Division, National Institute of Standards and Technology, Boulder, CO, US

ABSTRACT

Available sources of data for the effective thermal conductivity (ETC) of dry and fluid (gas, water, and oil)–saturated sandstones at high temperatures and high pressures have been collected and evaluated. A comprehensive review of the different contact (divided-bar, needle-probe, hot-wire, guarded hot plate, etc.) and contact-free (optical scanning, laser flash) measurement techniques used to accurately measure of the thermal conductivity of sandstones from various regions of the world were provided. The uncertainties of each methods and their advantages and disadvantages were discussed. The review also included the authors own results on thermal conductivity measurements of dry and fluid-saturated sandstones from various locations of the Russia (basically from Dagestan, North Caucasus, Russia, from wells with various depth). The measurements of the λ, P, T relationships for the dry, gas-, water-, and oil-saturated sandstones with a

* E-mail: zumrud81@mail.ru.

† E-mail: ilmutdin@boulder.nist.gov.

‡ E-mail: azizagatov@yahoo.com.

various porosities in the temperature range from 273 to 573 K and at pressures up to 400 MPa were made using the steady-state guarded parallel plate apparatus. The estimated uncertainty in ETC measurements is 4 % (expanded uncertainties at the 95 % confidence level with a coverage factor of k=2). The effect of pressure, temperature, porosity, and fluid saturation on the ETC behavior of sandstones was studied experimentally and theoretically.

The comprehensive review various existing theoretical and empirical models for the prediction of ETC rocks materials were provided. We interpreted measured ETC data for gas-, water-, and oil-saturated sandstones using various theoretical models in order to check their accuracy, predictive capability, and applicability. The effect of saturating fluids, structure (size, shape, and distribution of the pores), porosity, and mineralogical composition on temperature and pressure dependences of the ETC of fluid-saturated sandstone was discussed.

Keywords: Thermal conductivity, sandstone, fluid-saturation, microstructure, heat transfer, high pressure, porous rocks

1. INTRODUCTION

1.1. Practical Applications of the ETC Measurements of Porous Rock Materials

Thermal conductivity is a key thermophysical property of rocks because it has a first-order control on the configuration of isotherms and the flow of heat within the basin (modeling of the thermal history of the Earth). Effective thermal conductivity (ETC) studies of the fluid-saturated porous materials are useful in a variety of industrial/geotechnical applications such as: 1. Petroleum and natural gas geology (thermal methods enhanced oil and gas recovery technology; petroleum and natural gas economics; origin and accumulation of petroleum and natural gas; petroleum and natural gas geochemistry; reservoir engineering; thermodynamics and phase behavior in porous media) [1-10]; 2. Utilization of hydrothermal energy [11-18] and underground thermal energy storage [19-22]; 3. Applications to geothermal problems, geothermal gradients and heat flux through the earth's crust (mathematical modeling), hydro-geological studies (rock mechanics/petrophysics, drilling and drilling fluids, well logging, testing and evaluation, fluid mechanics in porous media and multi-phase flow; reservoir simulation) [23-28]; 4. Civil engineering applications (construction of roads and buildings and in drilling of wells through permafrost zones) [29-45]; 5. Composite materials (heterogeneous materials, nuclear materials, solar energy, ceramic-metal composites, powders, packed bed with a stagnant fluid, thermonuclear fusion, granular materials, fibrous composites, cryogenic applications, multilayer insulation systems; foam materials) [46-58]; 6. Climate change [59]; 7. Management of microsystems [60, 61]; 8. Biomedical engineering [62]; 9. Food and computer technologies, etc. [63, 64]; All of these processes are requiring accurately knowledge of the ETC data of dry and fluid-saturated porous materials as a function of temperature, pressure, and porosity.

Energy conservation is an important part of any national energy strategy and energy conservation in underdeveloped countries with inadequate resources is even more important [35]. Using natural rock with a low thermal conductivity improves insulation of buildings by providing an energy efficient solution. Estimation of soil thermal receptivity is of great importance for various engineering where heat transfer takes place through the soil mass [30, 39, 41].

Sedimentary rocks are composed of a mixture of voids generally filled with water, oil, gas, and various minerals. The porosity can be changed by stress and air can be replaced by oil or water. Fluid-saturated porous rock is a heterogeneous mixture of numerous solid and fluid components [292-297]. Particular sandstone might be composed of quartz, feldspar, and calcite, and be saturated with oil/water, oil/gas, and salt/water mixtures. For liquid-saturated sandstones the ratio λ_f / λ_S is within 0.01 to 0.5, while for gas-saturated rocks λ_f / λ_S is about 0.001 [159,292-297].

The thermal conductivity of rocks is a key parameter also in geothermal investigations. It is important in heat flow density studies and in exploitation of geothermal resources, but also in basin modeling. Geothermal energy has received much attention in recent years as one of the renewal sources that can help relieve the energy crisis in the next decade. To evaluate the geothermal energy generation systems we must predict the amount of heat present and the rate at which it can be extracted. Amounts of heat and recovery rate depend on basic physical properties of the reservoir rocks and the fluids containing within them. The rate of heat extraction depends on the thermal conductivity and the fluid flow characteristics, *i.e.*, permeability and relative permeability, of the water and steam in the rocks. All these important basic thermophysical characteristics of the rock and fluids are functions of both the temperature and pressure of the reservoir system.

Combined with local thermal gradients, thermal conductivity is used in constraining heat flow estimates [13]. Determination of the thermal conductivity of rock material is essential to obtaining reliable values of the Earth's heat flux Q using the Fourier equation

$$Q = -\lambda(P,T) gradT \,, \tag{1}$$

where $\lambda(P,T)$ is the thermal conductivity of the Earth layers as a function of temperature and pressure. Accurate estimates of thermal conductivity are required for determining heat flow from temperature measurements in the Earth and for extrapolating near-surface temperature data to estimate thermal regimes at greater depths. Luchenbruch and Sass [26] used rock thermal conductivity data and temperature gradient in the crust to temperature estimates down to several tens of kilometers based on data from the upper few kilometers. The pressure and temperature dependence of thermal conductivity is the most critical factor for these estimates. Several factors must be considered in estimating formation thermal conductivity of the rock matrix and pore fluid, anisotropy, and the effects of removing the lithostatic overburden in the process of coring. Oil well logs are an important source of geothermal data for studying regional tectonics, reconstructing the evolution of sedimentary basins, and theorizing petroleum generation, migration, and accumulation. About half of the heat producing the geothermal gradient measured in oil wells comes from the upper mantle; the other half is generated by radioactive elements concentrated in granites. These sources are so deep that they can be regarded as uniform; so lateral variations in heat flow over oil field structures are due to lateral variations in ETC and to heat transferred by moving fluids, principally water. Geothermal anomalies exist over many oil fields [65]. In oil fields disconnected from ground water, the area distribution of heat flow arises solely from variation of thermal conductivity due to lateral changes of faces and rock porosity, which also are often controlled by local structure. Two of the key steps in oil exploration are to assess when hydrocarbon generation occurred and to determine the volume of hydrocarbons produced. These depend to a large extent on the temperature history of the potential source rocks, as hydrocarbon generation is typically modeled with an Arrhenius-type kinetic relationship between temperature and time [66]. Consequently, the initial part of this overall problem is one of determining the thermal history of the basin, a process often referred to as thermal maturity modeling [17]. Assuming that we know the heat flow history, then we can estimate the temperature of any sedimentary layer as it is buried over time provided we know the thermophysical properties (thermal conductivity) of the sediments. Gallagher et al. [14] examined how variation in the ETC estimates of the sedimentary column influence the reconstruction of the thermal history of a given well.

The thermal conductivity directly controls the temperature gradient, (dT/dz), therefore key variable in thermal modeling [28]. The temperature gradient as a result of conduction is described by Fourier's law as inversely proportional to the thermal conductivity for a given heat flow

$$\frac{dT}{dz} = \frac{Q}{\lambda}. \tag{2}$$

As one can see from Eq. (2), the geothermal gradient (dT/dz) is a function of heat flow (Q) and the thermal conductivity (λ) of material in which the geothermal gradient is measured. Although heat flow Q varies considerably between basins, it is usually constant for a specific locality in a basin. It is obviously that a decrease in thermal conductivity in a sedimentary section results in an increase in geothermal gradients, and vice versa. In the case of the 1-D, steady state approximation, the temperature at depth z from time t is given as

$$T[z(t)] = T_0(t) + Q(t) \int_0^{z(t)} \frac{dz'}{\lambda(z')}, \tag{3}$$

where $\Omega(z) = \int_0^{z(t)} \frac{dz'}{\lambda(z')}$ is the thermal resistance, $T_0(t)$ and $Q(t)$ are the surface temperature and heat flow, respectively, at time t and λ is the thermal conductivity. Thus, the ETC is a function of depth for a given time and therefore depends on factors such as lithology and porosity, temperature, pressure *etc*. Therefore, the reliable ETC data are requiring estimating the temperature history adequately. If the temperature $T(z)$ is plotted versus thermal resistance $\Omega(z)$ in the so-called "Bullard plot" [67], it will be linear if conditions are conductive and at steady state and there are no internal heat sources. Thus a linear regression allows derivation of the surface temperature $T_0(t)$ from the intercept with z=0 and of heat flow $Q(t)$ from the slope Eq. (3).

As one can see, to calculate the thermal resistance $\Omega(z)$ the thermal conductivity data are needed as a function of T and P. A decrease in thermal conductivity from 3 to 2 $W\cdot m^{-1}\cdot K^{-1}$

increase the thermal gradient from 20 to 30 °C·km^{-1} and rises the temperature at a depth of 3 km by 30 °C. Therefore, to accurate predict the thermal and hydrocarbon generation histories we have to quantify thermal conductivity for a particular study area.

The effect of thermal conductivity variation on subsurface temperature is studied by several authors (see, for example [68]). The estimation of terrestrial heat flow requires the measurement of geothermal gradient and of bulk rock thermal conductivity. Ljubimova et al. [69] proposed the method of prediction of the deep Earth temperatures on the bases of fluid-saturated thermal conductivity data of rocks at high temperatures and high pressures.

The ETC of dry and fluid –saturated porous media continues to be an important problem in heat transfer science [70, 71]. All heat transfer problems require for their solution knowledge of the ETC of porous materials. The influence of medium morphology is difficult, if not impossible, to investigate experimentally.

The modeling is the only practical approach to explore the dependency of ETC of porous media on the morphology of the porous media. Unfortunately, the ETC of dry and fluid-saturated porous rocks cannot be accurately predicted theoretically, due to their complicated physical and chemical structure. Therefore, the ETC measurements of the porous rock materials are also research interest. It is well known that ETC and other thermophysical properties of rock materials are affected by various factors, among which the most important temperature, pressure, such as mineralogical composition, porosity, fractional content of quartz, microstructure, stratification, distribution, orientation, size and shape of the components, and nature of pore fluid (Somerton [72]).

Various empirical and semiempirical equations have been proposed to relate the ETC to these parameters, and some attempts have been made to predict the conductivity theoretically. Therefore, the reported thermal conductivity datasets by various authors are may vary significantly by as much as a factor of (2 to 3) for any given rock type. Even within the same rock type ETC can vary over a considerable range. This could explain the published data discrepancy for the same type rocks.

The main purpose of this work are: (1) comprehensive analyses of the available sources of data for ETC of dry and fluid (gas, water, and oil)–saturated sandstones at high temperatures and high pressures; (2) comprehensive review of the different contact (divided-bar, needle-probe, hot-wire, guarded hot plate, *etc*) and contact-free (optical scanning, laser flash) measurement techniques of the ETC and their uncertainty analyses; (3) to provide an accurate experimental ETC data for geological porous fluid-saturated sandstones with porosities from (1 to 13) % at temperatures from (273 to 523) K and at pressures up to 500 MPa using a guarded parallel-plate method, which has been used previously for accurate measurements on other solids and rocks (dry and fluid-saturated) materials; (4) to study the effect of temperature and pressure on ETC behavior of dry- and fluid-saturated rocks (sandstones); (5) comprehensive review various existing theoretical and empirical models for the prediction of ETC rocks materials, interpretation and test of the measured ETC data for gas-, water-, and oil-saturated sandstones using various theoretical models in order to check their accuracy, predictive capability, and applicability; (6) to develop new models which are accurately taken into account the combined effect of temperature, pressure, and porosity on the ETC of fluid-saturated porous rocks.

1.2. Scientific Applications of the ETC Measurements of Porous Rock Materials

Thermal conductivity is also the key transport property of the solids (porous materials) for scientific applications. Thermal conductivity is the process of heat transfer by molecular motion and sometimes free flow of electrons through a solid, liquid, or gas medium from high to low temperature. Radiation heat transfer takes place in the form of electromagnetic waves that propagate even through a vacuum. Thermal conductivity in metals is dominated by the transport of free electrons whereas in nonmetallic solids, such as porous rock materials, it occurs via lattice vibrations (phonons). In porous media, the thermal resistance ($1/\lambda$) that originate due to the structures of pores and grain contacts determines the heat conductivity transfer behavior. Thermal conductivity is defined the capacity of a materials to conduct (transmit) heat. The coefficient of thermal conductivity, λ ($W \cdot m^{-1} \cdot K^{-1}$), is a measure of the rate Q (W) at which heat flows through a material. It is the coefficient of heat transfer across a steady-state temperature difference $\Delta T = T_2 - T_1$ over a distance $\Delta x = x_2 - x_1$, or

$$Q = \lambda\left(\frac{\Delta T}{\Delta x}\right) = \lambda grad\, T. \tag{4}$$

Thermal conductivity can be measured by transient heating of a material with a known heating power generated from a source and measuring the temperature changes with time. There is different geometry heat generating sources (cylindrical, plate; wire *etc.*, see below Figures 3, 10, 19, 29, and 30). Since many rocks are anisotropic, thermal conductivity is a tensor quantity and is dependent on direction. Heat flow is thus not necessarily exactly parallel to temperature gradient.

Thermal conductivity $\lambda(P,T)$ also related with another very important thermophysical (transport, thermodynamic, and acoustic) properties of porous rocks such as thermal diffusivity (a), density (ρ), heat capacity (C_P), $\lambda = a\rho C_P$; and sound velocity (V_S and V_P), $\lambda = (1/3)\bar{V}aK_T / \rho\gamma^2 T$, where $\bar{V} = (1/V_P^3 + 2/V_S^3)^{-1/3}$. Therefore, reliable ETC data can be used to calculate and test derived properties such as thermal diffusivity a, isobaric heat capacity C_P, density and mean sound velocity. The reliable thermal conductivity measurements can be also used to consistence test of independent measurements various thermophysical properties (λ, a, C_P, $\rho, \bar{V}$) of rock materials. Since, thermal expansivity coefficient (α) and heat capacity (C_P) theoretically related through the equation of state

$$\left(\frac{\partial C_P}{\partial P}\right)_T = -TV\left[\alpha^2 + \left(\frac{\partial \alpha}{\partial T}\right)_P\right], \tag{5}$$

theoretical calculations of thermal conductivity λ finally involve thermal expansivity α and heat capacity C_P.

The heat conductivity of solids providing by the lattice waves or phonons can be expressed as

$$\lambda_{lat} = \frac{1}{3} C_V v l, \tag{6}$$

where C_V is the lattice heat capacity at constant volume, v is the phonon velocity, and l is the average phonon mean free path. At high temperatures C_V and v are almost independent of T. However, l varies inversely as the absolute temperature T, so that $\lambda_{lat} \propto T^{-1}$ (Peierls [73]). The variation $\lambda_{lat} \propto T^{-1}$ results from a first-order perturbation treatment, which is valid for the weak interaction. With temperature increasing the phonon mean free path approaching to lower limit l_0, which can be comparable to the linear dimensions of the unit cell of the crystal lattice (Ziman [74], Spitzer [75]). Therefore, with temperature increasing, the lattice thermal conductivity is transforming from Eq. (5) to $\lambda_0 = \frac{1}{3} C_V v l_0$. Roufosse and Klemens [76] studied the law of the gradual temperature transformation of λ_{lat} from Eq. (5) to λ_0 (see below, sec. 6). They found that at temperatures $T<T_0$, where $T_0 = (Mv^2/k)/(101.2\gamma)$, the lattice thermal conductivity varied as

$$\lambda_{lat} = \frac{3k}{6^{2/3}\pi^{4/3} A a T} \propto \frac{1}{T}, \tag{7}$$

while at temperatures $T>T_0$, λ_{lat} varied as

$$\lambda_{lat} = \frac{B}{T}\left[\frac{2}{3}\left(\frac{T_0}{T}\right)^{1/2} + \frac{1}{3}\left(\frac{T_0}{T}\right)\right] \tag{8}$$

Equation (8) indicate that deviations from the inverse temperature dependence ($\lambda_{lat} \propto T^{-1}$) are small (1.14 times) up to $T \approx 2T_0$, while at $T \approx 3T_0$ the deviation is about 38 % (Roufosse and Klemens, [76]). For the complex crystals with N atoms per unit cell the temperature dependence λ_{lat} analogously to Eq. (8) with $T_0 = T_1 N^{-2/3}$. As the pressure is increasing, T_0 and T_1 tend to increase because with increasing pressure v increase, and γ decrease. This is very important result for geophysical purposes when high temperature is accompanied by high pressures.

The Grüneisen parameter γ_{Th} is very important to determine the pressure changes with heating the material at constant volume. Hofmeister [77] developed new theory which can explain the effect of pressure on ETC of solids. Her model for the lattice thermal conductivity λ_{lat} is based on phonon lifetimes. She is studied of the pressure and temperature dependence of transport properties from the Grüneisen parameter $\gamma_{\mathrm{Th}} = \alpha V K_{\mathrm{T}}^2 / C_{\mathrm{V}}$, isothermal bulk modulus $\chi_T = -V\left(\frac{\partial P}{\partial V}\right)_T$, and thermal expansivity, $\alpha = \frac{1}{V}\left(\frac{\partial V}{\partial T}\right)_P$. The derived equation is Hofmeister [77]

$$\lambda(T) = \lambda(298)\left(\frac{298}{T}\right)^a \exp\left[-(4\gamma_{Th} + 1/3)\int_{298}^{T}\alpha(\theta)d\theta\right]\left(1 + \frac{\chi_0'}{\chi_0}P\right), \tag{9}$$

with a=0.33 for silicates and a=0.9 for MgO, $\chi_0' = d\chi_T / dP \approx 4$ to 5 is constant, $\chi_0 \approx 135$ GPa (Honda and Yuen [78]). Grüneisen parameter γ_{Th} is essentially a measure of the fractional change of pressure on heating a material at constant volume, so that seismic velocity and thermal conductivity are related by means of the effect of pressure on the lattice constant. According to the Hofmeister model (Eq. 9) the temperature coefficient of the ETC, β_T, is related to the thermal expansivity coefficient α as

$$\alpha = -\frac{1}{(4\gamma_{\mathrm{Th}} + 1/3)}\left(\beta_T + \frac{a}{T}\right). \tag{10}$$

According this relation (10), the measured values of temperature coefficient of ETC, β_T, can be used to calculate the values of the thermal expansivity coefficient α, therefore to calculate the temperature dependence of the specific volume of the rocks at various pressures and temperatures, i.e., to develop equation of state $V(T,P)$. Hofmeister [77] also explained the pressure dependence of the thermal conductivity λ based on phonon lifetimes. The pressure coefficient derived by Hofmeister [77] is

$$\frac{\partial \ln \lambda}{\partial P} = \frac{1/3 + 4\gamma_{\mathrm{Th}}}{\chi_T}, \text{ or } \frac{\partial \ln \lambda}{\partial P} = K_T(1/3 + 4\gamma_{\mathrm{Th}}), \tag{11}$$

where $K_T = -\frac{1}{V}\left(\frac{\partial V}{\partial P}\right)_T$ is the isothermal compressibility. The values of pressure coefficient, $\frac{\partial \ln \lambda}{\partial P}$, derived from the ETC measurements can be used to estimate the values of χ_T and K_T. Therefore, reliable ETC data can be used to calculate the values of isothermal compressibility, K_T, and develop equation of state of solid materials using the relation (11).

1.3. Heat Transfer Mechanism in Fluid-Saturated Rocks

The mechanism of heat transfer in porous materials is complicated by the irregularity of the microstructure. In most fluid-saturated porous materials, heat is propagated by three processes: (1) thermal conductivity through the solid and liquid phases; (2) radiation across the pore; and (3) convection through the pore. For the small pore size the convection can be neglected and the radiation is not a factor at low temperatures (below 300 °C). The total ETC in multiphase materials depends on the thermal conductivity of each phase, the fractions of the phases, and the way in which the phases are distributed. The distribution of the phases includes theirs size, shape, orientation, continuity relative to the heat flow direction and gas pressure in the pore. There are a number of models for the prediction of the ETC which is based on theoretical studies (see below sect. 5.3.3).

The influence of pore fluid on rock thermal conductivity depends on porosity and texture: if porosity is low or porosity is concentrated in vertical fractures, the effect of pore fluids on the gross conductivity is small because the matrix conducts heat so much better than the pore fluid. However, if porosity is very high or porosity is concentrated in horizontal fractures, the gross thermal conductivity of rock will decrease because the conductivity of fluid is much lower than the rock matrix. Decrease of the ETC of the rocks will trap some extra heat in the sequences with high porosity, resulting in an increase of the geothermal gradients. The heat trapped in the rocks would cause the thermal expansion of the pore fluid and increase the fluid pressure. Thermal expansion of rocks is relatively small in magnitude; however, thermal expansion behavior may have significant effects on the structure of rocks. Because of the multi-mineral composition of rocks, heating causes micro-fracturing due the differential thermal expansion of mineral grains. Differences in thermal expansion characteristics of various minerals in the assemblage of mineral grains can cause structural damage upon heating the rock. Even a given mineral the coefficient of thermal expansion may be different in different crystallographic directions. These differences in thermal expansion results in stress concentrations at grain contact points, when a rock heated, leading to the possibility of fracture of individual mineral grains and disaggregating of the rock. Therefore, the effects of heating rocks are altering their properties.

ETC is a function of the thermal conductivities making up the rock, including the fluid present in the pore space. The thermal conductivity of quartz is sensitive temperature variations, and this effect is manifested in a reduction of the thermal conductivity of quartz rich sandstones with increasing temperature [79,80]. It is obviously that the greater part of the heat is conducted through the solid phase, because the thermal conductivity of the skeleton much higher than the conductivity of saturated fluid (gas or liquid phases). Therefore, it is understandable that increasing the proportion of solid phase per unit volume (decreasing the porosity) will result in an increase in the ETC of the material.

The heat flow through fluid-saturated porous rocks can be expressed as in Figure 1. Physically it is very clear that ETC of fluid saturated porous materials is function of the thermal conductivities of solid phase, λ_S, fluid phase, λ_f, and conductivity of the interferences (contact region), λ_{SS} (grain-grain) and λ_{Sf} (grain-fluid) interfaces. The conductivities of the solid and fluid phases are well known, while the nature and physical mechanism of the thermal resistance solid-solid, λ_{SS}, and solid-fluid, λ_{Sf}, interferences (grain-grain and grain-fluid contacts resistences) is still less understandable. Increasing the pressure improves the thermal contact between grains (λ_{SS}) and increases the ETC. The reason the gas saturated conductivities are much lower than those of water or oil saturated rocks is partly because of the low conductivity of gas and partly because in the vicinity of grain contacts the distance between grain surfaces is not much larger than the mean free path of gas molecules. This results in a decrease in the ETC of the gas, so that essentially no heat flows from the surface of one grain to the gas and then to the surface of the next grain even though near grain contacts the surfaces are extremely closes. Consequently, nearly all the heat must be transferred through the grain contacts. Improved heat transfer efficiencies at the grain contacts occur when the sample is saturated with liquid (oil or water, denser phase than gas, $\lambda_{SL} >> \lambda_{SG}$). Water and oil has a much higher thermal conductivity than gas, its molecules have a much shorter mean free path.

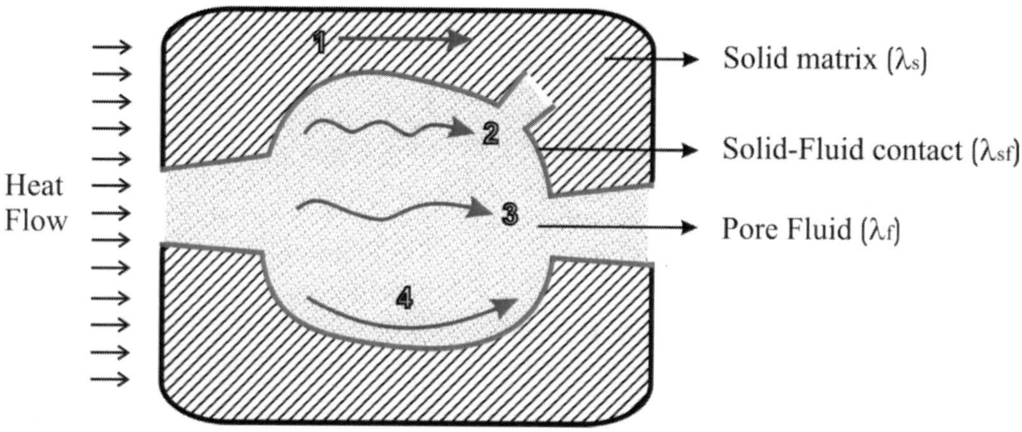

Figure 1. Heat transfer mechanism in the porous material containing a fluid in the pore space. 1. Conduction of heat through solid matrix; 2. Conduction of heat through saturated fluid inside the pores; 3. Radiation through the pore space; 4. Convection currents within the pore space.

Consequently, in the vicinity of grain contacts heat transfer from grain surface to liquid to grain surface becomes significant. The liquid conductivity is less than that of quartz but because the distance the heat travels through liquid phase is very small around the grain contacts, the intergranular heat transfer efficiency by this method is almost as great as if the heat flowed directly from one grain to another.

2. Experimental Techniques to Measure Thermal Conductivity of Porous Rocks

There are a lot of experimental techniques for the measurements of the thermal conductivity of dry and fluid-saturated porous rock materials. Each of these methods is suitable for a different group of rock materials depending on their characteristics. Blackweel and Spaffond [564] have summarized various laboratory techniques, including the divided-bar steady-state measurements and the needle-probe transient measurements. Although thermal conductivity is simple in concept it is a difficult quantity to accurate measure in practice.

All absolute methods of thermal conductivity measurements depend upon the knowledge of heat input and temperature gradient, and all are eventually limited, either by the inability to eliminate all unwanted thermal resistance, or by the inability to ensure that all the measured heat travels in the required direction and not elsewhere. Different methods were proposed by various authors to measure ETC of porous dry and fluid –saturated rocks material at laboratory conditions (see for example, [23, 81-84]). Detailed review various methods of ETC measurements were provided previously also by several authors [23, 70, 72, 83, 85-89]. Experimental methods for the measuring of thermal conductivity of rock materials can be divided into two groups: 1. Steady-state methods; and 2. Transient temperature (unsteady-state) methods. Steady–state methods are more frequently used than transient methods. Majority of laboratory measurements of the ETC of rocks were based on steady-state heat flow technique.

2.1. Steady-State Methods

The conventional methods of thermal conductivity measurement are steady-state methods which require the measurements of a steady heat flux and of the temperature gradient through test samples in the form of slabs, cylinders, or spheres (see Figure 2). In this method the temperature field and heat fluxes across the sample are kept constant, *i.e.*, the temperature field inside the sample changes in space only ($dT/dx \neq 0$), therefore for this method ($dT/dt=0$). Thus to measure the thermal conductivity using this method the temperature difference across the sample is require. Generally, the attainment of steady-state conditions requires many hours, and large thermal gradients may produce nonuniform saturation distributions. Transient methods are faster and less likely to move fluid. In order to heat the sample steadily, an almost perfect contact is required between the heat source and the samples. Thus, the samples have to be prepared with extreme care. Steady-state method of ETC measurements where the thermal conductivity is determined by Fourier's law

$$Q = -\lambda \frac{dT}{dz}, \quad (12)$$

where Q is the heat flow (W·m^{-2}), λ is the thermal conductivity (W·m^{-1}·K^{-1}), and $\frac{dT}{dz}$ is the temperature gradient normal to the isothermal surface, when there is a constant temperature gradient across the samples and the heat flow is stable. At constant heat flow rates Q, temperature gradient $\frac{dT}{dz}$ is inversely proportional to the thermal conductivity λ. As one can see from Eq. (12), thermal conductivity can be determined by measuring the heat flux Q and temperature gradient $\frac{dT}{dz}$ across the sample (temperature difference between two known positions).

Accurate measurements of these two parameters (Q and $\frac{dT}{dz}$) depend on the method and cause uncertainty in the thermal conductivity data. The disadvantages of steady-state methods are [90]: 1. Several hours are needed for an individual test to reach an acceptable accuracy level.

The time required to run these test too long; 2. Long time is necessary to reach steady-state condition. This is can be effected on the changes of the thermal properties of the sample, especially for uniform distribution of the moisture within a sample; 3. Unaccounted heat losses are difficult to prevent; 4. Problem is in establishing good thermal contact between the heating and cooling elements and rock samples.

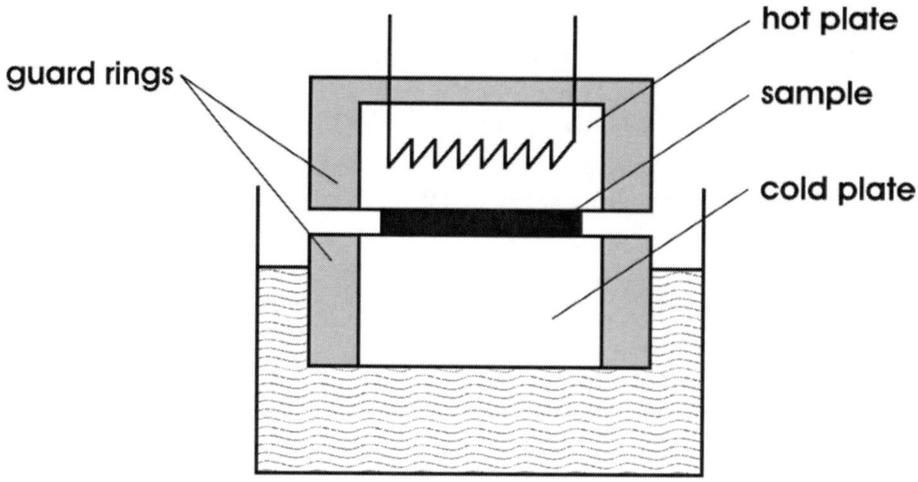

Figure 2. Schematic representation of the principle of a guarded plate thermal conductivity cell (steady-state method).

2.2. Transient Methods

Transient methods of thermal conductivity measurements are one of the frequently using method due to availability of fast data acquisition and measuring computer systems (see for example, [92, 523-526]). In this method heat is generating inside (except Laser Flash method) the sample and temperature rise of the heating source is measured with time. The temperature field inside the sample changes in time ($dT/dt \neq 0$) and in space (across the sample, $dT/dx \neq 0$). Practically, the temperature field inside the sample is produced by passing an electrical current through a line or plane electrical resistance embedded in the sample or by a non-contact method through electromagnetic waves (laser flash). A thermometer, that can be unified with the heat source or can be placed apart from the heat source, measures the temperature changes caused by the heat source.

Usually in many transient techniques, the heating source also serves as a resistance thermometer. In this method the temperature of the sample and the heating source are stabilized with the surrounding temperature. Then a small disturbance in temperature is applied by applying a heat pulse, *i.e.*, by passing a small current through the heating source for a short time. After that the change in the temperature of the heating source itself, the thermal properties of the surrounding sample can be calculated according to the type (theoretical model) of the heating source (sensor) used. The temperature difference is kept less than 2 K and sample size depends on the size and the theoretical model of the heat source used.

Unsteady-state (transient) methods are usually determined thermal diffusivity ($a = \lambda / \rho C_P$ -determine the rate of heat propagation in transient-state processes or governs rate of spread of temperature disturbances) rather than thermal conductivity directly. Therefore, in this method is possible to measure multiple thermal properties in a single run. This is other advantage of the transient method over steady-state methods. In transient methods values of thermal diffusivity are generally measured and then these data used to calculate the thermal conductivity. This method based on the measurement of the sample temperature as a function of time. The sample heated by a known line heat source. The main differences among various versions of the transient methods are the shape of the heating sources. The shape of the heat sources may be a point, a line, a plane or a cylinder. The most popular heat source configuration is the line as shown in Figure 3. For example, Ljubimova [91] used a square and flat heater sources to measure the diffusivity of rocks.

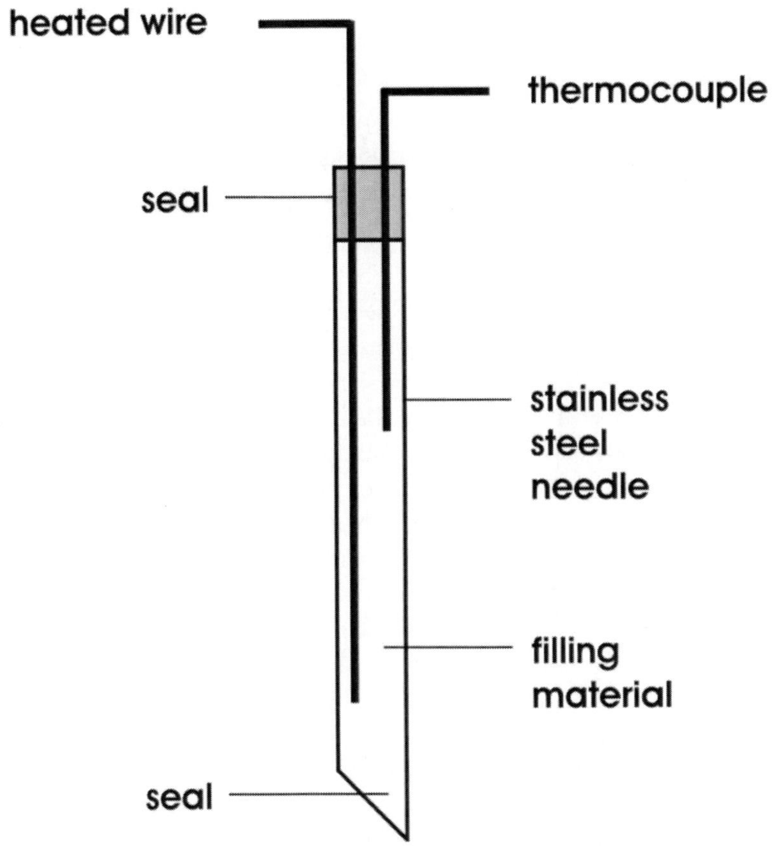

Figure 3. Schematic representation of the principle of heat- probe (transient method).

A cylindrical heat probe was used by Sepaskhah and Boersma [259], while Maqsood and Kamran [313, 527, 528, 578] used a transient plane source. But, in this case the process of placing the heater inside the sample is much less convenient and more disruptive of the sample than the simple needle probe. The cutting or drilling required for these methods seems to be just as extensive as the preparation for the divided bar, with no less potential for damage to the rock fabric.

The theory of the transient methods is based on an ideal model. An ideal model approximates that an infinite line (or plane) heat source of zero heat capacity is embedded in an infinitely extended in all directions (x,y,z) isotropic medium. In practice an infinite medium and heat source are not possible, therefore by applying boundary conditions to the ideal model a semi-infinite ($x>0$, $-\infty<x<\infty$, $-\infty<z<\infty$) medium is considered. The working equation of the heat source depends on its shape. Assuming that the line heat source is an infinity medium, and that the heat flow is radial, the temperature T at a point very close to the source, after a time τ will be

$$T = \beta + (Q / 4\pi\lambda) ln(\tau), \tag{13}$$

where β is the constant, and Q is the heat flow per unit length of the line source. The temperature difference ($\Delta T = T_2 - T_1$) after times τ_2 and τ_1 of heating will be

$$\Delta T = (Q / 4\pi\lambda) ln(\tau_2 / \tau_1). \tag{14}$$

Therefore, the thermal conductivity can be obtained from the slope $(Q / 4\pi\lambda)$ of a semilog plot of T versus τ. Transient hot wire method provide thermal conductivity λ more accurately (6 %) than thermal diffusivity a (30 %) [92]. This method is much faster (short measurement times) than steady-state methods but the results less accurate and relatively small samples. Density and heat capacity are needed to calculate the thermal conductivity from thermal diffusivity measurement. Most transient methods require lengthy and involved sample preparation. Thus, transient techniques have several advantages in compare with steady-state methods: 1. Experimental proceed relatively quickly; 2. semi-infinite domain is a convenient alternative to adiabatic boundary conditions; and 3. large thermal time constants are actually a benefit because they ensure that temperature changes are slow enough to be captured by typical data acquisition equipment. The typical uncertainty of the ETC measurements for this method is within 5-6 %. Middleton [93] detailed described transient method of measuring thermal conductivity of rocks.

Abid [526] used transient hot-bridge (THB) heat source (sensor) for the thermal conductivity measurements of moist porous material. This technique is based on the working principle of transient hot-wire (THW) method. The thermal conductivity and thermal diffusivity of a material are calculated from a single experiment. The basic idea of the method is to create an inhomogeneous temperature field inside the sample by constant heating and then monitoring the corresponding temperature rise of the heting source (sensor). The working equation of the transient hot wire technique is [523] (in term of voltage)

$$\Delta V(t) = V(t) - V_0 \approx \frac{\alpha V_0^2 I}{4\pi\lambda L}\left[\ln(t) + \ln\left(\frac{4a}{Cr^2}\right)\right], \tag{15}$$

The output voltage of the sensor is linear function of $\ln(t)$ with slope m and intercept n, $\Delta V(t) = m\ln(t) + n$, where slope of the output signal is $m = \frac{\alpha V_0^2 I}{4\pi\lambda L}$, and intercept is $n = m\ln\left(\frac{4a}{Cr^2}\right)$, $C = \exp(\gamma)$=1.781. From these equations the thermal conductivity and thermal diffusivity can be calculated using a single experiment as

$$\lambda = \frac{\alpha V_0^2 I}{4\pi m L} \text{ and } a = \frac{r^2}{2.25}\exp\left(\frac{n}{m}\right). \tag{16}$$

Transient hot –wire technique is the most widely used to measure the thermal conductivity of liquids and gases [529-531]. This technique is also frequently using for solid materials with uncertainties less than 1 % [538, 539]. Hot-disk [119, 120, 525, 540, 541], hot-strip [542, 543], or hot-bridge [524] techniques are more commonly applied for solids. In this case hot-wire is embedded in a groove between two equally sized sample halves. Unfortunately, fixing of hot-wire in solid samples is causing a problem of thermal contact resistance. Therefore, it is usually preferred to use transient hot-strip [532] method (THS) for the measurements on solid materials. On the other hand, THW technique is the best for thermal conductivity measurements on wetting fluids and gases because of the small contact resistance. A detailed review of the transient hot-wire cell used for the thermal conductivity measurements in liquids and gases is given in [531]. THS method can be used for the simultaneous measurements of thermal conductivity and thermal diffusivity of liquids and solids. The ideal model of a THS is based on a metal strip of infinite length and infinitesimal thickness embedded in an isotropic and homogeneous medium [525]. The strip is used as a continuous heat source and a resistance thermometer simultaneously. The output signal is obtained in the form of a voltage drop across the two ends of the sensor. This signal is the measure of the thermal conductivity and thermal diffusivity of the sample. The temperature distribution in the metal strip due to the heat flux produced per unit area is given by equation [525]

$$T(y,t)=\frac{q}{4a\sqrt{\pi}}\int_0^{\sigma}\left[\operatorname{erfc}\left(\frac{y-d}{\sigma}\right)-\operatorname{erfc}\left(\frac{y+d}{\sigma}\right)\right]dy, \tag{17}$$

where d is the half width of the strip (heat source) and $\sigma=2\sqrt{at}$. The total resistance of the strip can be represented as a function of temperature as

$$R(T)=R_0\left[1+\frac{\alpha}{2d}\int_{-d}^{d}T(y,t)dy\right], \tag{18}$$

where R_0 is the electrical resistance of the strip at 0 °C and α is the temperature coefficient of resistance of the metal strip. The average temperature of the strip can be obtained by integrating equation (17) over the width, which yields the working equation of THS technique

$$\overline{T}=\frac{q}{4\pi\lambda}\left[3+\ln(t)+\ln\left(\frac{a}{e^{\gamma}d^2}\right)\right]=m\ln(t)+n. \tag{19}$$

Now the thermal conductivity and thermal diffusivity of the test specimen can be calculated from the slope m and intercept as follows

$$\lambda=\frac{q}{4\pi m}\text{ and } a=d^2\exp\left(\frac{n}{m}-3+\gamma\right). \tag{20}$$

The relative uncertainty for 95 % confidence level for THS method are estimated by Hammerschmidt [532] is 5 % for thermal conductivity and 22 % for thermal diffusivity. Due to its very small electrical resistance the output signal of a strips very low [526], *i.e.*, the signal-to –noise ratio is poor. The end-effect (the non-uniform temperature profile along a heated strip or wire), is increase the uncertainty of the measurements. Therefore, this method also cannot be used to accurate measure thermal conductivity of porous rock materials.

Gustafsson [119] proposed a new transient plane source (TPS) method for the measurement of thermal conductivity and thermal diffusivity for a wide range of materials. The shape of the TPS sensor is made in the form of a 10 μm thick Nickel spiral covered on both sides with a 25 μm thick insulating Kapton layer. An electrical pulse is applied to heat the sensor for a short time. The resistance of the sensor changes with time, which can be calculated as

$$R(T)=R_0\left[1+\alpha\left(\Delta\overline{T}(\tau)\right)\right], \tag{21}$$

where R_0 is the initial resistance of the TPS sensor before applying current, α is the temperature coefficient of resistance and $\Delta\overline{T}(\tau)$ is the average value of the time-dependent temperature rise of the sensor, which is depend on output power of the sensor, thermal conductivity and thermal diffusivity of the surrounded test material and the design characteristics of the sensor. Average temperature $\Delta\overline{T}(\tau)$ can also be expressed as a function of only one dimensionless parameter [525] $\tau=\sqrt{\frac{ta}{r^2}}$, where r is the radius of the sensor, a is the thermal diffusivity of the sample and t is the time. The temperature response of the TPS sensor as a function of τ is given as $T(\tau)=\frac{P_0}{\sqrt{\pi^3}r\lambda}D(\tau)$, where P_0 is the total output power and $D(\tau)$ is the characteristic time-dependent function defined as

$$D(\tau)=\frac{1}{\left[m(m+1)\right]^2}\int_0^\tau\frac{1}{s^2}ds\left[\sum_{n=1}^{m}n\sum_{k=1}^{m}k\exp\left(-\frac{n^2+k^2}{4m^2s^2}\right)I_0\left(\frac{nk}{2m^2s^2}\right)\right], \tag{22}$$

where m is the number of concentric circles in the sensor (the optimal values of m>10 and τ >0.1 [526]). Due to non-linear implicit output temperature response, it is difficult to find the best linear part of the curve, therefore to find best value of the thermal conductivity and thermal diffusivity.

The simple experimental method determination of the field of the sample is pulse transient technique [149, 150, 533]. This method is based on the determination of maximum temperature rise in a sample at a specific distance from the heat source. Sample is divided into three parts. A planar heat source is sandwiched between sample I and II. Sensor is heated by passing an electrical pulse. A thermometer is located in between parts II and III of the sample, and records the temperature response due to the applied heat pulse. Theoretical model of the PTM assumes a planar heat source with infinitesimal thickness embedded in an infinite body. The temperature response at point x after a time t can be represented as [81],

$$T(x,t)=\frac{Q}{4c\rho\sqrt{\pi at}}\exp\left(-\frac{x^2}{4at}\right). \tag{23}$$

To reduce the thermal contact resistance between the samples, thermal paste and a very thin thermocouple is used. The equation for thermal diffusivity and heat capacity calculations from the temperature response (Eq. 23) can be derived from maximum of the function (23) at a point $x=h$ as $a=\frac{h^2}{2t_m}$ and $c=\frac{Q}{\sqrt{2\pi\exp(1)}}\rho hT_m$, where h is the thickness of the sample, t_m is the time at which temperature response reaches its maximum T_m, Q is the applied pulse energy, ρ is the density of the sample. The thermal conductivity can be calculated as $\lambda=\rho ac=\frac{hQ}{2\sqrt{2\pi et_m}T_m}$. The uncertainty of the method is 6 % to 15 % (for thermal conductivity), 3 % to 10 % (for thermal diffusivity), and 3 % to 5 % (for heat capacity) [149, 150, 533]. The disadvantages of the PTM are that, it needs a high pulse input, which establishes a large temperature difference at the two measuring ends of the sample. At high temperature pulse the moisture inside a porous material starts evaporating and the thermal conductivity itself depend on temperature. These factors are considerable affecting on the reliability of the results.

In order to measure the thermal conductivity of fluid saturated porous materials most suitable method is transient hot-bridge method (THB) [526, 577]. The THB sensor is made of a 7.5 μm thin printed circuit foil of Nickel which is sandwiched between two polyimide sheets each having of 27.5 μm. The output signal (V_{THB}) of the sensor is obtained by taking a difference of temperature rises between inner and outer strips $\Delta T=T_{inn}-T_{out}=T(D_1)-T(D_2)$. The output signal of the THB sensor is proportional to ΔT as $V_{THB}=0.5R_{eff}I_B\alpha\Delta T$, where α is the temperature coefficient of electrical resistance, R_{eff} is the electrical resistance and I_B is the applied current to the ridge circuit. The mean temperature rise of a line heat source as a function of dimensionless time τ is [524]

$$\overline{T}(\tau)=\frac{q}{\sqrt{4\pi}L\lambda}f(\tau),$$

where

$$f(\tau)=\tau erf\left(\tau^{-1}\right)-\frac{\tau^2}{\sqrt{4\pi}}\left[1-\exp\left(-\tau^{-2}\right)\right]-\frac{1}{\sqrt{4\pi}}Ei\left(-\tau^{-2}\right), \tag{24}$$

where $\tau=\frac{\sqrt{4at}}{D}$. Here $q=I_B^2R_{eff}/L$ is the heat flux per unit length of the embedded heat source strips, D is the width of the heating strip. After some mathematical manipulations (see details [526]), the fundamental relation for the basic operation of the THB sensor which represent the output signal is

$$V_{THB} = 0.5 R_{eff} I_B \alpha \frac{q}{\sqrt{4\pi L\lambda}} \left[f_{inn}(\tau) - f_{out}(\tau) \right]. \tag{25}$$

This is nonlinear equation and the thermal properties can be only numerically calculated. After applying first order approximations to Eq. (25), the mean temperature of the inner and outer strips can be obtained as [536]

$$\overline{T}_{out}(t) = \frac{q}{4\pi L\lambda} \ln\left(\frac{45at}{D^2} \right) \text{ and } \overline{T}_{inn}(t) = \frac{2q}{4\pi L\lambda} \ln\left(\frac{45at}{4D^2} \right). \tag{26}$$

Therefore, the final output signal of the sensor can be calculated as

$$V_{THB} = \frac{R_{eff}^2 I_B^3 \alpha}{8\pi L\lambda} \left[\ln(t) - \ln\left(\frac{45a}{16D^2} \right) \right]. \tag{27}$$

This equation is linear with logarithm of time. The slope of the plot of V_{THB} against $\ln(t)$ is m and intercept is n, i.e., the thermal conductivity is $\lambda = \frac{R_{eff}^2 I_B^3 \alpha}{8\pi Lm}$ and the thermal diffusivity is $a = 2.82\exp\left(\frac{n}{m} \right)$. The relative uncertainty in thermal conductivity and thermal diffusivity measurements are 3 % and 9 %, respectively. These numbers are not including the systematic uncertainties like contact resistance. The advantages of the THB method are [526]: (1) time required for a measurement is very short (1 to 2 minutes); (2) maximum temperature rise is 2 K; (3) constant penetration depth of heat for different samples; (4) low uncertainty range (3 % thermal conductivity and 10 % for thermal diffusivity); (5) design compensates the end-effect; and (6) minimizes the contact resistance problem. Kiyohashi and Deguchi [110] also successfully used hot-wire method to measure ETC fluid-saturated rocks.

2.3. Contact Methods

Most of the available ETC measurements for dry and fluid-saturated porous rocks were performed using the conventional (classical) contact methods (divided-bar, needle-probe, guarded hot plate). In these methods a heat source and thermocouples directly contact the sample. Therefore, all conventional contact methods include uncertainty due to contact thermal resistance between the heat source (heater) and thermocouples and the measuring sample at interface. Very important source of error which always present and cannot be removed completely at thermal conductivity measurements is the contact resistance at the sample-equipment interface (Beck [94]). Heat transfer across the boundary between heater and the sample depends on the properties of this contact. If a relatively high conductivity material is used in the sample-heater interface, its thermal resistance can be made quite small relative to the thermal resistance, its thermal resistance of the sample and the effect ignored. Water is a good fluid to use since its thermal conductivity is relative high and in many cases the requirement is for the conductivity of water saturated rocks. However, if air or oil are used, then large errors can results, especially if the rock is coarse granite and porous.

In contact methods the good contact between the samples and the heater or thermocouples is important to prevent contact resistance which is very important source of the experimental uncertainty. A smooth and well –prepared sample surface that exactly fits the sample cell is particularly important. Different types of waxes and contact-fluids (vaseline and glycerin, for example [168,220,526,560,561]) were used by various authors to improve the sample contact with the heater and thermocouples during the measurements (see also, for example, MidttØmme and Roaldset [95]).

For conventional contact methods of ETC measurements the differences between reported data reached up to 20 % and more, even at room temperature (Ross et al. [96]). Systematic discrepancy between various data measured with conventional contact methods arise from contact resistance and differential thermal expansion (Parker et al. [84]). Surface roughness creates "gap" through which phonons cannot propagate Hofmeister et al. [89] (see Figure 4). Differential thermal expansion between thermocouples and sample can create the additional thermal resistance [89].

The processes occurring during heat transfer at the contact are very complex [574,575]. Very little is known about the mechanism of heat transfer at the contact between two bodies. Figure 4 schematically depicts the contact between two bodies (heater and sample). The contact between two bodies (see Figure 4) contains surface of direct contact (Figure 4a) and free space (Figure 4b) forming a contact layer. In Figure 4a heat flux lines at the points of direct contact are depicted schematically as a narrows. The thermal conductivity of the contact layer is much lower than that of the bodies. The properties of contacts can be affected by varying the pressure on the body perpendicular to the contact surface. This is other way to improve the contact resistance between the heater and sample (see below). By changing the pressure the number and area of the direct contact surface is increasing, while the free spaces in the contact will decrease as the pressure on the body increases.

The maximum pressure is depending on the mechanical properties of the body and the pressure can be increased up to plastic deformation of the body. The contact between two bodies consists of direct contact region between two surfaces and of spaces filled with the contact layer-the medium. It is very difficult accurately describe the properties of such a contact.

The heat flux from body to another through a planar contact unit can be estimated as $q = \alpha(T_2 - T_1)$, where T_1 and T_2 are the temperatures of the surface in contact, α is the thermal contact conductance (or its inverse value -contact thermal resistance). The contact thermal resistance is a function of the pressure on the bodies perpendicular to the contact surface. The pretreatment of the contact surfaces is characterizing by the mean roughness factor N_{mean} (depends on materials). The detailed discussion of the effect of contact thermal resistance on the thermal conductivity measurements is provided by Kubicar [533].

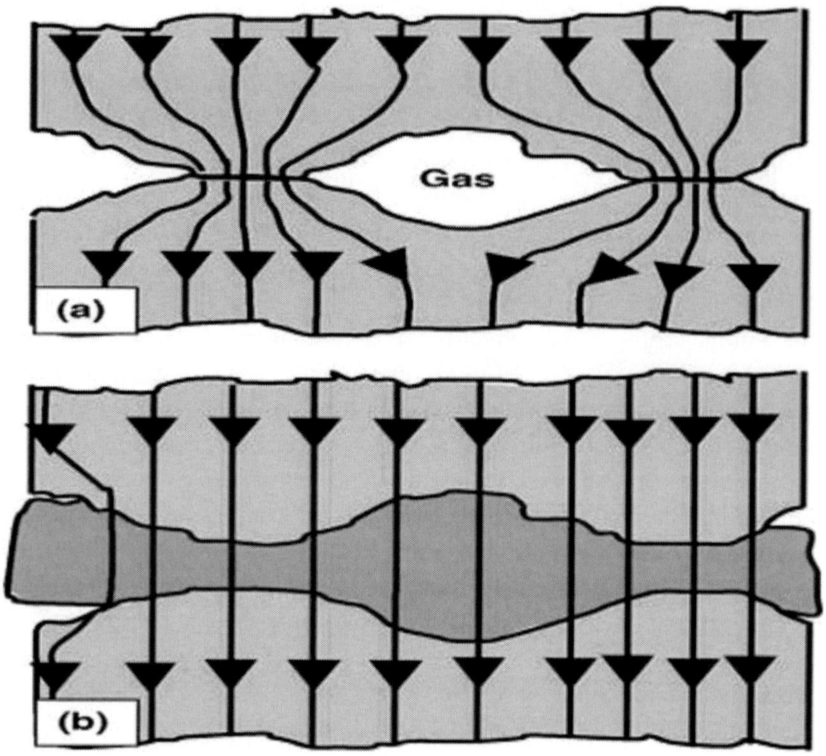

Figure 4. Schematic representation of the effect of contact thermal resistance between the heater and the sample on the heat transfer (flow). (a) contact space filled with gas (air); (b) contact space filled with contact well heat transfer material.

Most contact areas consist of more than 90 % air voids (free spaces in the contacts usually contain gas, see Figure 4), which represent a significant resistance to heat transfers since air is not a very effective heat conductor. Heat transfer across the contact occurs through the mechanism of radiation, convection and conduction. There are different ways to reduce the contact thermal resistance: surface roughness and flatness (Li et al. [97]), surface irregularities, "wariness", contact pressure (Dumont et al. [98]). Thermal interface materials (thermal greases, for example) can be used to fill the air gaps between the two surfaces (heater and sample). Strack et al. [99] improved the contact between heat source and the sample to heat the sample at steady-state condition by increasing the input power. Contact methods are take long measurements time and radiative heat losses from the surface (Parker et al. [84]).

Most measurements of thermal conductivity of rock materials have been performed using two conventional contact techniques such as the needle-probe and divided-bar methods. There are two differences between needle-probe and divided-bar methods (Pribnow et al. [100]): 1. The needle- probe method is a transient while the divided–bar is steady-state; 2. The divided–bar measurements yield thermal conductivity in the direction parallel to the divided –bar axis, whereas the needle-probe method yield thermal conductivity in a plane perpendicular to the needle-probe axis. Needle-probe and divided-bar methods were discussed in detail in several review publications (see for example [72, 101-103]).

2.3.1. Needle-Probe Method

In contrast to steady-state techniques (divided-bar, for example, see below), transient methods are fast, therefore less likely to produce nonuniform saturation distribution in the case of partially saturated samples [93]. For a needle-probe, the length *L* can be assumed to be infinite and the problem is reduced to two dimensions. The heat generated from looped wire is $Q=2I^2R/L$, where *R* is the resistance of the loop wire, *R*/*L* is the resistance of the needle per unit length. At any time *t* after heating has started, the temperature *T* is related to the thermal conductivity λ by

$$T=(Q/4\pi\lambda)ln(t)+C, \tag{28}$$

where *Q* is the heat input per unit length and unit time, *C* is a constant. This method assumes that the measurement curve is linear and ignores the imperfections of the experiment expressed in the constant *C*. The correct choice of a time interval is difficult. At the early stage of heating, the source temperature is affected by the contact resistance between the source and the surrounding material. At the later stage of heating, boundary effects of the finite length of the source affect the measurement. This method (needle-probe) widely used for soft rocks (see rock types classified by Mongelli [104]). Some sandstone is too hard to inject a hypodermic needle. According to Mongelli [104] the accuracy of the method is 2-3 %. A lot of experimental ETC data for rocks materials were determined using the needle-probe method (see below, sect. 3).

The needle-probe method is a transient method originally developed by Von Herzen and Maxwell [105]. In this method a probe, which consists of a heating wire (the sample heated by a known line heat source) and a thermistor, is inserted into the sample. The temperature of the sample is recorded as a function of time. From the temperature-time curve ($T-\tau$ curve, thermogram, Eq. 28) the thermal conductivity of the sample can be determined. This method is simpler and faster than the divided-bar method and requires less sample material. For these reasons it is used more widely than other methods. This method as all other contact methods suffers from contact thermal resistance problems (see above, sec. 2.3). In needle-probe method the thermocouples attachment involves drilling a hole in the center of the sample and inserting a smaller –diameter wire. Inserting the probe into the rock material disturbs the arrangement of the material grains and tends to minimize grain to probe contact. For consolidated rocks it is necessary to drill a hole to insert the probe. In this case good contact between the probe and the rock must be established. As was stated by Sass et al. [23, 106] the uncertainty in thermal conductivity measurements with the needle-probe method is within 10 %, although some cases this is may increase to 20 %. Transient hot-wire technique was used by many authors (see for example, [42, 100, 107-114, 190, 231-234, 241, 242, 247, 250, 257, 259, 262, 264, 305, and 556]) to measure ETC of the Earth materials (dry and fluid –saturated rocks and rock forming materials).

The needle-probe is based on the theory of a line source in an infinite medium. In practice, it is a needle-probe embedded in and flushes with the surface of a material of very low thermal conductivity (half-space line source, Pribnow and Sass [115]). The needle-probe (Vosteen and Schellschmidt [111]) consist of a cylindrical probe of 2 mm in diameter and 70 mm in length. A thermometer at the center of the needle probe records the probe temperature. The probe length can be assumed to be infinite. The needle-probe is embedded in a cylindrical block of transparent plastic such as Plexiglas. The heat propagation into the sample is about 6 to 12 mm [116] depending on the thermal conductivity of the material. Suleiman [117 and 118], Gustaffsson [119 and 120], and Maqsood and Kamran [313] used hot-plane source transient method to measure of the thermal conductivity of dry and partially saturated solid materials.

The advantage of the method is the possibility to reduce the contact resistance between the sample and the sensor. To insure good thermal contact, in the experimental arrangement the sensor is clamped between the sample halves that consist of two identical disc-shaped pieces (diameter is about 7-10 cm, thickness is about 2.0 to 2.5 cm). For this method the thermal conductivity can be obtained from $\Delta T(\tau)$ as

$$\Delta T(\tau) = P_0\left(\pi^{3/2} r\lambda\right)^{-1} D(\tau), \tag{29}$$

where P_0 is the total output power; λ is the thermal conductivity of the sample; and r is the radius of the sensor. The function $D(\tau)$ is the theoretical expression of the time dependent temperature increase, which describes the conducting pattern of a disk-shaped sensor, assuming that the disk consist of a number m of concentric ring sources [123]. The time dependence of the TPS element during the transient recording can be expressed as

$$R(t) = R_0\left[1 + \alpha\Delta T(\tau)\right] \text{ or } \Delta T(\tau) = \frac{1}{\alpha}\left[\frac{R(t)}{R_0} - 1\right] \tag{30}$$

The fitting of the theoretical values of $\Delta T(\tau)$ Eq. (30) to experimental we can calculate all of the desired information. The construction of the measuring cell insures that the heat will not reach the outer boundary of the sample. The short measuring times (seconds) and small temperature gradient across the sample is preventing a non-uniform saturation distribution in the case of partially saturated samples [117]. The accuracy of the method is within 3 % to 5 %.

Krishnaiah et al. [121] developed thermal probe to measure thermal resistivity of various materials. It represents a line-heat source of heat input per unit length, Q, of constant strength, in an infinite homogeneous medium, initially at uniform temperature. The probe, which is 95 mm long and 6 mm in diameter, consists of a nichrome heater wire. The space between the heating element and the cooper tube is filled with thermal epoxy, which provides excellent thermal conduction and acts as an electrically insulated material. The temperature of the probe was measured with Type –*T* thermocouple which is attached to its surface. However, the probe cannot be inserted in the rock sample without drilling a hole; therefore, this is causing the "contact resistance" effect due to the formation of air gap between the probe and the hole. Keeping this in view, cylindrical cement bricks have been used for calibrating the probe. In needle-probe method used by Xu et al. [112] the thermocouples attachment involves drilling a hole in the center of the sample and inserting a smaller –diameter wire. Vosteen and Schellschmidt [111] measured thermal conductivity of rock samples (dry and water-saturated) at an ambient temperature (about 25 °C) with needle-probe technique. Pribnow [113] and Popov et al. [114] also used needle-probe method to measure thermal conductivity of rock materials.

2.3.2. Divided-Bar Method

As was mentioned above, the majority of routine measurements (see for example, [14, 23, 106, 100, 111, 122-124, 127-129, 168, 171, 174, 220, 222, 223, 263, 268, 269, 560, and 561]) of thermal conductivity rock materials are still made by using the simple conventional steady-state techniques such as the divided-bar method. This has the advantages of reasonable accuracy, simplicity of construction and operation, and an equilibrium time that is compatible with routine production. Basically the divided bar consist of a cylinder of material of well-known thermal conductivity that is divided in the middle so that a piece of rock of the same diameter can be inserted. The ends of the bar are held at different constant temperatures, so that a steady heat flow occurs along the bar. The temperature is measured at each end of the bar and also on each side of the rock sample. Then, assuming perfect insulating on the sides of the bar (using guarded ring), the temperature differences over each section is directly proportional to the thermal resistance of that section. It is not necessary to know the amount of heat flowing along the bar since the method is comparative. The perfect insulation of the sides of the bar is impossible to achieve and any insulation is inconvenient when processing large numbers of samples, there is always some heat loss to or from the side of the bar. Usually the length of the bar is small compared with its diameter. Good measurement principles require that the thermal resistance of each half-bar should be of the same order of magnitude as of the samples being measured. The basic working equation of the divided-bar can be written as

$$\frac{\Delta T_2}{\Delta T_1 + \Delta T_3} = \frac{R_2 + d_S / \lambda_S}{R_1 + R_3 + d_g / \lambda_g}, \tag{31}$$

where ΔT_1, ΔT_2, and ΔT_3 are the temperature drops across zones 1,2, and 3, respectively; R_1, R_2, and R_3 are the unwanted thermal resistances; λ_g and λ_S are the thermal conductivities of the reference discs and of the sample, respectively; d_S is the thickness of the sample and d_g is the total thickness of the two reference discs. The relation (31) can be simplified to

$$d_g \frac{\Delta T_2}{\Delta T_1 + \Delta T_3} = d_S \frac{\lambda_g}{\lambda_S} + R\lambda_g, \tag{32}$$

where $R \sim R_2 - (R_1 + R_3)\frac{d_S \lambda_g}{d_g \lambda_S}$. The ratio $\frac{\lambda_g}{\lambda_S}$, for the set of discs of different thicknesses, can be obtained as the linear slope of $d_g \frac{\Delta T_2}{\Delta T_1 + \Delta T_3}$ plotted against d_S.

These methods involve the simultaneous measurements of steady state heat flux and temperature gradient through test samples (in the form of bars) for which the solution to the differential equation of heat conduction is readily available. This method has some disadvantages such as long time required to attain equilibrium; edge or end-effects that may increase due to lack of good thermal insulation at the sample boundaries. Also, if any moisture is present in the sample, non-uniform moisture distribution may arise when the temperature difference across the sample is maintained for a long time. This is may lead to conductivity measurements that depend upon the sample size and the magnitude of the applied temperature difference [117].

The divided-bar (Beck [101 and 127]) is a steady-state comparative method to measure thermal conductivity of solid rock samples or a cylindrical cell (Sass et al. [23]), containing water-saturated material. The basic concept of this method is to compare the unknown thermal conductivity of a rock sample to the known thermal conductivity of reference material. The sample and reference materials are shaped into discs with a diameter of 50 mm and approximately the same thickness (10 mm) [111]. The reference material usually has conductivity comparable to that of rock samples. The heat flows parallel to the divided –bar axis (stack) from a heater *A* (high temperature level) to a second heater *B* (low temperature level) through two reference disks and the sandwiched rock sample. The radial heat flow can be neglected because guard heaters and insulating material minimize it. A thermal compound is used to reduce the contact resistance between the reference and the rock samples. The temperature drop across the rock sample is compared to temperature drop across the reference sample of well-known thermal conductivity. Assuming the heat flow through the system (stack) to be constant, the unknown thermal conductivity of the rock sample can be calculated. Different designs of the divided-bar apparatus have been developed (see below, sec. 3) by various authors. This method was accepted as the most accurate [123, 125, and 126] and was recommended by the IHFC to be applied for rock samples (Beck [101 and 127]). For this method very important provide very good contact between the sample and measuring equipment (heater and thermocouples) to prevent the contact thermal resistance which is causing large systematic uncertainty. Due to the increased inter-granular contact resistance within the dry rock samples, thermal conductivity at ambient conditions for this method is usually lower than that determined by the needle-probe on the water-saturated rocks.

For hard rocks (see classification of the rocks type, Mongelli [104]), the divided-bar technique is often used with an accuracy of 4 % [104]. Some sandstone are too fragile for the high-grade sample preparation required by the divided-bar method. Mixtures of the crushed sandstone and water can be used in combination with a divided-bar apparatus (Sass et al. [23]). Schilling [130], Höfer and Schilling [131], and Gibert et al. [132] used method (RHTM, remote heating thermocouple measurement) in which heat is applied remotely (flux from a filament), but the sample is in contact with thermocouples for determining heat flow (single-physical contact method). The difference between measurements by the method and quartz data by (Höfer and Schiling [131]) is 5 % obtained using conventional methods which include radiative transfer and contact resistance effects. Single contact method was used also by Cahill et al. [133]. This technique has been used at pressures up to 0.8 GPa (Chen et al. [134]), but not Earth materials.

2.3.3. Guarded Hot-Plate Method

One of the examples of the widely used steady-state method is guarded hot-plate method (GHP). This method is used for the measurement of thermal conductivity of glasses, polymers, ceramics and thermal insulations materials. Application of the method for liquids and gases detailed described in the [529-531, and 537]. A steady-state heat flow through the homogeneous material at constant temperature is a necessary condition in this method. A GHP apparatus consist of a cold plate, a hot plate, guard heaters, samples and insulation.

Some GHP consist of two cold plates. Hot plate is surrounded by guard heaters and by thermal insulation to make sure that the heat released in the hot plate is passed only from the sample. In this case heat loses from the hot side may be minimized and accuracy in the results may be increased. In this method the main sources of the uncertainties are: (1) heat loses and (2) temperature difference measurement. The main heat loses from the heater and the sample to the surroundings. The sample (A) in cylindrical form is placed between the upper hot plate (B) and the lower cold plate (C) and is insulated from its lateral surface (F). The electrical heater is located inside the hot plate and generated heat. Two guard heaters (guard plate D and guard ring E) are mounted above the sample. Push rod (H) provides good thermal contacts between the sample and the heater. Heat flows from hot-plate side to cold –plate and establishes a temperature difference, ΔT, across the sample after achieving the steady-state condition. The thermal conductivity of the sample can be calculated by knowing the heat flux through the specimen and the temperature difference across its two faces using the following simple working equation

$$\lambda = \frac{Qd}{S(T_2 - T_1)} = \frac{qd}{(T_2 - T_1)}, \tag{33}$$

where $q=Q/S$ is the heat flux through the sample, d is the thickness of the sample, S is the cross-section area of the sample, T_2 and T_1 are the temperatures of the hot- and cold-sides of the sample, respectively. The measured thermal conductivities are correspondend to the mean temperature $T_m=(T_2+T_1)/2$. The steady–state methods basically used to measure of the thermal conductivity dry porous materials, although the method can be successfully used for measurements of the fluid-saturated materials under pressure (see below sect. 3).

Rausch et al. [548] developed a new guarded parallel-plate instrument for the measurement of the thermal conductivity of fluids and solids. Besides flexibility regarding sample material and thickness, the instrument design is aimed at the minimization of parasitic heat flows due to convection, radiation, or heat leakages. The uncertainty of the thermal conductivity in this method is 3 %. Zarr [549] developed guarded-hot-plate apparatus for measuring the thermal conductivity of insulation materials as well as other solids and composites in a standardized way. Disturbing heat flows due to convection, radiation, and heat leakages are minimized by a sophisticated design of the guard system of the instrument as well as by adequate evaluation methods. The measuring cell consists of two major parts. The lower part contains the cooling plate (146 mm diameter). Fans are installed to improve the convective heat removal from the fins with the help of ambient air. The main component of the upper part of the instrument is the heating plate (85.94 mm diameter) at room temperature. The system comprises a guard ring in which the balanced heating plate is immersed as well as a guard plate located above the heating plate. Another guard ring surrounding the above components minimizes effects of the ambient conditions on the temperature control system. The parallel plates as well as the components of the guard system are made of cooper, ensuring an approximately homogeneous temperature distribution within these components. The cooper surfaces which are in contact with the sample are polished and coated with a thin chrome layer. Thus, the emissivity of these surfaces and the radiative contribution to the total heat flow are minimized. In order to minimize the heat leakage from the heating plate to the guard system due to residual temperature difference the guard components were separated from each other and from the heating plate by insulating air gaps with a width of 1mm. The heat leakage location was tracked by modeling the heat distribution within the apparatus .The thermal resistances at all interfaces between the heating plate and the guard system were taken into account. A maximum sample layer thickness is 10 mm and can be adjusted.

2.4. Contact- Free Methods

2.4.1. Optical Scanning. Laser-Flash Analysis (LFA)

It is difficult to accurately measure the ETC of sandstones and other rock materials using the divided–bar and needle–probe methods (contact methods). The uncertainty in ETC measurements in this case reached up to 15 % and more (with 2 % reproducibility, 5 % calibration uncertainty). The samples for steady-state method of thermal conductivity measurements have to be prepared very carefully. For sandstones, the required grade of polish is almost impossible to achieve. In this case variable-state approaches have to be used.

As well known the measurements of heat flow involve some portion of the applied heat being carried from the heater across the material to the sensor by photons (light) and some portion being moved from atom-to-atom within the solid via phonons (vibrations) [77,89, and 138]. Photon transfer is not a diffusional process. Other most widely and frequently used method for the thermal diffusivity measurements is laser flash technique (LFM). In this method, heat flow across a sample is recorded as a function of time. Arrival times of the photons and phonons are easily distinguished because of a factor of 100000 differences in speeds. The mathematical models [143] were developed to quantitatively remove radiative transfer effects. Metal and graphite coatings are also used to reduce photon flux [240]. As was found in works [89, 572, and 573] contact methods (see above sec. 2.3) underestimated thermal diffusivity of rock materials at 298 K due to contact heat losses by 10 % per thermal contact, but at high temperatures the thermal diffusivity has been overestimated by a factor of 5. LFM method can be used for wide temperature (from -100 to 3000 °C) ranges [84 and 135]. The LFM is based on the heating of a sample from one side (front) using a laser pulse and measuring the temperature response at the other side (rear). The sample is usually in the form of a disc with a diameter of about 6 to 25.2 mm and thickness of about 2 mm. If the sample is under adiabatic conditions and a short laser pulse is applied then, the thermal diffusivity can be calculated using following relation [84 and 135]

$$a = 0.139\frac{d}{t_{1/2}} \tag{33a}$$

The main sources of uncertainty of the thermal diffusivity measurements in LFM are the finite resolution of the data acquisition system and the heat capacity of the pyrometer itself. If the experiments are repeated at each temperature and average values of the measurements are taken the total uncertainty is about 2 % [534]. The advantage of the method is that, it can measure thermal diffusivity of the sample without having any physical contact of the heat source with the sample (no contact resistance). Thermal diffusivity measurements can be performed only by knowing relative temperature change as a function of time. Therefore, the relative uncertainty of the measurements is between 2 % to 5 % [535]. Laser flash method is suitable to measure only thermal diffusivity.

Parker and Jenkins [135] used an electromagnetic heat source and a variable-state method. This is allowed to prevent contact resistance between the heat source and the sample, therefore increase the accuracy of the heat flux (measurements) which transferring through the sample. The method was originally developed by Parker et al. [84]. In this method the front face of a sample is heated by a short laser pulse. The resulting temperature rise on the rear face is measured as a function of time (thermogram) with an infrared detector. Various mathematical algorithms have been developed to analyses to correct the effect of the pulse length (shape and energy) of the laser and heat loss [136 and 137]. The advantages of this contact-free method are: (1) easy sample preparation; (2) fast measurement times; (3) small sample dimensions; and (4) high accuracy. The laser pulse used for heating the front face of the sample has a significant effect on the results. LFA method detailed described by Whittington et al. [138]. Details of the application of the method for Earth materials are reported by several authors [139-141]. Main components of the method are a controlled atmosphere furnace, a high-energy pulsed laser, and an IR detector (Braüer et al. [142]). A sample in the form of a small slab with parallel faces is held in temperature in the furnace while emissions from the top of the slab are monitored remotely with the IR detector. Additional heat is remotely supplied to the bottom of the slab by a pulse from an IR laser. Afterwards the variation of the rear-face temperature with respect to time t is considered. The normalized rear-face thermogram with respect to its maximum is a function of both diffusivity and heat losses. Parker et al. [84] showed that the half-rise time $t_{1/2}$ (is the time required for the back surface to reach half of its maximum temperature rise, T_m) is a function of diffusivity (see Eq. 33a), where d is thickness. A heat from the pulse diffuses from the bottom to the top of the sample; the increase in emissions is recorded by the IR detector. Because emissions are directly related to temperature (T^4), the detector response is known as a temperature –time curve (thermogram). Top and bottom surfaces of the sample are graphite coated to absorb laser light, thereby shielding the detector, and to enhance intensity of the emissions. Neither thermocouples nor the heater contact the sample, and neither power input nor temperature needs to be quantified. Equations used for data analysis requires the pulse width to be significantly shorter than the time heat takes to cross the sample. Because the rise in sample temperature associated with the pulse is small, 4 K, D is approximately constant during data acquisition, and the T dependence of D is determined by varying furnace temperature.

This method permits to separate of the lattice component from unwanted radiative transfer. Phonon and photon heat transfer are visually discernible in the temperature –time curves and are extracted from the mathematical model by Mehling et al. [143] and Hofmann et al. [144]. Sources of experimental uncertainty in laser-flash measurements are implicit in simple models for temperature –time curves arising from phonon transport alone (opaque samples, or very low temperatures). Opaque and soft materials lack radiative transfer and have good thermal contact, allowing calibration against results from conventional methods. The uncertainty in lattice thermal conductivity measurements by this method is 2%.

Hanley et al. [145] applied laser flash method to measure thermal conductivity of eight representative rocks. This figure also showed the characteristic temperature response of the rear face of the test sample. The values of T_m is typically 2 or 3 °C. Basically, the sources of uncertainty are related to the flash technique due to non-ideal boundary conditions and measurement errors. The non-ideal boundary conditions include finite pulse time effects, nonuniform heating, heat losses and nonhomogeneous material while measurement uncertainties include the uncertainties associated with monitoring the temperature history of the rear face of the sample and determining the effective sample thickness. Nonuniform heating might be caused due to a nonuniform heat source, laser or nonuniform absorption by a nonhomogeneous sample. To prevent nonuniform absorption of energy at the front surface of the sample, thin metallic coating can be deposited on the sample [145]. Rock is nonhomogeneous material, therefore may experience nonuniform heating at the front surface and may contain "heat paths" in various directions through the sample due to voids or large grains with high thermal conductivity. When measuring the temperature history at the rear face of a nonhomogeneous sample the response may vary from point to point due to the properties of the particular grains. To minimize the nonhomogeneous conduction effect, the sample size can be made larger. The temperature of the rear face was recorded using an infrared detector (contact-free) rather than a thermocouple. The rear face of sample can be coated with vacuum deposited metallic layers chromium or iron to prevent in depth viewing of the sample by the optical detector. This will provide the erroneous temperature measurements due to change in the effective sample thickness. The experimental apparatus developed by Hanley et al. [145] consists of four parts: 1. Thermal energy pulse source (laser with 2 cm diameter); 2. Sample holder and furnace for ambient temperature control; 3. Temperature rise detector; and 4. Data acquisition system. The effect of finite pulse times has been studied by Taylor [146 and 147].

Strack et al. [148] determined thermal conductivity of sandstones by measuring the thermal diffusivity, the density, and the heat capacity using a laser flash technique. A CO_2 laser (550 W) operating in a short pulsed mode (2.5 ms) is used as a heat source. Unpolished penny-sized samples of tight sandstones are irradiated on one side and the temperature is measured on the opposite side. The temperature was recorded with a digital data acquisition system. Carbon black is used to give the samples a uniform absorption (reflection losses are negligible). About 20 samples per hour can be measured with the experimental set-up. In this method, the highly energetic heat pulse hits the sample at t=0 and then a heat wave travels through the sample before radiation or convection heat losses can take place. The sides of the sample can be considered to be insulated from the surrounding medium. This method can reduce the measurement time by several orders of magnitude. A disadvantage of the method is that the method requires transducers and recorders with fast response times. The sample-to-transducer contact is improved by using thermal grease as an interface material. A thermocouple is used as a thermal transducer.

Usually thermocouples have a long response time. Strack et al. [148] used a thermocouple with a teflon-coated tip and a time constant of 5 ms. The tip of the thermocouple is larger than the grain size of the samples and the entire teflon tip is flexible and bends easily. Therefore, the heat contact is not very repeatable. This contact is the main source for inaccuracy of the thermal conductivity measurements. This method was applied for dry and salt-water saturated sandstones with various porosities. The large scattering in the thermal conductivity values of the dry samples was found. This is caused by air convection within the samples. The measurement repeatability is within 2 %. The entire system is calibrated with 5 %. Both errors are less than 15 % scatter due to rock changes among the subsamples.

Kubičár et al. [149] used pulse transient method to measure thermal conductivity of dry and water-saturated sandstone at low temperatures (1.5 and -10 °C). The sample consists of three parts (I, II, III). A plane heat source is clamped between the first and second parts. One junction of a thermocouple is placed between the second and third parts, where it measures the temperature response to the heat pulse. The ideal model of the pulse transient method yields the following relation for the transient temperature $T(h,t)$ at the thermocouple junction

$$T(h,t) = \frac{Q}{C_P \rho \sqrt{\pi a t}} exp\left(-\frac{h^2}{4at}\right), \tag{34}$$

where $Q = RI^2 t_0$ (≈ 79 kJ·m^{-2}) is the energy of the heat pulse released per unit area, R is the electrical resistance of the unit area of the heat source, t_0 is the width of the heat pulse, t is the time, and h (≈ 12.4 mm) is the distance between heat source and the thermocouple junction. Equation (34) is valid for a non-limited body in which an ideal heat source produces a heat pulse in the form of a Dirac delta pulse. In a real experiment the sample is limited and a real width of the heat pulse t_0 is used. The working equations for the calculation of the thermophysical properties for this method are: heat capacity, $C_P = Qf_C / \sqrt{2\pi e} \rho h T_m$; thermal diffusivity, $a = h^2 / 2t_m f_a$; and thermal conductivity, $\lambda = a\rho C_P = Qhf_C / 2\sqrt{2\pi e} t_m T_m f_a$, where T_m (≈ 1 K) is the maximum temperature response at the time of t_m (≈ 76.8 s), f_a (≈ 0.9633) and f_C (≈ 1.001) are the correction factors [150] due to deviation from the ideal model. The overall uncertainty in measurements of thermal diffusivity, heat capacity, and thermal conductivity are 5.88 %, 3.97 %, and 3.91 %, respectively.

Schatz and Simmons [151 and 152], Zaug et al. [153], and Chai et al. [154] also developed a time-varying quasi-steady-state contact-free technique that is contact –free for thermal conductivity measurements. Contact–free method for pressure studies was developed by Pangilinan et al. [155]. To our knowledge, so far contact-free method has not been applied to measure ETC of saturated porous materials. Optical scanning technique was used to measure ETC rocks materials by other authors, see for example [162, 163, 195-202, 256-258, 260, 261, and 303].

2.4.2. Radiative Heat Transfer

Any conductive heat transfer is accompanied by simultaneous radiative transfer. The correction for thermal radiation depends upon whether or not the sample absorbs radiation. Radiation originated from the sample, which can be a multiple of the radiation emitted from the plates. If the sample is entirely transparent, the conductive and radiative heat fluxes are additive and independent. In this case, the correction is simple and usually negligible. When the sample absorbs and re-emits radiation (partially transparent), the problem is more complicated, since the radiative and conductive fluxes are coupled. The semitransparent behavior of the samples can affect the thermal conductivity measurement technique. The interaction of radiative and conductive transfer can be reduced in the LFA method using reflective coating, especially for materials whose optical thickness is low. Radiative transfer can also affect the thermograms.

At lower temperatures the heat transfer in most rock materials dominated by phonons (lattice vibrations) (Hofmeister [77 and 156]). At elevated temperatures radiative heat transfer (photons) may become more important (Shankland et al. [157]). The heat transfer by photons can be estimated as

$$\lambda_{rad} = \frac{16}{3} k_B n^2 l_{rad}(T) T^3, \tag{35}$$

where n is the refractive index; $l_{rad}(T) = 1/\varepsilon(T)$ is the temperature–dependent mean free path length for a photon, and ε is the absorption coefficient. The temperature dependences of heat transfer by phonons and photons are different. In first estimation, the heat transfer by phonons decreases with temperature increasing as $\lambda_{lat} \propto T^{-1}$ (Peierls [73]) (three –phonon processes, higher-order phonon processes is neglects). This is valid if the mean free path length is longer than interatomic distances. For four-phonon processes and $T>\theta$ (Debye temperature) $\lambda_{lat} \propto \left(BT + B_1 T^2\right)^{-1}$(Klemens [158]) and $\lambda_{lat} \propto (BT)^{-1} + \left(B_1 T^2\right)^{-1}$ by Ziman [74]. Because the mean free path of the phonons cannot become less than the unit cell, the law $\lambda_{lat} \propto T^{-1}$ is no longer valid at high temperatures ($T>\theta$) and $\lambda_{lat} = (B/T)\left[(2/3)(B_1/T)^{1/2} + (1/3)(T/B_1)\right]$ (Roufosse and Klemens, [76]). The relationship $\lambda_{lat} \propto T^{-1}$ is valid only for structurally perfect isotropic single crystals. However, many rocks are composed of mixtures of highly disordered crystals of different compositions, therefore, the thermal conductivity of rocks tends to decrease more slowly than $\lambda_{lat} \propto T^{-1}$ ($\lambda_{lat} \propto T^{-m}$, where $0<m<1$). Klemens [158] concluded that m=0.5 for temperatures near or above θ. We found that the values of exponent m for some type of rocks varying within 0.3 to 0.4 (Abdulagatova et al. [159]). The heat transfer by phonons increases with increasing temperature as $\propto T^3$ (if neglects the temperature dependency of absorbance). Different type of the radiative heat transfer contribute to thermal conductivity is possible (Höfer and Schilling [131]).

To minimize the radiation emission, the plate surfaces should be highly polished and protected against oxidation. The radiative heat transfer between the plates (guarded plate method) is (for small temperature differences ΔT)

$$Q_{rad} = 4\pi R^2 \sigma T_L^3 \left(\frac{1}{\varepsilon_U} + \frac{1}{\varepsilon_L} - 1 \right) \Delta T, \tag{36}$$

where R is the radius of the plates, T_L the lower plate temperature, ε_L and ε_U are the emissivities of the upper and lower plates, respectively. To minimize the heat transfer by radiation, a solid material such as stainless-steel (1X18H9T) of low emissivity can be used for the plates.

3. MEASUREMENTS OF THE ETC OF ROCKS MATERIALS (SANDSTONE): LITERATURE REVIEW

Models require reliable ETC measurements for validation. Most of the available ETC measurements for rock materials have been made at room temperature and at atmospheric pressure. ETC of dry and fluid–saturated porous rocks and rock forming materials at high temperatures and high pressures were measured by various authors, see for example [11, 23, 72, 79, 82, 108, 109, 115, 127, 160-222, 250, 263, 301, and 305]. All measurements were performed by various methods (see above): (1) need-probe (uncertainty 2-3 %); (2) divided – bar (uncertainty 4 %); (3) guarded hot plate (4-5 %); (4) hot-wire (uncertainty is about 4-5 %); (5) optical scanning (uncertainty 2 %). Below we provided review of the reported ETC data for dry and fluid –saturated porous rocks materials.

Clark [168 and 220] measured the effect of compressing and water content on the ETC of rocks. The measurements were performed with the apparatus, which consist of vertical press, which provide apply the stress in (along the) vertical direction. The measuring thermal conductivity cell consists of 5 cylinders with diameter of 1.5 inch. The middle cylinder was used as a heat source, which is inserted into the sample under study. The temperature gradient along the sample was measured with thermocouple set on the distance close to the sample surface. The reliable contact between the heater and the sample was provided by using the contact materials such as vaseline, glycerin, and thixotropic *etc*. Side heat lost was measured using the sample of Pyrex glass with well-known thermal conductivity. The experiment results showed that the compressing of the sample slightly affected on the thermal conductivity if the sample porosity is small. For sedimentary rocks with low porosity, marble and sandstone, the effect of compressing is almost the same as simultaneously effects of pressure and fluid-saturation. For rocks with large porosity the effect of water is considerably than compressing.

Horai [107 and 108], Horai and Susaki [109], and Cho et al. [559] used transient hot wire technique to measure ETC of the Earth materials (rocks and rock forming materials). In the works [107, 108, and 109] the samples were of cylindrical forms (24 mm long and 12 mm in diameter), containing a heater of 0.1 mm thick chromel wire along their axis and a thermocouple at the center. The validity of the method and their accuracy was confirmed by measuring the thermal conductivity of standard materials such as quartz. Rapid increase of the ETC of the porous rock samples was found below 2 kbar which related with the closure of microcracks in the sample and slight increase above 2 kbar should be intrinsic to the material.

Asaad [166] measured ETC of three sandstones saturated with gas and liquids. The results were used to develop empirical prediction models for ETC. Three sandstones with different porosity and compositions saturated with gases (air, CO_2, He) and with oil, n-decane, ethanol, glycerin, and water. Measurements also were performed when vacuum in the pores. The uncertainty in the measurements is 10 %. Various empirical prediction models were proposed by the author on the bases of the thermal conductivity of components. Measurements were performed with steady-state method using the guarded hot plate method. Heat flowed through the sample parallel to its axis and side losses were prevented by using a guard ring. The sample was separated from this guard ring by a small annular air gap.

Özbek [177] measured ETC of porous rocks saturated with different liquids using divided-bar technique. The uncertainty of the measurements is within 5 %. This uncertainty do not included other uncertainties such as thickness of the sample, leaking of the fluid, uncertainty related with saturation. Therefore, total uncertainty of the measurements is considerable higher than 5 %. The results of the measurements were used to develop prediction models for ETC on the bases of permittivity, thermal conductivity of matrix, fluid saturation and quartz content. Brigaud and Vasseur [123] and Brigaud et al. [124] reported ETC of water- and air-saturated sedimentary minerals. These data were used to test geometric means models for ETC.

The prediction uncertainty is 10 % for water-saturated sample and 20 % for air-saturated sample. The measurements were made with the method of divided–bar. Reproducibility of the method is better than 2 % and interlaboratory agreement is better than 5 % (Chapman et al. [452]).

Anand et al. [222] used steady -state method to measure of dry and saturated with solution sandstones (marine sample). They studied the effect of pressure, temperature, and porosity on the ETC. Main part of the thermal conductivity experimental apparatus consist of 2 reference samples with well-known thermal conductivity and third sample with unknown thermal conductivity which ETC require to measure. The sample was saturated with silicone oil and n-hexane. The measurements were made for 6 sandstones. The maximum uncertainty in ETC measurements is 5 %. The authors developed correlation equation for prediction ETC on the bases density, porosity, permittivity, matrix and saturated liquid thermal conductivities. They also found that the effect of pressure on the ETC very weak and can be predicted on the bases of compressibility data. They used thermal conductivity apparatus like developed by Keese [223]. The linear relation between the ETC and the compressibility of sandstones is found by Anand [171].

Sugawara and Yoshizawa [224] measured ETC for three porosity materials (glass balls with different diameters, rubber plates with cylindrical pores and layer of glass plates) with various porosity and saturated with air and water. They used comparative method (steady state) to measure ETC. The values of the ETC were derived as a function of temperature. The porosity was used as a parameter. The method was used also for sandstones saturated with air and water. The same authors (Sugawara and Yoshizawa [225]) measured ETC of solid-liquid two-phase systems. The measurements were made for 13 sandstones with porosities of 9.7 % and 42.3 % and 6 building materials (fireproof bricks) with porosities of 15 % and 38 %. For the measurements were used standards materials together with the sample under study. The thermal conductivity was calculated as $\lambda = \lambda_0(\Delta T_0 / d_0)(d / \Delta T)$, where λ and λ_0 are the thermal conductivity of the sample under study and standard sample, ΔT_0 and ΔT are the temperature difference between the two samples (standard and under study, respectively), d and d_0 are the thickness of the samples (standard and under study), respectively. The authors studied the ETC as a function of porosity and saturated fluid. ETC of fluid-saturated porous sandstones have been measured also by several authors [79, 168, 220, 226, and 227].

Khan [228], Edmondson [229], and Woodside and Messmer [82 and 161] studied the effect of pressure on the ETC of rocks. They showed that the ETC increases with an increase in effective stress of the rock. This should be expected since increasing the pressure improves the thermal contact between grains and increases the over-all density of the rock and, consequently, the ETC. When good thermal contact is established, the change in ETC with added pressure is generally small. It is obviouse that changes in ETC are related with the bulk compressibility of porous rock. Edmondson [229] showed that the ETC of Berea, Bandera, and Boise sandstones increase by 7.8, 9.5, and 12.3 %/7 MPa, respectively in the pressure range from 6 to 25 MPa. The effect of pore pressure is to reduce the effective stress on the rock. Reduction in fluid pore pressure results in increased effective stress on the rock and thus an increase in ETC. Very important the phase behavior of the pore fluids. Reduction of the pore pressure may result in the vaporization of some of the liquid components and this can cause a large reduction in ETC.

Ljubimova et al. [180 and 181] reported the data for the ETC of dry, water-, and oil – saturated rocks in the temperature range from 20 to 200 °C and at pressures up to 1 kbar. They used the steady-state method proposed by Bridgman [229]. The results were presented as a linear function of pressure. Somerton [230] used comparative-type apparatus similar to that described by Zierfuss and van der Vliet [227] (see below) to measure the ETC of the several unconsolidated oil sands samples (sandstones, silty sand, siltstone, shale, limestone) fully saturated with oil and water. This is a steady-state method. In this method the steady-state heat flow characteristics of the test samples are compared to those of a crown glass standard. Measurements of the ETC were performed as a function of porosity and water saturation. Water-saturated sands were found to have ETC 6-8 times greater than values for the same sand packs air-saturated. Linear temperature dependence of the ETC is found. The effect of changes in pressure on the ETC of liquid–saturated unconsolidated sands is small and for practical purposes can be ignored.

Pribnow et al. [205] measured the ETC of the water-saturated rocks (two selected gneiss samples from 3380 m and one amphibolite core samples from 3858 m depth) at temperatures from 25 to 300 °C and at pressures up to 34 MPa with needle-probe method. They found that ETC of water-saturated amphibolite decreases with temperature at a rate approximately 40 % less than the rate for dry amphibolite, and the ETC of water-saturated gneiss decreases at rate about 20 % less than the rate for dry gneiss. The thermal cracking is the primary cause of more rapid decrease in dry ETC with temperature. The effects of thermal cracking were also observed in the water-saturated samples but the results are very small, less than 3 %. A dry sample from the same core section has been measured by Seipold [201]. Kiyohashi and Deduchi [110] measured the ETC of fluid-saturated rocks (sandstone-4 different specimens; zeolite-3 different specimens; tuff; cristobalite; perlite-2 different specimens; bentonite, china clay, kaolinite, and pagodite) with transient hot-wire comparative method at 20 °C. The porosity of the rocks was covered from 5 % to 51 %. The measurements were made as a function of water-saturation. The results were used to develop three-phase geometric-mean models. The models are expressed the measured values of the ETC with an accuracy of 22 %.

Zierfuss and van der Vliet [227] developed the apparatus to laboratory measurements of thermal conductivity of sedimentary rocks. They measured ETC of a number of sandstone and limestone core samples. The cores were entirely or partly saturated with air, water, and kerosene. Measurements were made with a steady-state comparative method. In this method heat is flow with constant rate through a pile consisting of a cylindrical rock sample, and a standard sample (crown glass cylinder) of well-known thermal conductivity. Evaporation of water and kerosene was prevented by wrapping the sample in plastic sheets. The uncertainty in the ETC measurements is about 4.8 %. Troschke and Burkhardt [231 and 232] used needle-probe technique to measure ETC of the sedimentary rocks saturated with water. Measurements were made as a function of fraction of water. The results were compared with prediction models (geometric mean, layer, and dispersion models). The geometric model is the most satisfactory described experimental ETC for water-saturated rocks.

The thermal conductivity of eighteen dried, moist, air- and water-saturated porous rocks was studied by Deguchi et al. [233 and 234]. Measurements were made with a transient hot-wire comparison method. The results showed that the ETC all of the rocks increased with increasing the water content in pores. The ETC of water-saturated rocks is 1.2 to 2.5 times higher than that of the dry rocks. The ETC of zeolite rocks is more affected by the degree of water-saturating. For moist rocks the measurements were made at temperature of 20 °C and 60 % relative humidity.

Bayuk et al. [235] measured ETC of water- and oil-saturated rocks. The results were used to develop the method to calculate of the ETC of oil-, water-, and gas-saturated sandstones. Popov et al. [203] measured the ETC of cuttings. The results were used to study of the correlations between rock ETC measured on cores and different logging data. About 600 cores were studied for the depth interval of 300-1200 m. The optical scanning method [114] (contact-free) was used to measure the ETC of two artificial samples (cylindrical shape with diameter of 3 cm and height 2 cm). The samples are mixtures of cuttings with plastic material epoxy resin and liquid sodium glass) whose thermal conductivity is known, 0.21 $W{\cdot}m^{-1}{\cdot}K^{-1}$ and 0.97 $W{\cdot}m^{-1}{\cdot}K^{-1}$, respectively. The size of the cuttings particles varies from 0.06 to 2.5 mm. This method also used to measure the ETC of rocks drilled by a horizontal well on core cuttings. The uncertainty in ETC measurements was 3-8 %. Popov and Remushkevich [204] reported ETC measurements for sedimentary rocks from cores recovered from Russian oil-gas fields: Povkhovskoe, Vat'-Eganskoe, Em-Egovskoe, Samotlorskoe *etc.* The sandstones (Kaliningrad group of fields) were collected from wells in a depth interval of 1637 to 2572 m. The rocks have quartz and occasionally quartz and calcite pisolitic cement. The measurements were made with the optical scanning technique (Popov et al. [114]) on 897 cores. Measured values of the ETC were used to estimate the anisotropy coefficient ($K = \lambda_{II} / \lambda_{\perp}$) for each air-dry and water-saturated samples. Almost all dry rocks were found to be anisotropy (K=1.03-1.38, $\lambda_{II} >> \lambda_{\perp}$) that can be explained by textural-structural peculiarities of rocks and oriented fracturing. For saturated rocks the thermal anisotropy coefficient is decrease up to 1.01 to 1.16. Increase in rock ETC at water-saturation is 36-115% for terrigenous rocks (porosity 11-22.5 %) and 13-20 5 for limestones (8 % porosity).

Keese [223] experimentally studied the relationships between ETC and air permeability, formation resistivity factor, median grain size, and grain size distribution. The oil sands were measured both air and brine saturated in order to estimate the dependence of ETC on fluid phase composition. Natural oil sands obtained from California oil fields, Ottawa and Del Monte sands were used to study the ETC. The grain size was from 0.0052 to 0.0285 inch. Porosity of the samples was within 31 % to 44 %, depth from 598 to 2796 m. The measurements were made with a steady-state comparator type apparatus. The unknown sample is placed in series in a stack with one or more standards of known conductivity. A temperature difference between the top and bottom of the stack and a guard heater on the side to prevent radial heat loss ensure linear heat flow through the stack. The central part of the thermal conductivity apparatus is the stack. It consists of three holders of 3.735 inches outside diameter, 2 inches inside diameter, and 0.62 inches thick. Two of these holders contain standards of known conductivity and the third (center) holder contains the sample to be studied. Copper plates with copper constantan thermocouples are attached to the top and bottom of each holder by nylon screws. The cooper disc (0.119 inch thick and 1.99 inch in diameter) is inserted between the sample and thermocouple plate. A temperature gradient is produced across the stack with a heat source at the top and a heat sink at the bottom. The heat source and sink consist of circulating baths of silicone oil. The temperature difference across the stack was 15 °C and a mean temperature in the unknown sample of 52 °C. The stack is packed in a low conductivity ceramic fiber and the pack surrounded by a cylinder 0.5 inch thick and 5.75 inch inside diameter on which guard heaters are mounted. The differential thermocouples were used to ensure that the guard heaters are at the mean sample temperature. Axial stress is applied to the stack with a hydrostatic press (2.1-2.4 MPa). The sample was saturated in a vacuum saturation apparatus with brine (7500 ppm NaCl). The measurements were made for oil and quartz sands. The porosity dependence of the ETC was studied with fluid saturation as a parameter. The maximum uncertainty in ETC measurements is 5 %. The results were compared with predictions of Woodside and Messmer [82], Kunii and Smith [236], de Vries [237], and Krupiczka [238].

Gruescu et al. [49] measured the ETC of clay using the dual-probe heat-pulse technique. This method is based on using two parallel probes separated by a distance *d*; the first one contains a heater and the other a temperature sensor. With the dual-probe device inserted in the clay, a heat pulse is applied to the heater and the temperature at the sensor probe is recorded as a function of time. The ETC can be further determined from these data. The details of the method described by Sammartino et al. [239] and André and Degiovanni [240]. A transient line-probe method for the measurements of the thermal conductivity of heterogeneous materials like oil shale was used by Prats and O'Brien [241] and Tihen et al. [242]. Problems of thermal contact resistance are severe with this technique. According to Tihen et al. [242], perfect contact between the shale and the heating source is not attainable; therefore the data must be corrected for the error which imperfect contact introduces.

Nottenburg et al. [243] developed an alternative thermal comparator method for the measurements of the ETC of oil shale. The measurements were carried out in the temperature range from 25 °C to 400 °C. The measurements were made as a function of organic content and temperature. The essence if the technique rests in a comparison on the ability of the experimental sample to transport heat and that of a known reference material to conduct the same amount of heat. The test sample is sandwiched between two reference samples and a thermal gradient is impressed across the stack arrangement. This method is similar to the guarded hot-plate method developed by Woodside [244] although it is considerably faster. Various versions of this technique were reported in the literature (Mirkovich [245] and Powel [246]). The results were correlated as a quadratic function of temperature and shale grade.

Thermal conductivity of core samples from Eromanga/Cooper Basin has been measured by Gallagher [14] with a divided-bar apparatus. The measurements were made with the samples dry and saturated with water. The results were compared with prediction models. The measurements were performed for three type rocks-sandstones, siltstone sand shale with effective matrix thermal conductivities of 6.0, 3.5, and 2.5 $W{\cdot}m^{-1}{\cdot}K^{-1}$, respectively. The porosity measured samples is within 0 to 33 %. The effects of temperature and pressure on ETC were not examined in this study. The method is similar to that described by Sass et al. [27]. The upper and lower ends of the bar were kept at 34 °C and 18 °C, respectively. The radial heat loss from the sample was minimized by lagging the bar with cotton wool and Styrofoam (Jessop [129]). To reduce the effect of the contact resistance (imperfect thermal contact between the sample and the adjacent cooper disc) the sample was polished flat to 0.03 mm. A contact fluid was not used to avoid the possibility of the fluid or grease migrating into the sample. In all cases an axial pressure of 1.5 to 3 MPa was applied axially to the divided-bar to improve the thermal contact. The system was calibrated with quartz within 5 %. The repeatability of the method is 5 %. Messmer [247] experimentally studied the effect of temperature and of water and oil saturation on the ETC of a variety of sandstones by using the probe method. The ETC of dry sandstones with no overburden pressure was found to decrease up to a temperature 540 °C which attributed to the decrease in the thermal conductivity of quartz with increase in temperature. The effect of saturation, both water and oil, was most pronounced in the low saturation regions. 80 % of the increase in thermal conductivity occurred during the first 20 % of saturation. A cylindrical sample originally at a uniform temperature is heated at a constant rate by a line heat source, and the temperature rise measured with time. The thermal conductivity of the tested material may be calculated from the known power input and the temperature-time variation record. The theoretical bases of the method were developed by Carslaw and Jaeger [81], and some limitation on the method by Blackwell [248]. The thermal conductivity for this method can be calculated from the slope $(Q/4\pi\lambda)$ of the straight line of T versus $\ln t$. The measurements were made for sandstones of about 90 % or more quartz varying in porosity from 3 % to 60 %. The sandstone samples were saturated with C_{10}-C_{12} aliphatic hydrocarbons or tap water. The reduction in ETC with temperature of all samples reflects their quartz contents. The significant hysteresis in conductivity with decreasing temperature is probably caused by the irreversible changes in this clay under the temperature conditions. The large effect for the small percentage of clay present indicates that the clay is located at the quartz grain boundaries. The measurements by Messmer [247] are agreed in trend with that of other authors (Birch and Clark [79] and Somerton and Boozer [250], for example). The ratio of the pore space filled with a fluid to the total pore space is described as the saturation of that fluid. In thermal conduction through rocks, the matrix conduction will be modified by both the saturating fluid conductivity and its location in the rock. Measurements were made for seven oil saturated sandstone samples of porosities ranging from 3 to 59 %. For all samples the ETC increase most rapidly in the first several per cent of saturation with an almost linear increase thereafter. In general, about 80 % of the change in ETC occurs in the first 15 % of saturation increase. At low saturations the liquid will accumulate about the grain contacts preferentially and improving the contact between the grains. The ETC of water-saturated samples is much higher than oil saturated samples due to greater conductivity of water.

Morinl and Silva [250] directly measured the thermal conductivity of four ocean sediments (two biogenic oozes and two clays) as a function of porosity, temperature and pressure. Measurements were made with the standard needle -probe method in the temperature range from 22 °C to 220 °C and at pressures between 2 and 62 MPa. Since sediments consist of primarily water, with porosities ranging from 70 to 88 %, and since the effect of pressure upon the thermal conductivity of the solid matrix is minimal, any variation in sediment thermal conductivity with pressure is expected to reflect the pressure effects upon the liquid phase alone. The measured ETC of sediments as a function of pressure is just like pure water thermal conductivity behavior. The slopes of the $\lambda - P$ lines are approaching the slope of the data for pure water as sediment porosity increasing. The pressure corrected thermal conductivity for sediments is

$$\lambda(P) = \lambda(P = 1atm) + 1.72\times10^{-5}\left(cal\cdot cm^{-1}\cdot s^{-1}\cdot C^{-1}\right)\left(\frac{depth}{1000m}\right). \quad (37)$$

Thus a sediment sample obtained from a water depth of 2500 m will require that a value of 4.3 $\times10^{-5}$ cal·cm^{-1}·s^{-1}·C^{-1} be added to the laboratory measured thermal conductivity. The temperature dependency of the ETC for sediment samples was presented as

$$log\,\lambda = (aT + c)\phi + b, \quad (38)$$

where *a*, *b*, and *c* are the empirical constants.

Ratcliffe [251] and MacDonald and Simmons [252] studied the variation in sediment thermal conductivity with temperature. Singh et al. [253 and 254] used transient method with a line heat source employed to measure the ETC of dune sand and brick sand samples with porosities of 0.42 and 0.52, respectively. The dune sand was moistened with distilled water up to 25 % by weight. Measurements were performed also with tertiary and alcohol saturated (23.3 %). In the case of brick sand the percentage of water was 47.5 %. The authors are claimed that the uncertainty of the method is 5 %. The probe used for this study was thin needle of 1 mm diameter and 13 cm in length. Verma et al. [255] measured the ETC of three-phase (powder of iron – aluminum-air; CuO- aluminum-air, Al_2O_3-ZrO_2-air) systems by a differential temperature sensors method.

Hratmann et al. [256] are reported ETC data for shaly sandstone and marls (porosity range 10 % to 30 %, from borehole with 570 m and 810 m depth). Measurements were made using the optical scanning technique (see for example [114, 257, and 258]). The anisotropy of the ETC was 3-4 %, therefore can be assumed that the rock is isotropic. They also measured density and sonic velocity using multi-sensor core-logger. The goal of this work was to measure various properties to predict ETC from petrophysical properties which can be measured in situ. A cylindrical heat probe technique was used by Sepaskhah and Boersma [259] to measure ETC of loamy sand, loam and silty clay at several water contents and temperatures. The results were compared with predicting model by de Vries [237]. The water content is ranged from 0 to 30 %. Moyne et al. [260] measured the effect of water content on the ETC of porous media (light concrete, bed of glass spheres of 170 μm diameter, composite material, and highly compacted clay) using flash method. The principle of the method (see also above section 2.4.1) is to submit the "front" face of a cylindrical flat specimen to a thermal pulse of very short duration –which can be assumed to be a Dirac-shape time pulse-and to record the temperature variation on the "rear" face. Comparison of the experimental normalized thermogram with theoretical normalized thermogram is lead to the identification of the effective thermal diffusivity of the media.

Zeng et al. [261] used transient hot-wire technique to measure of the thermal conductivity of aerogel opacified with the fine carbon particles. The uncertainty in ETC measurements is about 4 %. Huenges et al. [262] reported ETC measurements for water-saturated rocks (500 core samples from KTB-well). Measurements were made with transient heat-flow method. The apparatus is very similar to previous version of this technique as described by Sass et al. [106]. The core surface was prepared with a thixotropic fluid to improve the thermal contact between the heat source and sample. The uncertainty in measured values of ETC is about 5 %. To study the effect of water saturation on the ETC, the measurements were made on two typical rock samples. The measurements were performed with nitrogen and water saturated samples. The difference between nitrogen and water saturated samples ETC was 0.4 $W{\cdot}m^{-1}{\cdot}K^{-1}$. Gomaa [174] developed high temperature thermal conductivity apparatus. The main part of the apparatus is stack containing the test sample sandwiched between two disc-shaped standard materials. The thermal conductivity of standard materials is known as a function of temperature. The standard materials have the same diameter as that of the sample but their thickness is one half of the sample thickness. Each of the sample and the two standard materials is sealed in its holder by stainless plates in which thermocouples are mounted at the center so that the temperature difference across the sample and the standard materials could be measured. Linear steady-state heat flow in the stack is established by the use two electrical heaters mounted in large copper blocks at both ends of the stack and guard heating system. The temperature drop across the whole stack is of the order of 20-25 °C and radiation heat transfer is negligible. The stack and its heaters are mounted in a loading frame which allows application of controlled axial stresses. Fluid seal in the sample is established by the use of O-rings mounted between the sample holder and its stainless steel plates. Axial stresses up to 270 $kg{\cdot}cm^{-2}$ and 135 $kg{\cdot}cm^{-2}$ pore pressure may be applied. Fired lava was used for the standards. Its thermal conductivity is quite close to that of common sandstones. Temperatures at the different locations in the stack as well as differential temperature across the sample and two standards are recorded in a multipoint strip chart recorder. ETC is calculated from the temperature distribution data at steady state and the known thermal conductivity of the standards. The measurements were made for consolidated sandstones (5 cm in diameter and 3.2 cm thick) and unconsolidated sand packs. Their faces were flat, smooth and parallel with a diamond laping wheel. The samples was washed and then dried in vacuum at 100 °C for at least 3 h. After that they was saturated with 100 % of the test liquid by using vacuum (below 100 micron of Hg column) saturation equipment specially designed for this purpose. The saturating liquid was then allowed to flow the sample container until the sample was completely covered with the liquid and then the vacuum was released and the samples were immersed in the liquid for at least the next 3 h. A weight check on the saturated samples showed 100 % saturation. Liquids used were Brine (5000 pp KCl aqueous solution), n-heptane, and solvent. The measurements were made for four different samples of porous media, Boise, Berea, and Bandera sandstones (consolidated materials) and Ottawa sand (unconsolidated material). These samples were measured in dry and partially and fully saturated conditions.

Surma and Geraud [257] measured the ETC of fluid saturated granite in two perpendicular directions. They used two methods of measurements of the thermal conductivity: classical method line source (LS) and optical scanning (OS). In the LS method a needle-probe embedded in and flushes with the surface of a material of very low thermal conductivity (Plexiglas) (see details Huenges et al. [262]). The uncertainty of the method is 5 %. The length of the thermal source is 5 cm. Thermal conductivity was measured in two directions to determine the possible anisotropy. Second method OS is based on scanning a sample surface with a focused, mobile, and continuously operated constant heat source in combination with a temperature sensor. The heat source and the temperature sensor move with the same speed and to a constant distance to each other (see Popov et al. [114 and 258]). The size of the heating spot is 1 mm in diameter. The uncertainty in thermal conductivity measurements is less than 2 %. The porosity of the samples was between 0.3 and 10 %. No scale effect on the studied volume was found in the experiment (from cm^3 to dm^3). Two samples were tested from depth of 1630.5 m and 1432.8 m. OS method is defining local thermal conductivity values. First sample showed the quartz cementation. Quartz –sealed fractures have a high thermal conductivity in contrast to feldspars. The derived values of thermal conductivity were used to estimate the anisotropy $K = \lambda_{par} / \lambda_{per}$ and inhomogeneity $\varepsilon = (\lambda_{max} - \lambda_{min}) / \lambda_{aver}$ factors. The average value λ_{aver} of the OS is 3.39 $Wm^{-1}K^{-1}$ comparable to the value of 3.23 $Wm^{-1}K^{-1}$ obtained with the LS method. The value of K is around 1, which shows no anisotropy. The inhomogeneity factor determines the thermal conductivity heterogeneity which can be related to the mineralogical or structural heterogeneity of the sample. The inhomogeneity factor is varied between 0.375 and 0.515.

Buntebarth [263] studied thermal properties (thermal conductivity and diffusivity) of rocks (metabasite and gneiss) from the pilot well of the KTB drilling site at temperatures from 50 to 200 °C and at pressures from 2 to 60 MPa. Measured values were used to calculate the heat capacity using density data. Metabasite demonstrate a low thermal conductivity and a negligible anisotropic behavior. Measurements were made with heat flow meter technique. The one-dimensional heat flow is determined while passing through a disk with a known thermal conductivity and establishing a temperature gradient along it. In each of the three boundaries, i.e., between heat source, specimen, heat flow meter (reference probe) and heat sink, the temperature is measured using three shielded thermocouples of 0.5 mm in diameter in each boundary. The diameter of the bar is 50 mm and thickness of the reference probe is 23 mm. The specimen may have a thickness of 5 to 10 mm. After reaching the stationary state, the thermal conductivity is determined. More than 100 specimens were measured. The effect of pressure and temperature on the ETC was analyzed. Vosteen and Schellschmidt [111] studied the temperature dependence of the ETC for dry rock samples from 0 to 500 °C by using a divided-bar device. Measurements were made for 118 rock samples from 26 locations along the TRANSALP-profile. Measurements were made for dry and water saturated rocks. Porosity varies for the sedimentary rock samples from 0.9 % to 5.7 % and for magmatic and metamorphic rocks porosity ranges from 1 % to 4.7 %. All rocks samples were cut to obtain two discs: parallel to the plane of bedding and perpendicular to the plane of bedding. The anisotropy factor was determined using measured values of the ETC in parallel and normal directions. The anisotropy factor for measured samples is varied from 0.9 to 1.6. High anisotropy factor was explained as an alternation of strong and weak conductive layers (e.g., quartz –or feldspare-mica-layers) or a high content of oriented minerals with strong anisotropy (e.g., mica). At 25 °C the measurements were made for dry and water-saturated samples by using a needle-probe technique. At high temperatures (up to 500 °C) the measurements were made with a divided-bar device. For this method the sample preparation is difficult and time consuming. The needle-probe method is a transient while the divided-bar is steady –state. The needle probe (or half-space line source) is based on the theory of a line source in an infinite medium. The needle-probe consists of a cylindrical probe of 2 mm in diameter and 70 mm in length. A thermometer at the center of the needle- probe recorded the probe temperature. The probe length can be assumed to be infinite. The working length of the sample is about 6-12 mm. The uncertainty of the method is discussed by Grubbe et al. [264]. The divided-bar (Beck [101 and 127]) is a steady-state comparative method to determine thermal conductivity of rock samples containing water-saturated material. The basic concept of this method (see above section 2.3.2) is to compare the unknown thermal conductivity of a rock sample to the known thermal conductivity of reference material. The sample and reference (pyroceram which has thermal conductivity comparable to that of rock samples) materials are shaped into discs with diameter of 50 mm and approximately the same thickness (10 mm). The heat flows parallel to the divided bar axis (stack) from a heater (high temperature level) to a second heater (low temperature level) through two reference disks and the sandwiched rock sample. The radial heat flow can be neglected because guard heaters and insulating material minimized it. A thermal compound is used to reduce the contact resistance between the reference and the rock sample. The temperature drop across the rock sample is compared to that across the reference sample of known thermal conductivity. Assuming the heat flow through the system (stack) to be constant, the unknown thermal conductivity of the rock sample can be calculated.

Dobrynin [265] measured some physical properties (porosity, permeability, resistivity, density and velocity of elastic waves) of consolidated sandstones under pressure. Homogeneous quartz sandstones were chosen for the measurements: (1) Torpedo sandstone (20.2 % porosity) from Kansas and (2) the Medina sandstone (8.7 % porosity) from Ohio. Each sandstone samples containe about 5 % clay minerals, consisting mostly of kaolinite and chlorite. One cylindrical sample 2 inch in diameter and 5 inch in length was cut from the each sandstone and then saturated in a vacuum with a 3N solution of NaCl. Changes of resistivity under pressure were studied for sandstones with 100 % water saturation. Measurements were made at overburden pressure ranged from 0 to 138 MPa. The internal pore pressure was 0.1 and 12.4 MPa. The uncertainty in resistivity and compressibility measurements was 1-2 %.

Aichlmayr and Kulacki [70 and 266] developed an experimental technique for measuring the ETC of fluid saturated porous media. The method based on the transient heating of a semi-infinite cylinder by a constant heat flux at the boundary. This method is avoids determining effective thermal diffusivity and quantifying the boundary heat flux. The uncertainty in the ETC measurements is about 15 %. The method used to measure ETC of water-glass, air-glass, and steel –air systems. The solid –fluid conductivity rations for these systems are 1.08, 25.7 and 2400, respectively. The porosity of the glass sample is 36.5 % and for steel sample is 40.3 %. The test cell consists of a cylinder that is 22.2 cm in diameter and 40.6 cm long. The chamber is heated from above by a thin foil heater sandwiched between brass and acrylic disks. A guard heater of identical type is located between the acrylic disk and the aluminum cover plate. The power dissipated in the guard heater is adjusted until the temperatures at opposite faces of the acrylic disk are equal i.e., zero heat flux. This procedure ensures that the heat flux applied to the medium is constant. The solid phase of the fluid-solid system consist of 3 mm glass beads and 14 mm steel ball bearings. Saturating fluids include air and water. The relative uncertainty in the ETC measurements is within 14 to 16 %. The measured values of the ETC were compared with the values reported by Ofuchi and Kunii [267] for glass spheres with diameters between 1.15 to 12.1 mm and steel balls having diameters between 3.09 and 10.9 mm. Saturating fluids were water, helium, carbon dioxide, and air.

King [268] used a divided–bar apparatus to measure the ETC of two water-saturated sandstones and a limestones at permafrost temperatures (from -13 to 5 °C) and at 0.7 MPa. Pribnow et al. [100] measured the ETC of 15 whole –round sediment samples from Ocean Drilling Program between 17 and 440 m below the seafloor by using the divided -bar technique. Measurements were made in the temperature range from 5 to 60 °C. For low porosity samples (if room temperature thermal conductivity is λ_{rt} =1.6 W·m^{-1}·K^{-1}) the ETC decrease with temperature; for high porosity sample (if λ_{rt} =0.8 W·m^{-1}·K^{-1}) the ETC increase with temperature; and for moderate porosity (is λ_{rt} =1.2 W·m^{-1}·K^{-1}) the ETC almost does not change with temperature. This behavior is the result from a positive temperature coefficient for seawater ($\lambda \propto T$) and a negative coefficient for rock matrix ($\lambda \propto T^{-1}$). The measurements were made for horizontal λ_{hor} and vertical λ_{ver} components of the ETC independently and determine a mean anisotropy ($\lambda_{hor}/\lambda_{ver}$ =1.2). These data are very useful to determine the thermophysical characteristics of hydrothermal circulation in the upper oceanic crust as influenced by crustal topography, sediment cover, and permeability. Heat flow calculations are requiring the ETC data for the sediments. The measured ERTC data were used to improve the accuracy of heat flow values determined during drilling (Davis [189]) by applying anisotropy and temperature corrections for ETC. Pribnow et al. [100] also used the transient line source (needle probe) technique to measure more than 1000 ETC. The line source is a cylindrical probe, 1.5 mm in diameter and 70 mm long that contains a heating wire and a thermoresistor. After pushing the probe into the sediment core, the line source is heated with constant power. The thermal conductivity is calculated from the temperature response to the line source. This technique allows rapid thermal conductivity measurements. Davis and Seemann [269] also measured the anisotropic thermal conductivity of Pleistocene turbidity sediments of the northern Huann de Fuca Ridge using divided- bar apparatus. Measurements were made in three directions. The average anisotropy values is $\lambda_{hor}/\lambda_{ver}$ =1.25-1.26. They found that the anisotropy is almost independent of depth. 12 samples were measured at temperatures from 5 to 60°C. The temperature dependence of thermal conductivity ($\Delta\lambda / \Delta T$) is a linear function of thermal conductivity at room temperature

$$\frac{\Delta\lambda}{\Delta T} = A_0 + A_1\lambda_{rt}. \tag{39}$$

Therefore, the thermal conductivity at any temperature can be calculated based only on room temperature value λ_{rt} as

$$\lambda(T) = \lambda_{rt} + \frac{\Delta\lambda}{\Delta T}\Delta T = \lambda_{rt} + (A_0 + A_1\lambda_{rt})\Delta T. \tag{40}$$

Dorofeeva and Shapova [270] studied the ETC of sediments from underwater boreholes in Lake Baikal. The measurements were performed in the Laboratory by comparator of thermal conductivity (Antipov et al. [271]). In total about 2985 thermal conductivity measurements have been taken in three boreholes from a depth of 102-673 m. The results were compared with the values derived from borehole loggings: natural gamma ray, conductivity and resistivity. This is steady state method relative method. The uncertainty is within 3-5 %. The measurements were made without disturbing their integrity. The results of ETC measurement of the same samples with needle -probe method is higher by 30 %. The relationship between the ETC of Baikal bottom sediments and water content, resistivity and natural gamma rays was studied for one of the borehole (BDP-98). Water content was 40 %.

Sukharev and Sterlenko [272] and Sukharev et al. [160] experimentally studied the thermal properties of water, oil- and gas saturated rocks samples (sandstones). Measurements have been made at temperature of 25 °C and atmospheric pressure for 14 sandstone samples as a function of water and oil-saturations. Steady-state method was used to the measurements. The sharp increasing of the ETC with saturation was found in the range of 0 to 25-50 %. Between 50 and 75 % saturation the ETC slightly increases with saturation and between 75 and 100 % saturation the ETC againg increasing with saturation. The measured values of the ETC were used to determine the location of the oil filed.

In the works [110, 178, and 273] the authors reported ETC data partially saturated sandstones at temperature of 293 K. Kiyohashi et al. [110] developed prediction model for ETC of wet sandstones which predict own measurements within 22 %. The measurements were performed using hot-wire method. Seipold et al. [162, 163, 195-202, and 513] studied the ETC and thermal diffusivity of rocks at high temperatures and high pressures. The measurements were performed up to 1000 MPa. At pressures up to 500 MPa the linear pressure dependence of the ETC has been found. They used laser flash (impulse method) method at pressures up to 500 MPa and at room temperature. The method allow simultaneously measure ETC and thermal diffusivity of rocks. Kukkonen et al. [190] used transient method for the measurements of ETC and thermal diffusivity of porous rocks as a function of temperature and pressure. The measurements were made at temperatures from room temperature to 1150 K and at pressured up to 1000 MPa. They found that in this temperature range the ETC is changes within 12-20 %. Horai and Susaki [109] measured ETC of silicone rocks at temperatures from 300 K to 700 K and at pressures up to 1200 MPa with the uncertainty of 4 % to 5 % using steady-state method.

The same steady-state (guarded parallel plate technique, see below sect. 4) method was used by several research groups [274-297, and 522] to measure ETC of fluid (gas, water, and oil) saturated rocks with various porosities at temperatures from 275 K to 523 K and at pressures up to 500 MPa. Ljubimova et al. [298 and 299] and Maslennikov and Ganiev [300] reported thermal conductivity data for dry, wet, and oil-saturated rocks at high temperatures and high pressures by applying the steady-state method. The maximum uncertainty of the method is less 4 %, while the reproducibility is 1 %. Djamalova [301] measured thermal conductivity of 125 rock samples from various region of the Dagestan (located south Russian).

The measurements were made using steady-state method. The uncertainty of the measurements is within 5 to 10 %. The samples were thoroughly polished, provided good contact between the sample and heater, and cooler using vaseline, and therefore side heat loses were minimized. After polished the sample was coated (covered) with thin the layer of bakelite lacquer to prevent migration of the vaseline into the sample and leaking the water from the sample. Djamalova [301] also used divided-bar technique to measure thermal conductivity of sandstones. Steady –state condition was reached in approximately 4-6 hours. The values of the thermal conductivity were calculated using the working equation

$$\lambda = \frac{Qx}{S\Delta T}, \tag{41}$$

where $Q = I^2 U\tau$ is the heat input; x is the thickness of the sample; S is the area of the sample; ΔT is the temperature difference between the heater and cooler. This method is providing the measurements of the ETC of rocks at high pressures using the pressure creating and the pressure transferring units. The measurements were performed at pressures from 1 to 280 atm. Most measurements were made for sandstones with different porosities and quartz contents.

Buntebarth and Schopper [302] experimentally and theoretically studied the influence of fluids, solids and interactions between them on thermal properties of porous rocks. Recently, Fuchs and Förster [303] reported laboratory measurements of the thermal conductivity of Mesozoic sandstones from 8 wells of the Northeast German Basin. The measurements were made with optical scanning method. The porosity of the samples was within 16 to 30 %. Matrix thermal conductivity ranges from 3.4 to 7.4 W·m^{-1}·K^{-1}. The higher values reflect the large quartz content in sandstone. Based on the temperature log information the thermal conductivity was indirectly calculated for the entire borehole profiles.

The discrepancy between laboratory-measured and computed thermal conductivities in the two boreholes is in the order of 0.24 and 0.56 W·m^{-1}·K^{-1}. Cheng et al. [305] used transient-state apparatus for the in situ measurements of the thermal conductivity of dry, water-, and oil/bound-water saturated sandstones at temperatures from 293 to 523 K and at pressures up to 11 MPa. The accuracy of the data was checked on measurements standard sample of fused quartz.

To reduce the influence of the thermal resistance between the sample and the hot wire, some films or coatings were used. The measuring cell of the thermal conductivity apparatus developed by Nozad et al. [310 and 311] consists of an acrylic rectangular parallelepiped 10 cm wide, 10 cm deep and 20 cm tall. The thermal boundary conditions are applied by heat exchangers located at the ends of the measuring cell. Cooper plates constitute the thermal surface between circulating water and sample. The water streams originate in hot or cold temperature regulated reservoirs. The internal temperature profile is acquired by 8 thermocouples located at the centerline of the cell and sampled at 1 s intervals. To determine the thermal diffusivity (and thermal conductivity), Nozad et al. [310 and 311] plot $erf^{-1}(1-\Theta)$ versus $z/2\sqrt{t}$ and determine the effective thermal diffusivity (a) from the slope of the line, as $erf^{-1}(1-\Theta) = a^{-1/2}\left(z/2\sqrt{t}\right)$. The authors estimate the uncertainty in thermal conductivity measurements to be 5 %.

Deissler and Eian [308] and and Deissler and Boegli [309] used radial conduction to measure ETC of gas satuaretd portous media. The preliminary heater is located at the axis of the measuring cell and nichrome-wire heaters are wrapped around the ends to compansate for axial heat loss. A set of radial distributed thermocouples located at the midplane of the measuring cell acuire the radial temperature profile

$$T = T_1 - \frac{Q}{2\pi L \lambda_1}\left[\frac{2\lambda_1}{2\lambda_1 + c_1 T}\ln\left(\frac{r}{r_1}\right)\right]. \qquad (42)$$

This profile was used to determine the ETC. They plot the measured temperature profile versis the $\left[\frac{2\lambda_1}{2\lambda_1 + c_1 T}\ln\left(\frac{r}{r_1}\right)\right]$. The thermal conductivity from this method is $\lambda(T) = \lambda_1 + c_1 T$. The value of λ_1 was determined using the slope of the line. Yagi and Kunii [307] used steady-state radial conduction to measure the ETC of gas-saturated porous media. Locating of the heater at the center of the poorous media rather than at the end consideraable reduce heat losses. Guard heaters are located along the circumference and at the ends of the chamber for thermal management and to minimize heat loss. Thermocouples are positioned at several radial locations within the chamber to obtain internal temperature measurements. In order to determine the ETC following working equation for this method was used

$$\lambda = \frac{Q\ln(D_0/D_1)}{2\pi L(T_i - T_0)}, \tag{43}$$

where D_0 and D_1 are the outer and inner diameters of the annular chamber, respectively; T_0 and T_1 are the their temperatures. Hadley [34] employs axial conduction to measure the ETC of porous media. The sample disk is sandwiched between cooper and standard disks. Standrds disks are made of Dynasil 4000 glass and 304 stainless steel. The thermal conductivity of both very well known. The sample is heated by a nichrome wire embedded in an aluminium plate. He determines the ETC in a manner similar to that of Ofuchi and Kunii [267] (see below).

The heat flux through the entire assembly is determined from the temperature gradient across the standard materials. Then the ETC of the sample estimated from its surface temperatures and dimensions. The uncertainty in ETC measurements is about 10 %. The major contributions to the overall uncertainty are non-uniform contact resistance and thermistor drift. Ofuchi and Kunii [267] also used axial conduction in a cylidrical chamber to measure the ETC of porous media saturated with water, air, and CO_2. The measuring cell has an inside diamter of 20 cm and a height of 10 cm. The top is heated by steam and the bottom is cooled by water. A marble disc with well known thermal conductivity occupies the lower half of the measuring cell. The axial heat flux is determined by estimating the temperature gradient through it. Several thermocouples placed inside the porous media provide the axial temperature profile, which in turn, is used to determine the ETC.

Xie et al. [306] developed transient short-hot-wire (SHW) method to simultaneously measure ETC and diffusivity of porous materials. In this method, a platinum wire (8.8 mm in length and 48 μm in diameter) coated with an alumina layer (0.7μm thickness) served as both the heating source and the thermometer sensor. The authors used non-linear relation between the temperature rise and the logarithmic time. ETC of the sample was determined by fitting the experimental data with the calculated temperature rises. The uncertainty in ETC and diffusivity measurements by this method is 2 % and 7 %, respectively. They found linear relation between the ETC and density. Waite et al. [42] used Maxwell's needle probe method to measure ETC of four porous mixtures of quartz sand and methane gas hydrate. Measurements were made at low temperatures between -20 °C and 15 °C and at pressures from 3.5 to 27.6 MPa. Needle-probe was an epoxy-filled high-pressure tube with radius r of 0.79 mm containing a 4 kΩ thermistor and a heater wire loop with resistance of 335 Ω/m. The vessel was immersed in a temperature –controlled bath. The bath temperature drift was less than 0.18 °C/hr.The sample temperature measured by the thermistor every 2 second was recorded. The thermistor temperature T at time t after current is first supplied to the heater wire is given as

$$T(t) = \frac{q}{\lambda 4\pi}\left[\ln(t) + \ln\left(\frac{4k}{r^2}\right) - \gamma + \frac{2\lambda}{rC}\right], \tag{44}$$

where k is the thermal diffusivity, r is the probe radius (0.79 mm), and γ id Euler's constant; C is the thermal contact between the probe and sample. Thermal conductivity was calculated from the slope of the best-fit line relating T to $ln(t)$ for t satisfying $t >> \frac{r^2}{k}$.

Maqsood and Kamran [313] measured ETC and thermal diffusivity of porous consolidated air-saturated sandstones by the transient-plane source (TPS) technique over the temperature range from 280 to 330 K at ambient pressures. Measurements were made for five types of sandstone samples with porosities from 8 to 17 % taken from various positions above the baseline.

The results were interpreted in term of mixing laws and empirical models. Simple correlation between the ETC and density and porosity was proposed. The specimens had rectangular shapes of $0.043 \times 0.045 \times 0.025$ m^3. Each sample consists of two identical slabs. The surfaces of the samples were polished to provide good thermal contact with the transient plane source sensor and to minimize thermal contact resistance. This method uses a resistive element as both a heat source and temperature sensor. By supplying a constant current to the TPS sensor and by monitoring the subsequent voltage increase over a short period of time it is possible to get information about thermal conductivity and diffusivity of the material surrounding the heat source. The time-dependent resistance of the TPS during the transient recording was expressed as

$$R(t) = R_0\left[1 + \beta\Delta\overline{T}(\tau)\right], \tag{45}$$

where R_0 is the resistance of the TPS sensor before the start of transient recording; β is the temperature coefficient of resistivity of the TPS sensor; $\Delta\overline{T}(\tau)$ is the average value of the temperature increase of the TPS sensor during the transient recording; $\tau = (t/\theta)^{1/2}$ is the dimensionless variable; t is the time measured from the start of the transient heating; $\theta = d^2/a$ is the characteristic time; d is the radius of the disk; a is the thermal diffusivity. The thermal conductivity and diffusivity were measured as a function of temperature. The uncertainties in thermal conductivity and thermal diffusivity measurements are 6 and 8 %, respectively. Effect of compaction pressure and water content on the thermal conductivity of some natural clay was studied by Beziat et al. [514]. Morgan [515] has estimates of mantle thermal conductivity based on high-temperature measurements of mantle xenolith conductivities.

Abid [526] used THB technique to measure thermal conductivity of Sander sandstone (specimen is taken from te place Sander Schifsandsteinbruch "Hermannsberg, Sand Main", Haßberge, Germany). This method take short measurement time, maximum temperature rise up to 2 K, constant penetration depth, low uncertainty range and compensation of the end-effect problem. Two samples of sandstone in the form of rectangular blocks with dimensions of 100×60×20mm^3 were studied.

The samples dried at temperature of 105 °C at least for 24 hours. The water saturated state is reached by immersing the dry specimen into water for 95 hours and reweighed until the mass change between two consecutive weighing is less than 0.3 %. Free saturation porosity is determined using Archimedes principle by the mass changes between dry and water –saturated specimens.

In this method not all the pores can be filled, therefore porosity derived from this method was usually lower than other methods. Total porosity of the samples was 19.32 % for dry and 18.63 % for wetted stone. The method used to measure of the thermal conductivity was described above in the sec. 2.2. The temperature dependence of the thermal conductivity of sandstone was measured in the temperature range from -20 °C to 40 °C. The measurements also were made by filling the porous sandstone with six different liquids (water, alcohol, toluene, water+alcohol mixtures) and gases (nitrogen, helium, argon, Freon -12) with various thermal conductivities. Abid [526] studied the effect of saturated fluids on the overall effective thermal conductivity of the sandstone. The results of the measurements were used to develop of the empirical model for the determination of effective thermal conductivity of porous materials in terms of easily measurable parameters such as porosity, thermal conductivity of the solid matrix, *i.e.*, mineral compositions and the thermal conductivity of the saturated fluids.

The measured results showed that ETC of sandstone is inversely proportional to the temperature. The anomaly temperature behavior of the thermal conductivity of water – saturated sandstone was found near the water freezing point. Abid [526] also tested the contact resistance problem experimentally. He put successive loads of 1 kg on the top of the upper sample half to make a better thermal contact between sensor and the sample halves. The results showed that the difference in the values of thermal properties for different runs remained within the experimental uncertainty 3 % for thermal conductivity and 9 % for thermal diffusivity.

Gegenhuber and Schoen [550] measured the thermal conductivity of sandstones (from different geological units of Austria) using transient method. In this method the authors used probe plane (10×10) cm and needle length with 7 cm (Plexiglas cylinder with diameter of 9 cm), boundary effects were negligible. The line source and the sample were fixed by a constant pressure of 15 bar. The thermal conductivity was calculated directly from the heating curve.

The authors also measured wave velocity and presented two models that describe the correlation between thermal conductivity and compressional wave velocity for various types of rocks. Novikov [554] and Vertogradskii et al. [555] measured high porous fluid (heavy oil, water, air, ethylene glycol, kerosene, working solutions) petroleum) saturated quartizided sandstones (250 samples) at high pressures. In these works the authors developed new apparatus for high pressure thermal conductivity measurements for fluid saturated rocks materials. The measurements were performed using optical scanning technique. The results of the measurements were interpreted using the model developed by Lichtencker et al. [479 and 483] and Asaad [166]. The measurements were made at temperatures from 25 to 100 °C. The uncertainty of the thermal conductivity measurements less than 10 %, while for thermal diffusivity is within 15%.

Nikolaev et al. [560 and 561] used divided-bar technique to measure of the thermal conductivity of dry and fluid –saturated (oil, water, air) sandstones (quartz -95 % and 15 % clay). They used various contact fluids (glycerin) to improve the thermal contact between the sample and heater (to reduce the effect of the contact resistance). Babaev et al. [563] reported thermal conductivity of 208 sandstone samples with various porosities (from 4.1 to 24.3 %), densities (from 2120 to 2980 kg·m^{-3}), and compositions (clay content from 3.3 to 39.5 %) from the deep (1747 to 4080 m). The measured thermal conductivity values are lied within from 0.38 to 2.7 W·m^{-1}·K^{-1}. The measured values of thermal conductivity were interpreted as a function of density, porosity, *etc*. The sandstone with high thermal conductivity of 2.7 W·m^{-1}·K^{-1} was observed for sandstone with low porosity of 6.5 %, high density of 2530 kg·m^{-3} and low permeability. Low thermal conductivity 0.38 W·m^{-1}·K^{-1} was found for the sandstone with high porosity of 17.1 % and low density of 2230 kg·m^{-3}. The uncertainty of the thermal conductivity measurements is within 3.5 to 18.5 %.

4. Guarded Parallel-Plate Method for Thermal Conductivity Measurements of the Dry and Fluid Saturated Sandstones

ETC of dry, water-, and oil-saturated sandstones has been measured by a guarded parallel-plate apparatus. It is an absolute, steady-state measurement device with an operational temperature range of 270 to 600 K and pressures up to 1000 MPa. The method (apparatus, procedure of measurements, and uncertainty assessment) has been described fully in our previous publications [274-297], thus only a brief review will be given here. A schematic diagram of the experimental thermal conductivity apparatus used in works [274-297] is shown in Figure 5. Details of the construction of the thermal conductivity cell are shown in Figure 6. It consists of a high-pressure chamber, a thermal - conductivity measuring cell, an air thermostat, a high precision temperature regulator, and a high - pressure liquid and gas compressors.

The temperature in the air thermostat was controlled automatically to within ±5 mK. In this method, thermal conductivity is obtained from simultaneous measurements of the steady-state heat flux Q and temperature gradient in the sample placed between the heating and cooling plates.

The good thermal contact between the heater and the sample was assured using thermal interface materials (wetting agent such as vaseline or glycerin). To reduce the effect of contact resistance (imperfect thermal contact between the sample and the adjacent bronze disc) the sample and heater surfaces were polished flat to within 0.05 mm and smooth. The spring (see Figure 5) was also used to crate contact pressure (3-6 MPa) to improve the contact between sample and heater.

The lateral heat-losses were estimated using the samples of Pyrex glass of well-known thermal conductivity. Two thermocouples were embedded in the center of the inner surface of the bronze disk (see Figure 6).

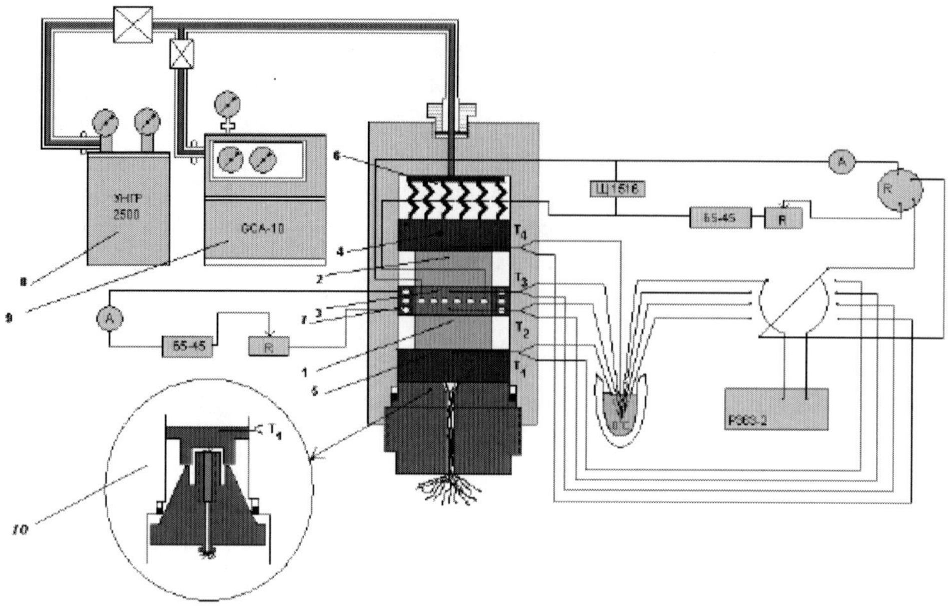

Figure 5. High temperature and high pressure apparatus for thermal conductivity measurements of rocks materials developed in [274-297]. 1and 2-samples; 3-heater; 4 and 5-coolers; 6- spring; 7-compansation heater; 8-high pressure oil **compressor** (УНГР-2500); 9-high pressure gas compressor (GCA-10); 10-obturator; 11 and 12 –bronze disks; 13- heater; 14-copper-constantan thermocouples; T_1, T_2, T_3, T_4 -thermocouples.

The heater was located between these thermocouples. Other two thermocouples were soldered to the body of the heater. The temperature difference (temperature gradient) and temperature of the chamber were measured with four copper - constantan thermocouples. The pressure was created with liquid and gas compressors (Unipress Type GCA-10, Poland). The pressure in the chamber has been measured with a manganin pressure transducer with an uncertainty of 0.25 %. The high-pressure chamber is located in the air thermostat.

Working equation of the method:

The thermal conductivity λ of the specimen was deduced from the relation

$$\lambda = \frac{Q - Q_{los}}{\frac{S_1}{h_1}\Delta T_1 + \frac{S_2}{h_2}\Delta T_2}, \tag{46}$$

where $Q = Q_1 + Q_2$ is the heat flow transferred from the heater to the upper and lower specimens; $Q_1 = \lambda S_1 / h_1 \Delta T_1$ and $Q_2 = \lambda S_2 / h_2 \Delta T_2$ are the heat flows transferred by conduction through the lower and upper specimens, respectively; Q_{los} is the heat losses through the lateral surface of the samples; S_1 and S_2 are the cross-sectional areas of the specimens that heat flows through; h_1 and h_2 are the height of the samples; and ΔT_1 and ΔT_2 are the temperature differences across the samples thickness. The thermal conductivity was obtained from the measured quantities: Q, Q_{los}, ΔT_1, ΔT_2, S_1, S_2, h_1, and h_2.

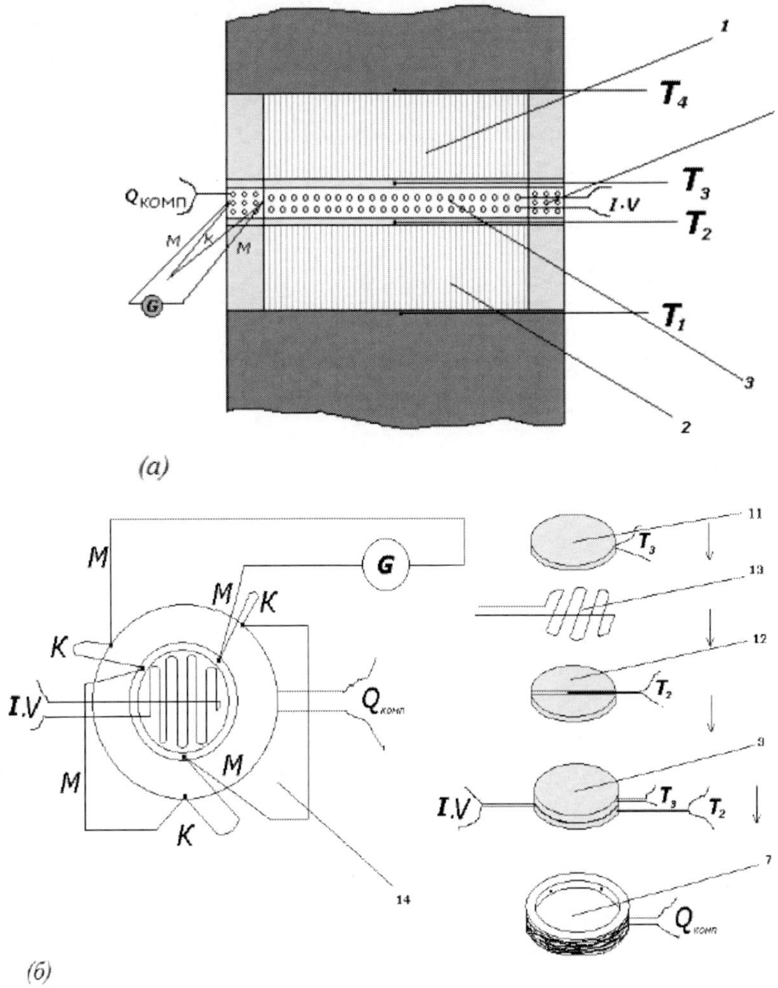

Figure 6. Construction of the measuring thermal conductivity cell (*a*) and compensation heater (*b*)

The heat flow Q from the heater was distributed between the two samples Q_1 and Q_2. The values of Q were corrected by a specimens lateral loss factor Q_{los}. The values of Q_{los} was estimated from the relation

$$Q_{los} = \lambda_m \Delta T \frac{2\pi h}{\ln(d/D)}, \tag{47}$$

where d=12 mm and D=22 mm are the diameter of the sample and the inner diameter of the high-pressure chamber, respectively; $h = h_1 + h_2$ is the height of the samples; λ_m is the thermal conductivity of the media of transmitted pressure (oil), and $\Delta T = T_m - T_C$ is the temperature difference between average values $T_m = (T_2 + T_1)/2$ or $T_m = (T_3 + T_4)/2$ of the temperatures (T_2 and T_3) of lateral surface of the specimens and temperatures (T_1 and T_4) of the inner surface of the high-pressure chamber; $T_C = T_1 = T_4$. The values of the temperature difference were almost constant, $\Delta T \approx 1$ to 2 K. The heat losses by conduction along the electrical leads and heating are negligibly small (Abdulagatov et al. [292-297]).

Uncertainty of the thermal conductivity measurements:

The maximum relative uncertainty of ETC measurements associated with measured quantities can be estimated from the equation

$$\delta\lambda = \frac{\delta Q}{1-\dfrac{Q_{los}}{Q}} + \frac{\delta Q_{los}}{\dfrac{Q_{los}}{Q}-1} + \frac{\delta S_1 + \delta h_1 + \delta(\Delta T_1)}{1+\dfrac{S_2}{S_1}\dfrac{h_1}{h_2}\dfrac{\Delta T_2}{\Delta T_1}} + \frac{\delta S_2 + \delta h_2 + \delta(\Delta T_2)}{1+\dfrac{S_1}{S_2}\dfrac{h_2}{h_1}\dfrac{\Delta T_1}{\Delta T_2}}. \quad (48)$$

Measurement uncertainties exist in the measured quantities contained in working Eq. (46) (Q, Q_{los}, ΔT_1, ΔT_2, $S_{1,2}$ and $h_{1,2}$) and the uncertainties of the temperature, T, and pressure, P, measurements. The uncertainty in all of the measured quantities was δQ=0.57 %; δQ_{los}=0.2 %; Q=0.28 W; Q_{los}=0.02 W; ΔT_1=2 K; ΔT_2=1.5 K; $\delta S_{1,2}$=0.33 %; $\delta h_{1,2}$=0.33 %; $\delta(\Delta T_{1,2})$=0.1 %. The propagation of uncertainties related to the uncertainties of pressure, temperature, and sample's height are (0.03, 0.01, and 0.002) %, respectively. Heat losses through the side surface of the specimens were 3.5 % of the total amount of heat supplied to the specimens. Therefore, from the uncertainty of the measured quantities and the corrections mentioned above, the total combined expanded uncertainty in the thermal conductivity measurement at the 95 % confidence level with a coverage factor of k=2 less than 4 %. This value of the uncertainty is not including the uncertainty due to contact resistance and radiative heat transfer.

Therefore, the uncertainty of the ETC data obtained with the method is higher than 4 %. Unfortunately, due to lack of the characteristic optical properties (refractive index and absorption coefficient) of the sample it is impossible quantitatively take into account the contribution of the radiative heat transfer for fluid saturated porous materials. Therefore, the thermal conductivity data for dry, water-, and oil-saturated sandstone did not include the corrections to radiation and contact resistance. However, it is possible to considerable minimizes their effect on the measured values of ETC by improving the contact between the heater and sample using thermal interface materials and contact pressure. According Hofmeister et al. [89] contact resistance with heaters and thermocouples, and possibly among constituent grains, leads to systematic and substantial underestimation of lattice thermal conductivity of 20 %. The relative version of the present contact method allowed considerable improvement of the accuracy of measurements by calibrating the thermal conductivity cell with the data derived from contact-free method for reference sample. In this case the uncertainty due to contact-resistance will include into the calibration parameter.

To check the reproducibility, the measurements at selected experimental temperature T and pressure P were repeated (5 to 10) times. The scatter of the experimental results did not exceed ±0.5 %. The measurements were made with various temperature differences ΔT_1 between (1 and 5) K.

To check and confirm the validity of the method and procedure of the measurements, the thermal conductivity measurements were made with standard (reference) materials (fused quartz) using the present apparatus. Fused quartz has been recommended as a standard material for the test and calibration of the thermal conductivity apparatus (see, for example, Devyatkov et al. [304]). The thermal conductivity of fused quartz has been used to calibrate apparatus for measuring the ETC of rock specimens by many authors using the various versions of the contact methods. Figure 7 shows the comparison between the present and reported thermal conductivity data for fused quartz. Excellent agreement within 1.0-2.5 % was found between the present data and the majority reported values by other authors. Unfortunately, to our knowledge, there are no reported thermal conductivity data for fused quartz obtained using the contact free methods to compare with the present data. This excellent agreement for fused quartz and quartz ceramic demonstrates the reliability and accuracy of the present measurements for porous rocks and correct operation of the instrument.

Recently [495] we also measured the thermal conductivity of ceramics (SiC-BeO) at high temperatures and high pressures to confirm the reliability and accuracy of the method and correct functioning of the thermal conductivity apparatus. Good agreement between our data and reported reference data for ceramics confirms the reliability and high accuracy of the measurements for rocks material and gives us an assurance that our instrument is functioning correctly. The method described above was used to measure of the ETC of dry and fluid (water, oil, gas) saturated sandstones at high temperatures and high pressures. The results are presented in the Appendix.

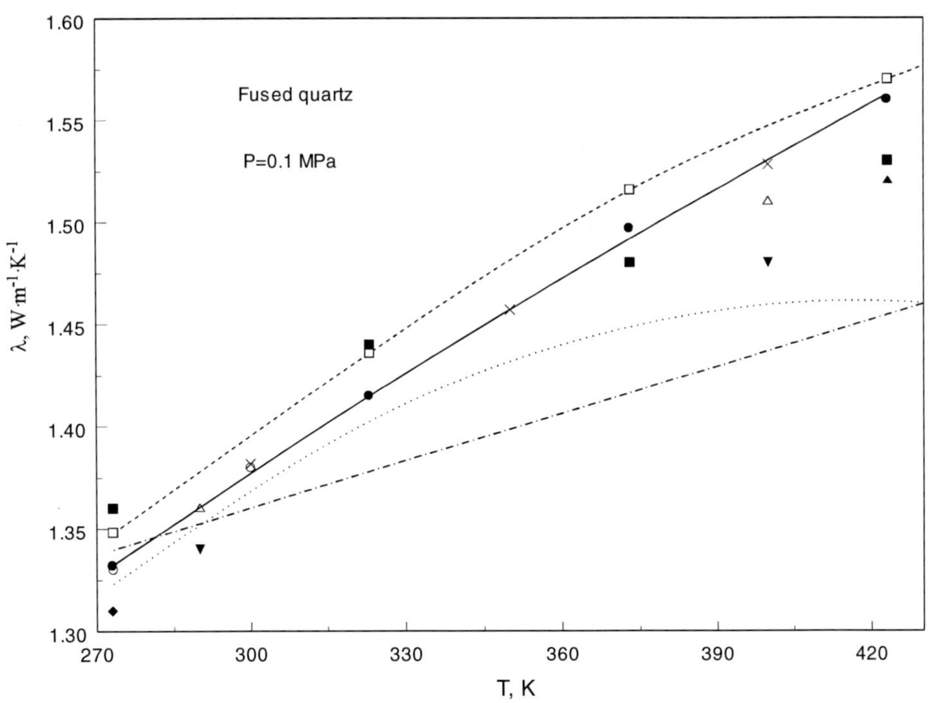

Figure 7. Comparison measured thermal conductivities of fused quartz with reported data. ●-Abdulagatov et al. [294]; ○- Touloukian [446] (recommended); ×-Devyatkova et al. [485]; □-Sugawara [482]; ▲-Camirand [486]; ■-Clauser and Huenges [348]; (- · · - · · -) - Kaye and Higgins [487]; (············) – Ratcliffe [251]; ▼- Xia Cheng et al. [305], Δ-Incropera et al. [488] (recommended); ♦-Horai and Susaki [109].

5. Effective Thermal Conductivity of Sandstones: Measurements and Modeling

5.1. Sources of Experimental ETC Data for Sandstones

The summary most of the reported ETC data sets for dry, air-, water-, and oil-saturated sandstones from various locations of the world are given in Table 1. Measured ETC of sandstones have been also presented in several handbooks [11, 88, 168, 516, and 522]. In this table, the method employed; the uncertainty claimed by the authors, and the temperature, pressure, and porosity ranges are presented together with the information about the samples (type of sandstones, saturation, and location). Table 1 provides the most useful collection of reliable ETC data for sandstones including high temperature and high pressure measurements. The Appendix includes only some selected data at high temperatures and high pressures which were measured in our Laboratory. To our knowledge this is good comprehensive compilations most of the reliable ETC data for dry and fluid-saturated sandstones reported by various Laboratories.

Some selected experimental ETC data as an example of the temperature, pressure, and porosity dependencies for several selected of fluid-saturated sandstones are given in Figures 8 to 11.

Table 1. Summary of experimental effective thermal conductivity data for dry and fluid saturated sandstone

Type of sandstone, samples, locality, saturation	Temperature, pressure, porosity	Method of measurements, uncertainty	Ref
Jacobsville and Freda sandstones, deep 457 and 1524 m,	303 K, 17.3 MPa	Modified divided-bar method	[516]
11 data points, fluid saturated sandstones	ambient conditions	transient method (needle-probe), divided-bar	[88, 129]
Karoo (7 data points) and Jacobsrille (8 data points) sandstones	ambient conditions	divided –bar method	[517]
Fluid saturated sandstones	ambient conditions	divided-bar method	[518]
447 data points, dry and fluid saturated sandstones	ambient conditions	divided-bar method	[520]
Consolidated sandstones	280-330 K, 0.1 MPa, porosity from 8 to 17 vol.%	transient-plane source method	[313]
Boise, Berea, and Bandera sandstones (consolidated), dry, partially, and fully saturated with 5000 pp KCl aqueous solution, n-heptane, and other solvents	27 MPa and at ambient temperatures	divided –bar method	[151]
Dry and water-saturated sandstones (Eromanga/Cooper Basin)	porosity from 0 to 33 %	divided-bar method	[14]
Quartz sandstones; grain size contg. 70-80% of size 0.16-0.20 mm. The QS contain SiO_2 97.7 and Fe_2O_3 0.20-0.27 wt.%.	at atmospheric pressure as a function of temperature	guarded parallel-plate method, uncertainty is 4 %	[319]
Dry, water-, and oil/bound-water saturated sandstones	293-523 K, 10-11 MPa, porosity 3 %, 5 %, 25 %, and 32 %.	transient-state apparatus for in-situ measurements	[305]

Type of sandstone, samples, locality, saturation	Temperature, pressure, porosity	Method of measurements, uncertainty	Ref
Silty sandstone, water-saturated and dry, alevretic sandstones,	porosity 26 % and 24 %, ambient conditions	regular thermal regime technique	[490]
From oil reservoirs, Carpathian Depression	porosity 4.9-10.2 %, ambient conditions	contact-free method	[320]
Natural Berea sandstone, saturated with gas (N_2O, air, He, H_2), liquid (n-heptane, water), major solid componentis quartz	at 298 K, 0.1 MPa, porosity 22 %	transient line heat source	[491]
Quartz-biotite, seliceous sandstones, deep 143-748 m, Krivoy Rog, Ukraine (10 data points)	286-297 K, 0.1 MPa	source method	[492]
East-European Platform, Russia (15 data points)	ambient conditions	divided-bar method	[493]
Water-, natural gas-, and oil-saturated sandstones	293-573 K, up to 150 MPa	guarded parallel-plate method, uncertainty is 4 %	[284]
13 sandstones saturated with air and water, with porosities of 9.7 % and 42.3 %	273-373 K, 0.1 MPa	comparative (steady state) method	[224, 225]
Dry and fluid (silicone oil, brine, and n-hexane) saturated sandstone (6 sample)	293-413 K, at 4 MPa	divided –bar, steady -state method, 5 % uncertainty	[222]
Gas, oil and water saturated sandstones (14 samples)	298 K, 0.1 MPa	steady-state method	[272, 160]
Sander sandstone dry and water-saturated	274.65 K and 263.15 K, porosity 14 %	pulse transient method, uncertainty 5.88 %	[149]
Berea sandstone	300 K, 0.1 MPa	laser flash method	[145]
Quartzite dry and salt-water saturated sandstone (tight sandstones)	ambient conditions, porosities from 1 to 20 %	laser pulse, uncertainty is 5 %	[148]
Clean and clayey sandstones	ambient conditions	steady-state comparative method (divided - bar), uncertainty is 4.8 %	[315]
Gas, oil, and water saturated sandstone	273-523 K, 0.1 to 400 MPa, porosity 7 %	guarded parallel-plate	[290]
Dry and fluid-saturated, Mordovo-karmalskii, Russia, 125 samples	ambient conditions	N/A	[494]
entirely or partly saturated with air, water, and kerosene (38 samples)	porosities from 4.4 to 48.5 %, ambient conditions	steady-state comparative method (divided - bar), uncertainty is 4.8 %	[227]
Water-saturated sandstones	at normal (298 K) room temperature and 0.1 MPa	divided-bar method, uncertainty is 2-4 % (corrected for thermal contact resistance).	[190]
Water-, air-, and oil- saturated (oil production zone)	305.4 K, 0.1 MPa, porosity 19.6 %, 40 %	steady-state method (comparative-type apparatus)	[230]
Gas-, oil- and water- saturated sandstone, Dagestan, Russia	open porosity 16.243 %, up to 400 MPa, from 273 K to 423 K	guarded parallel-plate method, uncertainty is 4 %	[274]
From deep 509-533 m, South Hungery (Mechek, Komlo, Zaboka)	298-310 K, 0.1 MPa	comparative method	[317]
Dry and fluid (gas, oil, water)-saturated sandstones, Soltanbek-Gasha (Dagestan, Russia)	from 273 K to 423 K, up to 400 MPa, porosity from 1.5 % to 20 %	guarded parallel-plate method, uncertainty is 4 %	[289]

Table 1. (Continued)

Type of sandstone, samples, locality, saturation	Temperature, pressure, porosity	Method of measurements, uncertainty	Ref
Tertiary sandstones, Uinta basin (Utah); clean, clay-reach	porosity up to 60 % for different mineralogical content	divided-bar and needle probe methods	[124]
Gas-, water-, and oil-saturated sandstones (Dagestan, Russia), from deep 2970-5106 m	porosity 12 %, 5 %, 16.2 %, from 273 to 423 K, up to 250 MPa	guarded parallel-plate method, uncertainty is 4 %	[275]
Seven dry, water, and oil saturated (C_{10} - C_{12} aliphatic hydrocarbons or tap water) sandstones, quartz content > 90%,	porosity from 3% to 60 %, up to 813 K, at 0.1 MPa	probe method (line heat source)	[247]
Shaly sandstone, borehole with 570 and 810 m depth	porosity 10 % to 30 %	optical scanning technique	[256]
Boise, Berea, and Bandera sandstones, consolidated sandstones, dry and fluid saturated (brine, aqueous KCl solution, heptane, solvent)	4.6 MPa, up to 406 K, porosity 20%, 22%, 29%, and 33.5 %	high temperature ETC apparatus, steady-state divided –bar	[174]
Water saturated sandstone	260 K to 278 K, at 0.7 MPa	divided–bar method	[268]
Partially saturated sandstones	293 K, 0.1 MPa	hot-wire method, uncertainty is 5 %	[178]
Dry and partially saturated sandstones (Berea, Kishu, Izumi, Tako sandstones), 4 samples	293 K, porosity from 5 % to 51 %	hot-wire method, uncertainty is 5 %	[110]
Partially saturated sandstones	293 K, 0.1 MPa	hot-wire method, uncertainty is 5 %	[273]
Oil-saturated sandstone	13 % porosity, 288-520 K, up to 250 MPa	guarded parallel plate technique, uncertainty is 4 %	[292]
Type of sandstone, samples, locality, saturation	Temperature, pressure, porosity	Method of measurements, uncertainty	Ref
Dry and salt water saturated	porosity 1.0% - 20 %, ambient temperatures, at 0.1 MPa	variable state approach (contact-free technique, electromagnetic heat source), uncertainty 5 %	[148]
Dry and fluid (air, CO_2, He, oil, n-decane, ethanol, glycerine, and water - saturated porous sandstones (3 natural sample)	294.3 K, at 0.1 MPa, porosity 12.6 %, 15.8 %, 22.0 %	uncertainty is 10 %, steady-state guarded hot plate method	[166]
fluid-saturated porous sandstones (Pennsylvania)	from 273 K to 673 K, 0.1 MPa	divided –bar method	[79]
Type of sandstone, samples, locality, saturation	Temperature, pressure, porosity	Method of measurements, uncertainty	Ref
Dry, helium-, water-saturated porous sandstones (Pennsylvania, Colorado)	318 K, up to 70 MPa	divided –bar method	[168, 316]
Fluid-saturated porous sandstones	ambient conditions	divided –bar method	[220]
Berea, Bandera, and Bakelite sandstones, fluid (water and kerosene)-saturated porous sandstones	from room temperature to 353 K and up 12.5 MPa, porosity are 18 % and 22 %	steady-state apparatus (divided-bar), uncertainty is more than 4.8 %	[227]
Gas, water-, oil-saturated sandstones	273-423 K, up to 400 MPa, porosity 5 %, 12 %, 16.2 %	guarded parallel-plate method, uncertainty is 4 %	[278]

Type of sandstone, samples, locality, saturation	Temperature, pressure, porosity	Method of measurements, uncertainty	Ref
Water-saturated porous sandstones (4 different samples)	at 20 °C, porosities from 5 % to 51 %	transient hot-wire comparative method	[189]
Fluid-saturated (air, kerosene) porous sandstones	porosity 3-29 %	guarded parallel-plate method, uncertainty is 4 %	[318]
fluid-saturated (gas, oil, water) porous sandstones	up to 300 MPa, 283-573 K, porosity 6.9 %, 10.5 %, 12.0 %, 13.9 %	guarded parallel-plate method, uncertainty is 4 %, repeatibility is 1 %	[280, 285]
Clean sandstones	Ambient conditions	calculated values	[421]
Dry sandstones, Dagestan Russia (125 samples)	Ambient temperature, up to 280 atm	divided – bar method, uncertainty is 5-10 %	[301]
Oil-content (12.5 %) sample	293-393K, 0.1 MPa	steady-state method,	[226]
Berea, Bandera, and Boise sandstones	6 to 25 MPa, ambient temperatures	NA	[229]
Kaliningrad (Russia) group of fields, from wells in a depth interval of 1637 m to 2572m. Quartz and occasionally quartz and calcite pisolitic cement. Dry and water-saturated (30 samples)	ambient conditions, porosity from 2.13 % to16.6 %	optical scanning technique	[204]
3 type rocks-sandstones, dry and saturated with water	porosity is within 0 to 33 %.	divided-bar apparatus, uncertainty is 5 %	[113]
Mesozoic sandstones from 8 wells of the Northeast German Basin	porosity from 16 % to 30 %.	optical scanning method	[303]
from different geological units of Austria	NA	Transient method (needle-probe)	[550]
high porous fluid (heavy oil, water, air, ethylene glycol, kerosene, working solutions) petroleum) saturated quartizided sandstones (250 samples)	25 to 100 °C	optical scanning technique, uncertainty is less than 10 %	[554, 555]
Sander sandstone (specimen is taken from te place Sander Schifsandstein-bruch "Hermannsberg, Sand Main", Haßberge, Germany), dry and fluid (water, alcohol, toluene, and mixtures)	Porosity is 19.32 % for dry and 18.63 % for wetted sample, -20 to 40 °C.	THB technique, uncertainty is 3 %	[526]
dry and fluid –saturated (oil, water, air) sandstones (quartz -95 % and 15 % clay)	NA	divided-bar technique	[560, 561]
208 sandstone samples with porosities from 4.1 to 24.3 %, densities from 2120 to 2980 kg·m^{-3}, and compositions (clay content from 3.3 to 39.5 %), from the deep (1747 to 4080 m).	NA	Uncertainty is within 3.5 % to 18.5 %	[563]

Figure 12 shows the plot of (λ_{eff}/λ_S) verses of (λ_{oil}/λ_S) and (λ_w/λ_S), where λ_{oil}, λ_w, and λ_S are the thermal conductivity of the saturated fluid (oil and water) and solid matrix, respectively for different pressures. This figure demonstrates the effect of conductivity of the saturated fluids (water and oil) on the ETC of solid-fluid system. As one can see, the plot of the (λ_{eff}/λ_S) verses of (λ_{flu}/λ_S) is almost linear, except at high pressures where slightly deviations from the linearity were found. These figures show the typical temperature, pressure, and porosity dependencies of ETC fluid-saturated porous rock materials. As one can see from Table 1, most of the reported ETC data for fluid-saturated sandstones were obtained by contact methods such as divided-bar [14, 79, 88, 113, 124, 129, 151, 160, 168, 174, 190, 220, 222-227, 230, 268, 272, 301, 315-317], guarded parallel-plate [166, 274, 278, 280, 284, 285, 289, 290, 292, 318, and 319], needle-probe [110, 178, 189, 247, 273, 305, and 323], and contact-free methods [145, 148, 149, 204, 256, 303, and 320] and consequently their nominal accuracy limited to 4-5 % for contact methods and 2-3 % for contact-free methods. The experimental techniques employed to measure ETC of sandstones were discussed above (see sec. 3).

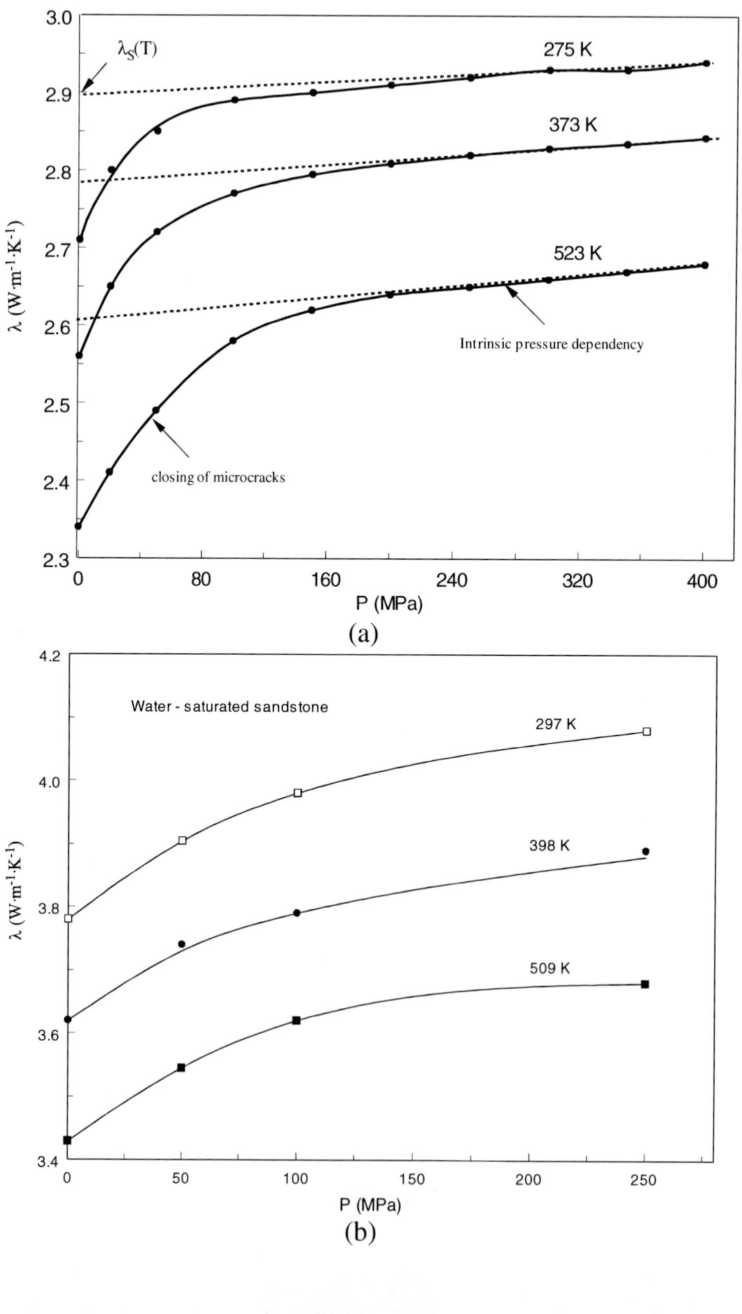

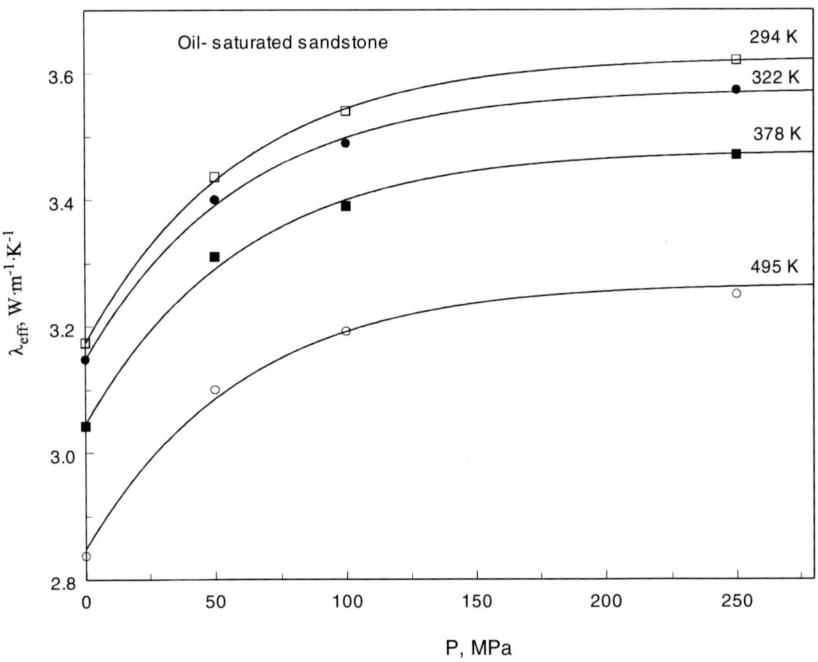

(c)

Figure 8. Experimental ETC of air-(a) (Intrinsic pressure dependency and closing of microcracks (reduction of pore space, compacted rock matrix)), water-(b), and oil-(c) saturated sandstones (Abdulagatov et al. [292, 293, 314]) as a function of pressure *P* along the selected isotherms together with values calculated from Eq. (232) (dashed lines) Dashed (a) lines are linear extrapolation to the zero pressure.

The following section will provide a detailed outline of the available theoretical and empirical prediction models for the temperature, pressure and porosity dependence of the ETC fluid-saturated porous rocks.

These models will be employed in sec. 7 to the some selected experimental ETC data for fluid –saturated sandstones in order to exam their accuracy, predictive capability, and applicability for fluid-saturated sandstones.

According Reiter and Tovar [404] summarizing of the ETC data for rocks, the ETC of sandstones varied in the range from 1.88 to 4.98 $W{\cdot}m^{-1}{\cdot}K^{-1}$. The possible reasons for disagreements are as follows:

(1) The constituents may have widely variable content [6];
(2) Conductivity, unlike other properties, depends on direction. ETC of multicomponent matrix is thus a function of the conductivities of the individual minerals, and the shape and orientation of the crystals;
(3) Crystals in a rock matrix may contain impurities and have a conductivity significantly different from that of pure monocrystalline specimens;
(4) Boundaries between individual crystals represent a thermal resistance. Depending on the nature of the boundary, and whether it has a filling of residue material, water or gas.

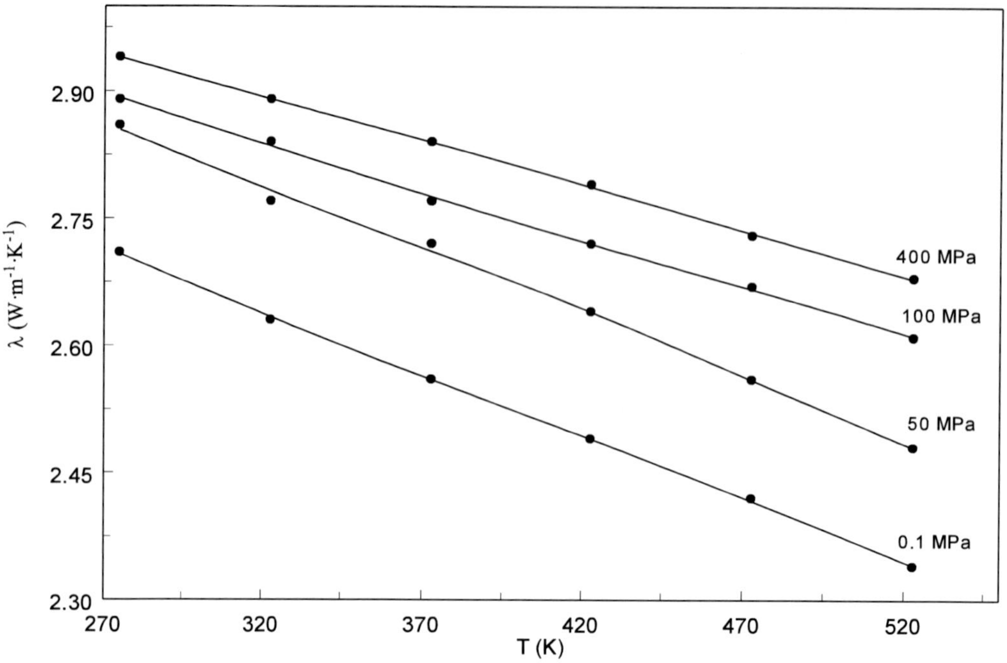
λ (W·m⁻¹·K⁻¹)
2.90
2.75
2.60
2.45
2.30
270
320
370
420
470
520
T (K)
400 MPa
100 MPa
50 MPa
0.1 MPa

(a)

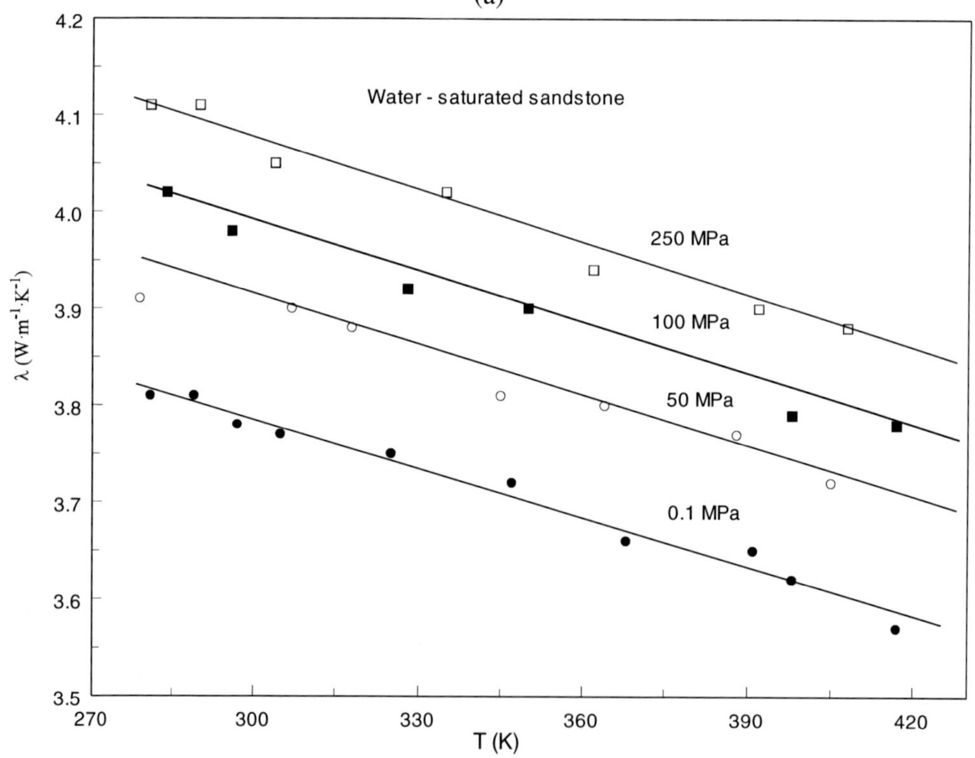
λ (W·m⁻¹·K⁻¹)
4.2
4.1
4.0
3.9
3.8
3.7
3.6
3.5
270
300
330
360
390
420
T (K)
Water - saturated sandstone
250 MPa
100 MPa
50 MPa
0.1 MPa

(b)

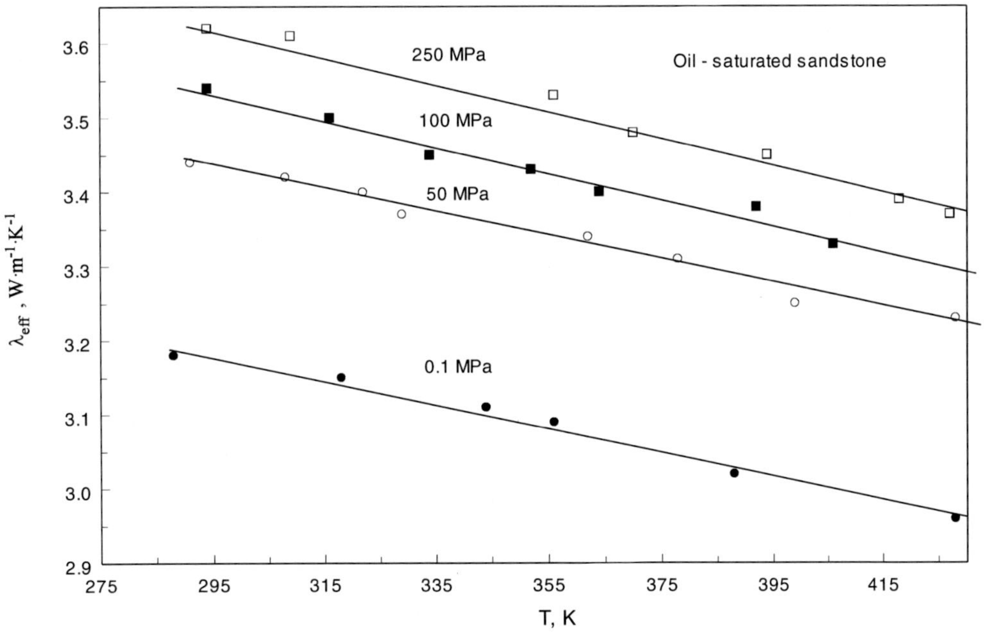

(c)

Figure 9. Experimental ETC of air-(a), water-(b), and oil-(c) saturated sandstones, porosity 13 %, (Abdulagatov et al. [292, 293, 314]) as a function of temperature *T* along the selected isobars together with values calculated from Eq. (232).

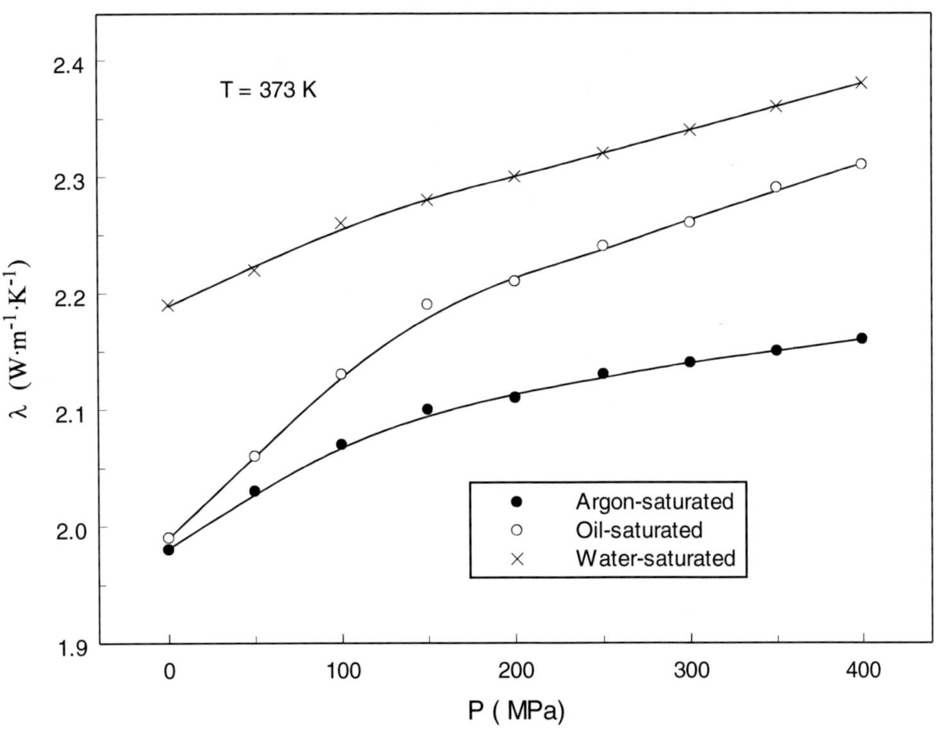

Figure 10. Comparison of the measured ETC of argon-, water-, and oil-saturated sandstones (porosity is 7 %) by Kurbanov [290] as a function of pressure *P* along the selected isotherm of 373 K.

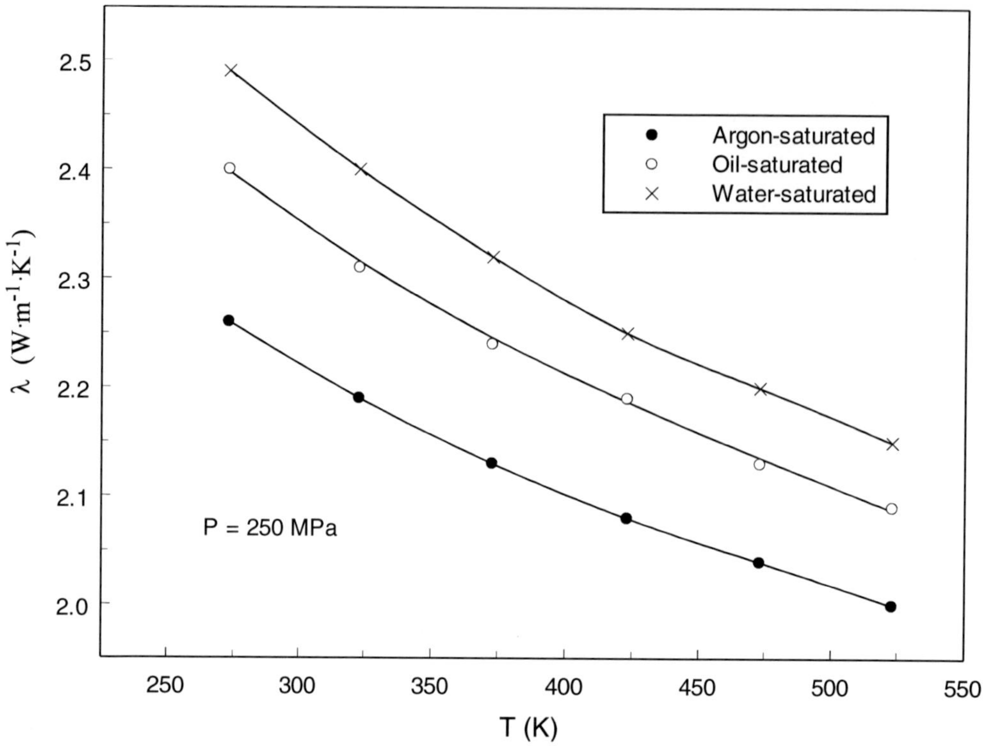

Figure 11. Comparison of the measured ETC of argon-, water-, and oil- saturated sandstones (porosity is 7 %) by Kurbanov [290] as a function of temperature T along the selected isobar of 250 MPa.

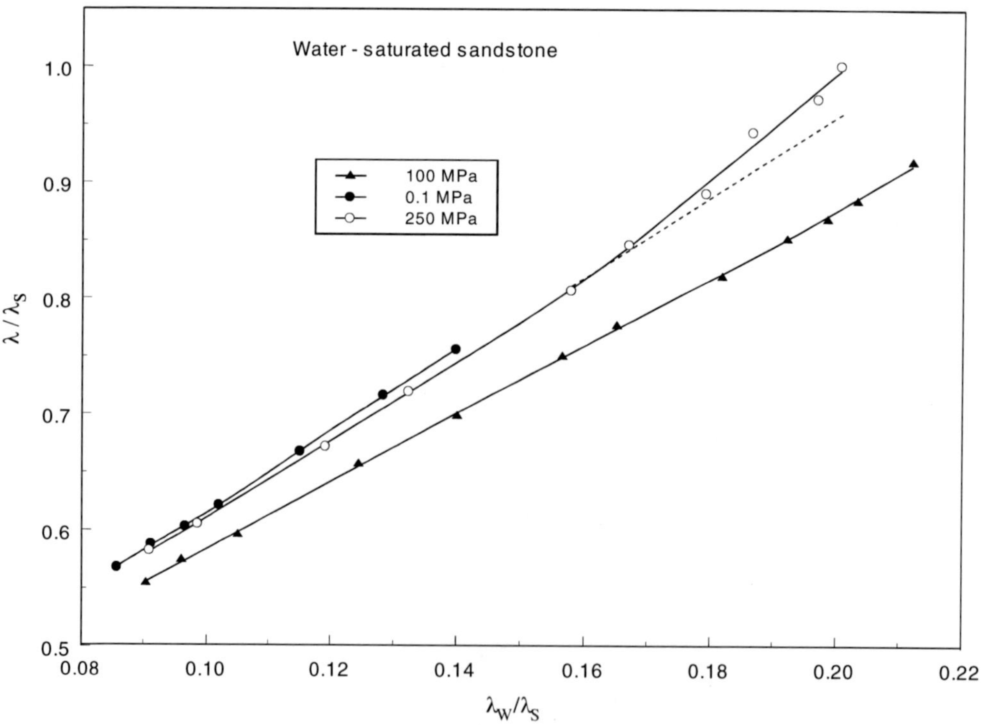

(a)

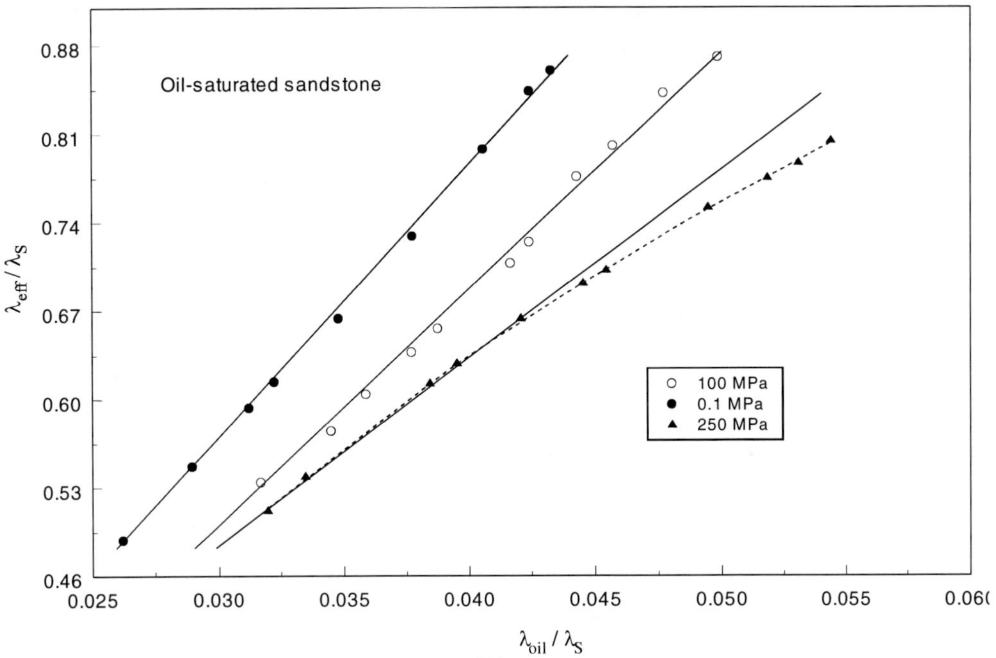

(b)

Figure 12. Normalized ETC of rock-water [314] (a) and rock-oil [292] (b) systems (λ / λ_S) as a function of normalized thermal conductivity of saturating water (λ_W / λ_S) and saturated oil ($\lambda_{oil} / \lambda_S$).

5.2. Theory and Modeling of the ETC: Porosity Dependency Models at Constant *T* and *P*

As was mentioned above (see sec. 2), the measurement of ETC from core material and by using the down-hole needle-probe are time consuming and costly. Other hand, since high-temperature and high-pressure measurements are difficult and since it is not possible to duplicate in the laboratory the conditions of the earth's interior, theoretical and prediction methods can be used to extrapolate laboratory measurements to high temperatures and high pressures (Schatz and Simmons [151]; Roufosse and Klemens [76]; Hofmeister [77]). It is, therefore, predictive models are very important. Lot of methods have been developed in the last decades to estimate the effective thermal conductivity of unsaturated porous rocks (see, for example, [1, 49, 57, 82, 161, 343-345]). Despite decades of experimental and theoretical work, predicting the ETC of fluid-saturated porous media continues to be one of the great unsolved problems in heat transfer science. The ETC depends on the geometry and arrangement of the phase-information which is generally not available from experiments.

For many engineering and scientific applications the relationship between physical bulk properties and porosity is needed. Porosity of a material is the most important parameter affecting its mechanical, thermal, and electrical properties, but its relationship to porosity is poorly understood. Increasing porosity leads in general to a decrease in any bulk property, but in detail, the microstructure of the granular material is important, i.e., the shape and spatial arrangement of the grains (texture) as well as the conditions in the contact area of the connecting grains (contact stiffness). Most models are only valid for low-porosity crystalline rocks and not considering different grain shape and textures (Schoen [321]). Keller et al. [322] proposed a simple model which is valid for the complete porosity range (see below).

The comprehensive review of the correlation, prediction, and estimation techniques for the ETC of fluid-saturated rocks are reported in the works by Singh [87], Ramey [12] and Aichlmayr and Kulacki [70] (see also [72, 323-329]. Here we will briefly review the basic theoretical models, correlation, and prediction techniques which were frequently used to predict the effect of porosity on the ETC of fluid-saturated porous materials and examine their applicability for sandstones ETC data. Various widely used theoretical models for the porosity dependence of the ETC of fluid-saturated porous materials will be considered in this section. In these models the ETC is a function of thermal conductivity of constitutes (thermal conductivity of matrix, λ_S, and saturated fluid, λ_f) and porosity, ϕ. ETC is most sensitive to interface geometry if the solid-fluid conductivity ratio is large. The predictive capability of the ETC models essentially depends on how they representing of the microstructure of the porous media. The physical limits for the ETC fluid-saturated rocks are very well established. ETC of small to modest solid-fluid conductivity ratio $\left(\lambda_S/\lambda_f\right)$ media can be predicted with reasonable accuracy, while our understanding of heat conduction in large solid–fluid conductivity ratio, $\left(\lambda_S/\lambda_f\right)\rightarrow\infty$, still is incomplete (Aichlmayr and Kulacki [70]). The prediction of the ETC fluid-saturated porous media requires a detailed description of the medium morphology, which in general, is not available. Based on the fact that a porous medium is composed of a solid matrix and pore spaces which may be filled with fluids, many researchers have developed the models for prediction of the ETC from porosity and solid and fluid thermal conductivity data (see below). In developing these models, it was basically assumed that the process of transferring heat in fluid-saturated porous media involves the following: (1) heat conduction through solid particles; (2) heat conduction through the contact region between solid particles; (3) heat conduction through pore fluids; (4) thermal radiation across the solid surface (can be neglected below 500 °C). It was also concluded that convection in porous media is negligible for the small and moderate pore size and only for those media with very large pore spaces convection is slightly effective. Therefore, ETC in porous media is a result of conduction in the solid, contact region, and fluid phases. Thus, in order to estimate the ETC of fluid-saturated porous rocks, it is essential to define precisely the contribution of each component (minerals, λ_S, and fluids, λ_f) to bulk conductivity. There are three types of models for expressing ETC of fluid-saturated rocks: (1) mixing-law type models (statistical mean equations) for porous minerals aggregates; (2) empirical models in which thermal conductivity is related to easily measure physical parameters and regression to lab data; and (3) theoretical models based on heat transfer in composite materials with simplified geometries. Because ETC of rocks is difficult to measure (especially fluid saturated at high pressures) and time consuming, there is a need to correlate ETC with other more easily measured physical properties such as porosity, density, electrical resistivity, and permeability, fractional content of quartz, or sonic velocity with a known value of ETC through fitting method (Anand [171]; Beck [176]; Somerton [72]; Strack et al. [148]).

Empirical models have their shortcomings in that the resulting models may be applicable only to the particular suite of rocks being investigated. Theoretical models are based on the mechanism of heat transfer applicable (heat transfer theory) to simplified geometries of the rock-fluid system. The difficulty with these models is the degree of simplification necessary to obtain a solution. There is still a lack of detailed knowledge of how the heat is transferred through sedimentary rocks, and in particular, at the transitions surfaces and interfaces at the grain-pore-fluid and grain-grain interfaces. Preferably, one would use a theoretical model to describe the physical nature of heat conduction, but sufficiently reliable models have not yet been developed, and empirical modifications of the equations are needed.

As was mentioned above, the main difficulties in developing theoretical models to predict the ETC for porous rocks are the complexities of the geometries of the rock structures. Due to irregularity of the microstructures, theoretical calculation of the ETC of porous materials, especially for fluid-saturated porous materials, is rather difficult and sometime impossible. Existing prediction methods are based on certain simplifications such as parallel cylinders, spheres dispersed, simple cubic in a conducting medium, etc., [36, 44, 71, 174, 177, 224, 236, 322, 327-329, 331-341]. Even with a well-defined microstructure, the problem remains complex due to the existence of the interface (grain-grain, grain-pore liquid) resistance. Semiempirical approach is the only practical way of predicting the ETC of porous materials. A large number of theoretical and semiempirical models have been developed for the prediction of the ETC of multiphase porous materials. Some selected and widely used theoretical models for the ETC porous materials as a function of temperature, porosity and thermal conductivity of solid matrix and pore fluid will be discussed below. Some of the models were developed for limited values of the porosity and for the various magnitude of the solid-fluid thermal conductivity ratio $\left(\lambda_S / \lambda_f\right)$. An extensive review of literature on ETC models for fluid-saturated and dry porous materials was provided in the works [70, 72, 238, 323-329, and 342].

5.2.1. Mixing-Law Models

Mixing-law models were used by many authors to estimate ETC of fluid-saturated rocks. These models are based on using various ways of averaging thermal conductivity of components with respect to their volume fractions, such as geometric mean model or fabric model. Such models have been discussed by several authors [72, 101, 115, 123, 167, 221, 323, 329, 346-348, and 526]. The basic assumption for mixing –law models is ETC of rocks may be expressed as a function of the thermal conductivity of each component constituting the rock, and of its relative proportions. The possible effect of structure is therefore neglected.

Different type mixing-law, theoretical, and empirical models were reviewed previously by various authors [70, 72, 177, 189, 238, 232, 236, 238, 267, 293, 323-326, 331, and 342]. Mixing-law model can be applied for porous mineral aggregates containing various fluid phase. These models do not take into account the structural characteristics of rocks, i.e., the possible effect of structure is therefore neglected, and therefore they are limited applicability. Mixing-law models were studied by many authors (see for example, [102, 350-360]) to estimate the ETC of fluid-saturated rocks. These models combine values of the thermal conductivity of the rocks solid matrix (λ_S) with the conductivity of the pore fluid (λ_f) on the basis of porosity, i.e., representing the porosity dependency of the ETC of fluid-saturated porous materials relative to components constituting the rock (λ_S and λ_f), other words, mixing-law models are based on using various ways of averaging thermal conductivity of components (λ_S and λ_f) with respect to their volume fractions, such as geometric mean model or fabric model. These models are of a general character and can be used for all sediment types. The geometric-mean model is another mixing-law model which also widely used to calculate the ETC of rocks. Although these models has no physical basis, the models have been successfully used by many authors and clearly simple (Woodside and Messmer [82 and 161], Drury and Jessop [361], and Sekiguchi [80], Brigaud and Vasseur [123]). This model also based on the rock thermal conductivity components and the rock porosity [72, 82, 107, 108, 123, 161, and 362]. For this model the ETC can be expressed as (geometric mean model)

$$\lambda = \lambda_S^{1-\phi} \lambda_f^{\phi} . \tag{49}$$

Some empirically modifications of these models have been proposed by various authors to improve the agreement between the calculated and measured values of the ETC fluid-rock systems (see below).

In order to derive the equation for ETC, the heat transfer through a porous medium can be considered at in a manner analogous to an electrical circuit (see Figure 40) made up of the following resistances: (a) resistance to conduction of heat through solid particles; (b) contact resistance between solid particles; (c) resistance to conduction through fluids in the pores; (d) convection in large size pores; (e) radiation across the solid surfaces at high temperatures. How these resistances contribute to the total resistance (or conductance) of a natural porous rock is the main problem of the prediction techniques of the ETC of fluid-saturated rocks. Three commonly used, basic mixing-law models are the arithmetic mean, geometric mean, and harmonic mean. The harmonic and arithmetic mean models

$$\lambda = (1-\phi)\lambda_S + \phi\lambda_f \text{ ,Arithmetic mean,} \tag{50}$$

$$\lambda = \left[\frac{(1-\phi)}{\lambda_S} + \frac{\phi}{\lambda_f} \right]^{-1} \text{ Harmonic mean,} \tag{51}$$

are based on parallel and series arrangements of the components relative to the direction of heat flow (see Figure 13). Figure 14 shows ETC as a function of porosity predicted from mixing-laws Eqs. (50), (51), and Maxwell models Eq. (52) (see below) together with Hashin and Shtrikman [363] bonds of ETC (physical limits of ETC). The values of ETC estimated by these models are assumed as the upper (λ_{max}) and lower (λ_{min}) limits of the ETC for a rocks of given composition. Hashin and Strickman [363] modified the ETC boundary of the mixing-law models. The geometric mean model gives Eq. (49) an intermediate value of the arithmetic and harmonic means (see Figure 14). This model has been widely used by different authors [23, 27, 82, 124, 161, 364-367, and 512]. Therefore, the geometric mean model is the most realistic of these models, as it's the only one that honors the fact that there are grain-to-grain contact paths through the material in the direction of lowest thermal conductivity. The main criticism of the geometric mean model is that it does not take into account the texture of the samples, and is therefore valid only for isotropic rocks. Buntebarth and Schopper [302] model at the cases α =0 and α >80 are correspond to arithmetic and harmonic mean models, which are the upper and lower ETC limits of the layered media. If α =0, pore fluid has perfect wetting condition and pores completely filled and ETC resistance within the rock is minimal, while the case α >80 corresponds to the opposite conditions. The model excellent describes the measured ETC data if the parameter properly select from the experimental data. Walsh and Decker [169] and Beck [176] are developed correlation equation for the estimating of the ETC of geological porous materials using three-phase geometric-mean model. The accuracy of the prediction values of the thermal conductivity was 22 %.

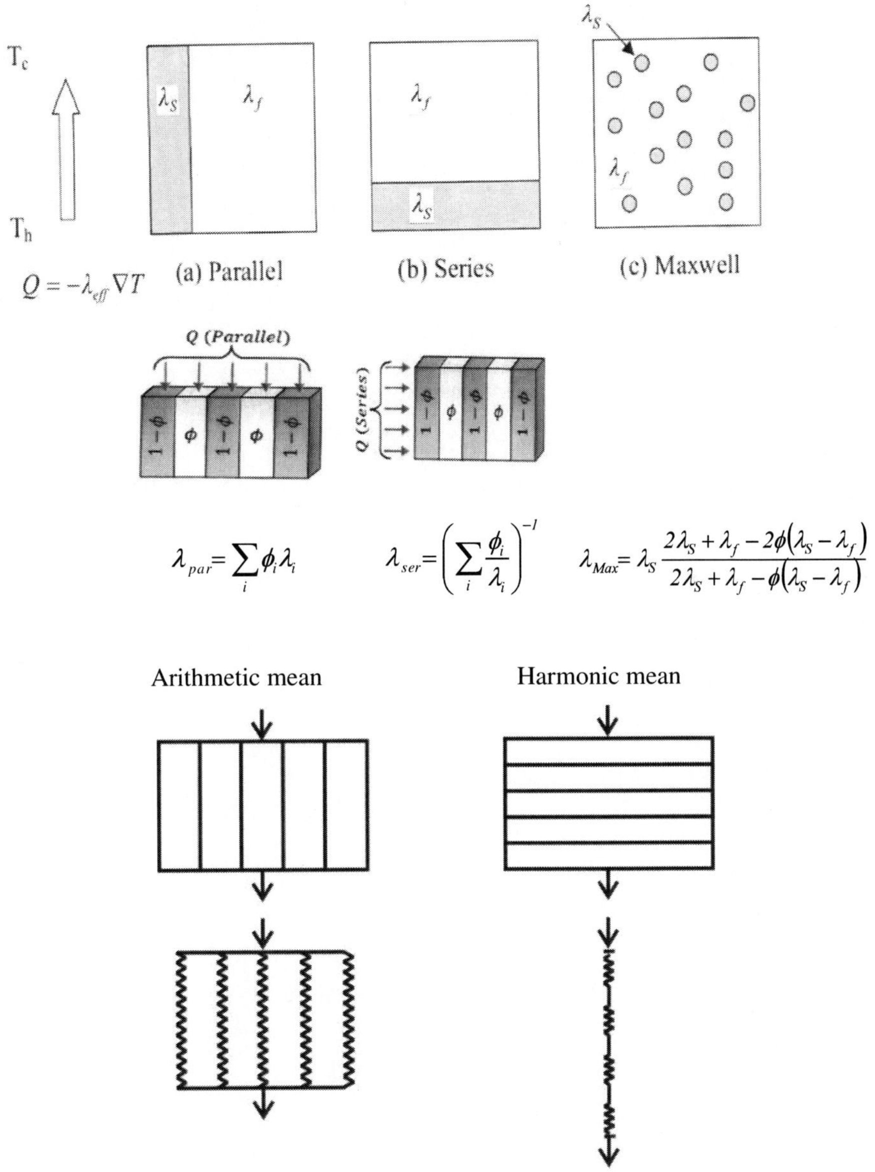

Figure 13. Schematic presentation (physical interpretation) of the arithmetic and harmonic mean ETC mixing-law model.

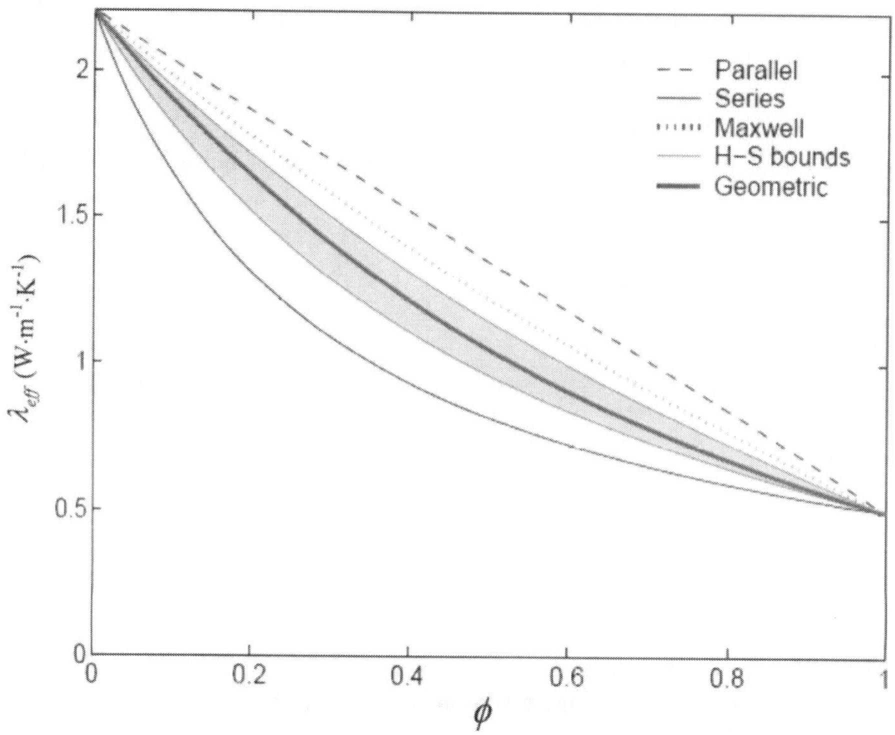

Figure 14. ETC as a function of porosity predicted from mixing-law and Maxwell models together with Hashin and Shtrikman [363] bonds of ETC (physical limits of ETC).

Brown [369] studied the relationship between the mixing models and the statistical properties of phase geometry. For this purposes, he is expands the expression for the relative dialectical constant in terms of the arithmetic mean. The analogous thermal conductivity expression is

$$\frac{\lambda}{\lambda_f} = \phi + (1-\phi)(\lambda_S / \lambda_f) - \frac{\phi(1-\phi)\left[1 - 2(\lambda_S / \lambda_f) + (\lambda_S / \lambda_f)^2\right]}{3\left[\phi + (1-\phi)(\lambda_S / \lambda_f)\right]}$$

in the first-order approximation and

$$\frac{\lambda}{\lambda_f} = \frac{1}{\phi + (1-\phi)/(\lambda_S / \lambda_f)} + \frac{2\phi(1-\phi)\left[1 - 1/(\lambda_S / \lambda_f)\right]}{3\left[\phi + (1-\phi)/(\lambda_S / \lambda_f)\right]}$$

in the second –order approximation based on the harmonic means

$$\frac{\lambda}{\lambda_f} = exp\left[(1-\phi)ln(\lambda_S / \lambda_f) + \frac{\phi(1-\phi)}{6}\left[ln(\lambda_S / \lambda_f)\right]^2\right].$$

The accuracy of these relations is limited

$$\frac{\lambda_S}{\lambda_f} < \frac{1+\phi}{\phi}.$$

5.2.2. Maxwell-Law Models

The Maxwell model is [332]

$$\frac{\lambda}{\lambda_f} = \left[\frac{2\phi + (3-2\phi)(\lambda_S/\lambda_f)}{3-\phi+\phi(\lambda_S/\lambda_f)}\right]. \tag{52}$$

This relation restricted to large porosities. When the values of (λ_S/λ_f) is large (gas and oil saturated rocks), these models give large different results. Maxwell's model is appropriate for the low porosity (ϕ <20 %). Maxwell considered dilute suspension of particles (see Figure 13c). Maxwell relation for the electrical resistance of a medium comprised of a random distribution of spherical particles in a continuous matrix can be used to estimate the ETC

$$R = R_2 \frac{2R_1 + R_2 + p(R_1 - R_2)}{2R_1 + R_2 - 2p(R_1 - R_2)}, \tag{53}$$

where R_1 is the sphere resistance, R_2 is the matrix resistance, p is the volume fraction of the spheres is small. The Maxwell's law for ETC followed from the relation for resistance by substituting $R = 1/\lambda$, $R_1 = 1/\lambda_S$, $R_2 = 1/\lambda_f$, and $p = 1-\phi$. Maxwell considers that the spheres do not interact, *i.e.*, the spheres radii are small relative to their separation distances. This model can be extended for interacting particles. If the solid contain random distributed spherical fluid-filled voids the ETC is $R = 1/\lambda$, $R_1 = 1/\lambda_f$, $R_2 = 1/\lambda_S$, and $p = \phi$

$$\frac{\lambda}{\lambda_f} = \frac{2(\lambda_S/\lambda_f)^2(1-\phi) + (1+2\phi)(\lambda_S/\lambda_f)}{(2+\phi)(\lambda_S/\lambda_f) + 1 - \phi}, \tag{54}$$

The Maxwell's relations, Eqs. (52) and (54), give the best estimate for the ETC when the geometry of the solid-fluid interface is unknown (see Figure 14). Although mixing-rules less stringent than Maxwell-law models, provide convenient predictions for the physical limits of the ETC. Maxwell model restricted to small porosity and at large values of (λ_S/λ_f) ratio becomes equal to Waff [339] model. Eucken [368] first modified Maxwell's equation. Therefore, Maxwell- model has a good physical basis, gives quite reliable results when the volume proportion (porosity) of one of the two components does not exceed ϕ =0.25 and conductivity ration λ_S/λ_f<10. Beck [176] reported the improved method of the ETC calculation for fluid-filled rocks. He used an empirical approach to extend the range of validity of the Maxwell relation up to ϕ =0.4. His equation, $\lambda = \lambda_S^{1-\phi}\lambda_f^{\phi}$, do not reflect the specifics of the pore geometry.

5.2.3. Theoretical Models

Most developed theoretical models based on the mechanism of heat transfer applicable to simplified geometries of the rock-fluid system or composite materials with simplified geometries. According to the relation for ETC of dispersions spheroids (Fricke [336]), the spherical pores are caused the minimum changes in thermal conductivity, while crack-like pores have the maximum possible effect. For dry sample r=0 or for air saturated sample r=0.008. The prediction capability of the Fricke model is very accurate (within 3 %).

Luo et al. [347] reported comprehensive review of the methods of ETC prediction for fluid saturated reservoir rocks (clay, carbonate, and sandstone) by using the fabric theory. They found that fabric model prediction of ETC much better than geometric mean model. The fabric theory predicts ETC of a rock based on a weighted average of the thermal conductivity of its mineral elements arranged in series and parallel relative to the direction of heat flow (Davis [269] and Somerton [72]). The basic elements of the model are solid rock matrix and the pore water. The ETC can be expressed as

$$\lambda = FSE\left[(1-\phi)\lambda_S + \phi\lambda_W\right] + (1-FSE)\left[\frac{(1-\phi)}{\lambda_S} + \frac{\phi}{\lambda_W}\right]^{-1}, \tag{55}$$

where FSE is the fraction of series elements, λ_S and λ_W are the thermal conductivity of the solids and water, respectively, and ϕ is the porosity. This equation is valid when specific geometrical effects (shapes of the pore spaces and solid grains) can be ignored and all pore spaces filled with water. If the element components are only in series (FSE=1), the upper bound of the ETC is $\lambda = (1-\phi)\lambda_S + \phi\lambda_W$ (arithmetic means or Ohm's law model, gives highest values of the ETC rock-fluid system of all the mixing-law models). This model is equivalent to parallel element arrangement of the components relative to the direction of heat flow. This model equivalent to a group of conductors in parallel (i.e., layered parallel to the direction of heat transfer). This is classic mixing rule give rise to a considerable error especially when $\lambda_S/\lambda_f >> 1$. If the element components are parallel (FSE=0), the lower bound of the ETC is $\lambda = \left[\frac{(1-\phi)}{\lambda_S} + \frac{\phi}{\lambda_W}\right]^{-1}$ (Wiener bounds or weighted harmonic means, gives lowest values of the ETC rock-fluid system of all the mixing law models, Chan and Jeffrey [221], Somerton [72]). This model equivalent to a group of conductors in series element arrangement (i.e., layered perpendicular to the direction of heat transfer). The models used quite successfully over limited ranges of porosity and conductivity. For half series and half parallel elements FSE=0.5. Luo et al. [347] proposed the modified form of the Eq. (55)

$$\lambda = FSE\left[\sum_{i=1}^{n1}(1-\phi)\xi_S^i\lambda_S^i + \sum_{i=1}^{n2}\phi\xi_f^i\lambda_f^i\right] + (1-FSE)\left[\sum_{i=1}^{n1}(1-\phi)\frac{\xi_S^i}{\lambda_S^i} + \sum_{i=1}^{n2}\phi\frac{\xi_f^i}{\lambda_f^i}\right]^{-1}, \tag{56}$$

where n1 and n2 are the number of mineral components of rock and types of fluids saturated in the pore space, λ_S^i is the thermal conductivity of the ith solid mineral, ξ_S^i is the volume fraction of the ith mineral in the total weight of minerals, λ_f^i is the thermal conductivity of the ith fluid type with saturation ξ_f^i. This is fabric model of the ETC of rocks. This modified fabric model also ignores the effect of the pressure on ETC. The effect of grain texture and sedimentation structure on the ETC can be taken into account by adjusting the FSE values for different categories of rocks. From this point of view, the fabric model is generally superior to the geometric mean model.

The problem of conduction through arrays of particles that touch at flatted surfaces is considered in detail by several authors [340, 370-372]. They estimated ETC of elastic spheres bounded by plane surfaces. Kaganer [370] estimate of the ETC of bed spherical particles in a vacuum. He assumed that particles touch at flattened regions resulting from elastic deformation. Chan and Tein [340] like Kaganer [370] assumed that particles contact at surfaces flattened by elastic deformation. They analytically studied the heat transfer through the solid phase of a packed bed of spheres bounded by two infinite plane surfaces of different temperatures. The prediction of the conductance is based on the constriction resistance for spheres in contact. Exact equations are derived for solid, hollow, and coated spheres and for several regular packing patterns of different void fractions. Rayleigh [373] has analyzed the case when the spheres of radius a and conductivity λ_S are arranged in a simple cubic array. He is analyzed regular arrays of cylindrical and spherical objects surrounded by a fluid. He is extended Maxwell's model to slightly higher particle concentrations for the case of an ordered cubic array of spherical particles of uniform size. Rayleigh's relation [373] is

$$\lambda = \lambda_f \left[\frac{1 - 2k\phi + kF}{1 + k\phi + kF} \right], \tag{57}$$

where $k = (\lambda_f - \lambda_S)/(\lambda_S + 2\lambda_f)$; $F = 1.65[(\lambda_S - \lambda_f)/(\lambda_S + 4\lambda_f/3)\phi^{10/3}]$. This model is rigid and artificial that it does not predict ETC of mixtures in practical case.

Meredith and Tobias [374] extended Rayleigh's model to higher concentrations as

$$\lambda = \lambda_f \left[\frac{1 - 2k\phi + 0.409k(6\lambda_f + 3\lambda_S/4\lambda_f + 3\lambda_S)\phi^{7/3} - 2.133\phi^{10/7}(3\lambda_f - 3\lambda_S/4\lambda_f + 3\lambda_S)}{1 + k\phi + 0.409k(6\lambda_f + 3\lambda_S/4\lambda_f + 3\lambda_S)\phi^{7/3} + 0.906k(3\lambda_f - 3\lambda_S/4\lambda_f + 3\lambda_S)\phi^{10/3}} \right]. \tag{58}$$

This relation was analyzed and improved by Sangani and Acrivos [375].

Kiyohashi [324] reviewed of studies of the ETC models of three-phase materials and their applicability for various porous materials. Kiyohashi and Deguchi [110] and Deguchi et al. [234] extended this model for three-phase solid-liquid system as

$$\lambda = \lambda_S^{\phi_s} \lambda_f^{\phi_f} \lambda_a^{\phi_a}, \tag{59}$$

where $\phi_S + \phi_f + \phi_a = 1$, λ_a is the thermal conductivity of air. For the practical used the Eq. (59) was modified to

$$\lambda = \left(\phi_S^{-1}\right)^{\phi_a/(\phi_a+\phi_W)} \left[\lambda_S\phi_S + \lambda_f\left(1-\phi_S\right)\right]^{\phi_S} \lambda_f^{\phi_f} \lambda_a^{\phi_a} \tag{60}$$

Equation (60) predicted ETC for water-saturated rocks within 22 %.

Brigaud and Vasseur [123] proposed simple relation for ratio of thermal conductivity measurements for water λ_{eW} and air λ_{ea} saturated rocks

$$log\,\frac{\lambda_{eW}}{\lambda_{ea}} = \phi\, log\,\frac{\lambda_W}{\lambda_a}, \tag{61}$$

Experimental results show good agreement with Eq. (61). Brigaud and Vasseur [123] also developed geometric mean type ETC model for two-phase media for various types of structure. Brigaud et al. [125] proposed that ETC measured in laboratory must be corrected for in-situ temperatures as

$$\lambda_{mT} = \lambda_{m20}\left[293/\left(273+T\right)\right], \tag{62}$$

where λ_{mT} is the matrix conductivity at temperature T, λ_{m20} is the matrix conductivity at 20 °C determined in the laboratory. This correction is especially significant in deep sedimentary basins.

The predictions of ETC from sonic travel time (Δt), form porosity neutron signal (ϕ_{HI}), and from the bulk density signal (ρ_b) are

$$\lambda = \lambda_{mT}^{[(\Delta t_W-\Delta t)/(\Delta t_W-\Delta t_m)]} \lambda_{mT}^{[(\Delta t-\Delta t_m)/(\Delta t_W-\Delta t_m)]}, \tag{63}$$

$$\lambda = \lambda_{mT}^{[1-(\phi_{HI}-HI_{SH}V_{SH})]} \lambda_{wT}^{[(\phi_{HI}-HI_{SH}V_{SH})]}, \tag{64}$$

$$\lambda = \lambda_{mT}^{[(\rho_b-\rho_w)/(\rho_m-\rho_w)]} \lambda_{wT}^{[(\rho_m-\rho_b)/(\rho_m-\rho_w)]}, \tag{65}$$

where λ_{WT} is the water conductivity value as a function of temperature.

Odelevskii [327] developed various approaches to estimate the ETC of two-phase porous systems. The following prediction models for the calculation of the ETC of fluid-saturated porous materials were proposed by Odelevskii [327]

$$\lambda = \frac{1}{4}\left[(3f-1)\lambda_s + (3\phi-1)\lambda_f\right] + \sqrt{\left[\frac{(3f-1)\lambda_s + (3\phi-1)\lambda_f}{4}\right]^2 + \frac{\lambda_s\lambda_f}{2}} \tag{66}$$

and

$$\lambda = \lambda_S \left[1 - \frac{3\phi(1 - \lambda_f / \lambda_S)}{3 - (1 - \phi)(1 - \lambda_f / \lambda_S)} \right], \tag{67}$$

where $f = 1 - \phi$. Equation (66) was derived for the multiphase porous media having nonstretched particles. He assumed that all of the phases are spherical in shape and have the same size. For fluid-saturated porous systems with ratio (λ_S / λ_f) between 30 and 100, the most appropriate thermal conductivity model proposed by Odelevskii [327] is

$$\lambda = \lambda_S \left[1 + \frac{2\phi(\lambda_f / \lambda_S - 1)}{(1 - \phi)(\lambda_f / \lambda_S - 1) - 2} \right]. \tag{68}$$

This relation is applicable only for porous media with open-pores. For gas–saturated rocks $(\lambda_S >> \lambda_f)$, Eq. (68) can be rewritten in a simpler form as

$$\lambda = \lambda_S \frac{(1 - \phi)}{(1 + \phi)}. \tag{69}$$

Mendel [328] developed following form an equation to predict the ETC of fluid-saturated rocks

$$\lambda = \lambda_S \left[1 - \frac{3\phi(\lambda_S - \lambda_f)}{2\lambda_S + \lambda_f} \right]. \tag{70}$$

This model was derived for heterogeneous systems with a disperse phase having the shape of a parallel cubic. This model is reduced to Budiansky [388] model at $(\lambda_S / \lambda_f) \to \infty$. Luo et al. [347] modification of this equation is

$$\lambda = \left[\Pi (\lambda_S^i)^{\xi_S^i} \right]^{1-\phi} \left[\Pi (\lambda_f^i)^{\xi_f^i} \right]^{\phi} \tag{71}$$

Brailsford and Major [376] have extended the Maxwell model to cover the full range of porosity. The model is given by

$$\lambda = (3\phi - 1)\lambda_S + (2 - 3\phi)\lambda_f + \left[\{(3\phi - 1)\lambda_S + (2 - 3\phi)\lambda_f\}^2 + 8\lambda_S \lambda_f \right]^{1/2} \tag{72}$$

In this model the constituent phases are mixed in a definite proportion for a two-phase system. This relation Eq. (72) was applied to the Sugawara and Yoshizawa [224] data.

Chan and Jeffrey [221] used Ohm's law model to develop ETC uniform-heat flux model as

$$\lambda = \lambda_f\left[\left(1-\phi^{1/3}+\phi\right)+\lambda_f\left(\phi^{1/3}+\phi\right)/\lambda_S\right]\left[\left(1-\phi^{1/3}\right)+\lambda_f\left(\phi^{1/3}\right)/\lambda_S\right]^{-1} \tag{73}$$

and parallel –isotherm model

$$\lambda = \lambda_f\left[\phi^{2/3}+\lambda_f\left(1-\phi^{2/3}\right)/\lambda_S\right]\left[\phi^{2/3}-\phi+\lambda_f\left(1-\phi^{2/3}+\phi\right)/\lambda_S\right]^{-1}. \tag{74}$$

These models were applied by Chan and Jeffrey [221] for water-filled low porosity (2 %) fractured granite. The model developed by Bruggeman [377] is

$$\left(1-\phi\right)=\frac{\lambda-\lambda_S}{\lambda_f-\lambda_S}\left[\frac{\lambda_f}{\lambda}\right]^{1/3} \text{ or } \lambda=\lambda_S^{(1-\phi)(1+k\phi)}\lambda_f^{\phi[1-k(1-\phi)]}, \tag{75}$$

where $k=\frac{3}{2}\frac{\lambda_S-\lambda_f}{\left(2\sqrt{\lambda_S}+\sqrt{\lambda_f}\right)\left(\sqrt{\lambda_S}+2\sqrt{\lambda_f}\right)}$. He used Maxwell's model for cylindrical particles. For gas-saturated porous rocks at temperatures of 275 K and 323 K, the values of k are about 0.305 and 0.295, respectively. This model is valid for the heterogeneous systems which constant of stretched ellipsoids and spherical particles. Waff [339] developed a self-consistent solution, for a spherical assemblage, similar to that of Bruggeman model [377]

$$\lambda=\left[\left(1+\phi\right)\lambda_f+2\left(1-\phi\right)\lambda_S\right]\left[2+\phi+\left(1-\phi\right)\lambda_f/\lambda_S\right]^{-1}. \tag{76}$$

Frey [378] and Dul'nev and Zarichniak [379] proposed geometric lattice models for the ETC with porosities above 60 %. Staicu et al. [51] are presented three different models for calculating of the ETC of composite materials: numerical methods; predictive analytical methods; and the bounds obtained by variational methods. Their prediction was compared with measured results for nuclear materials with various microstructures.

Ziman [74] proposed the model which yields the relation between ETC and porosity for fluid-saturated porous materials

$$\lambda=\frac{\lambda_S\lambda_f}{\lambda_S\phi+\lambda_f\left(1-\phi\right)}. \tag{77}$$

Using a differential effective medium approach, Revil [1] developed following relationship for the ETC of a fluid saturated porous material

$$\lambda=\frac{\lambda_f}{f}\left[fr+\frac{1}{2}\left(1-r\right)\left(1-r+\sqrt{\left(1-r\right)^2+4fr}\right)\right], \tag{78}$$

where $r = \frac{\lambda_S}{\lambda_f}$, $f = \phi^{m/(1-m)}$ is the thermal formation factor; m is the Archie's first exponent (or cementation exponent) ranging typically between 1.3 (for unconsolidated sands) to 2.0 (for consolidated rocks). Alishaev [380] and Alishev et al. [576] proposed the model for calculating of the ETC of rocks filled with gas, water, and oil. His model is

$$\frac{\lambda}{\lambda_S} = 1 - \beta^2 + \beta^2 \sqrt{\frac{\lambda_f}{(1-\beta)\lambda_f + \beta\lambda_S}}\text{, where } \beta = \sqrt[3]{2\phi}. \quad (79)$$

The Eq. (79) allows estimate of the ETC of fluid-saturated porous media (rock materials). At ϕ =0.5 the Eq. (79) reduced to the geometric mean $\lambda = \sqrt{\lambda_S \lambda_f}$ model. At λ_f=0 (vacuum in the pores or air), we have $\lambda = \lambda_S(1-\beta^2)$, therefore, the condition β=1 is define the thermal conductivity limit. If we consider the case when $\lambda_f << \lambda_S$ (gas saturated porous media or vacuum). Then, Eq. (79) for ETC can be simplified as

$$\lambda = \left(1 - \sqrt[3]{(\phi/\phi_0)^2}\right)\lambda_S + \sqrt{\lambda_S \lambda_f (\phi/\phi_0)}. \quad (80)$$

where ϕ_0 is the maximum possible value of the porosity of sample, $0 \le \phi \le \phi_0$. We applied this Eq. (80) for various type fluid saturated rocks. For most fluid-saturated sandstones the values of the structural parameter is ϕ_0=0.28, although the values of ϕ_0 strongly depend on structural and mineralogical characteristics of the porous media.

Yu et al. [333] are developed two unit cell models, the point-contact hollow spherical model and the surface –contact hollow cubic model. Expression for ETC based on the point-contact hollow spherical model is

$$\frac{\lambda_e}{\lambda_f} = 1 - \frac{\pi}{4} + \frac{\pi}{8}\left\{\begin{array}{l} A(1+A) + (1-A^2)\left(1+\sqrt{1-A^2}\right)\beta - \left(1 - A + 3\sqrt{1-A^2}\right)/\beta - \\ \ln\left[\left(1-\beta+\beta A\right)\left(1-\beta\sqrt{1-A^2}\right)^3\right]/\beta^2 + \beta^2\left(1-A^2\right)^2 \ln\left[\sqrt{\frac{1+A}{1-A}}\frac{1-\beta(1-A)}{1-\beta\sqrt{1-A^2}}\right]\end{array}\right\}, \quad (81)$$

where $A = \sqrt[3]{1 - \frac{6}{\pi}\left(\frac{1-\Phi}{1-\Phi_i}\right)}$, $\Phi = 1 - \frac{\pi}{6}(1-\Phi_i)\left[1 - \left(\frac{R}{R+h}\right)^3\right]$ total porosity of the unit cell, Φ_i is the porosity of the shell, R is the inner radius of hollow spheres, and h is the shell thickness, $\beta = 1 - \lambda_f/\lambda_C$, λ_f and λ_C are the thermal conductivity of the fluid and shell, respectively. The surface –contact hollow cubic model predicted ETC of fluid-saturated porous media as

$$\frac{\lambda_e}{\lambda_f} = \frac{(2-\gamma_b)\gamma_b\gamma_a^2}{1-(1-\beta)\gamma_a} + \frac{\gamma_a^2\left[(1-\gamma_b)^2 - \gamma_c^2\right]}{1-(1-\beta)\gamma_a\gamma_b} + \frac{2\gamma_a\gamma_c(1-\gamma_a)}{1-(1-\beta)\gamma_a\gamma_c} + \frac{\gamma_a^2\gamma_c^2}{\beta + (1-\beta)(1-\gamma_b)\gamma_a} + (1-\gamma_a)(1+\gamma_a-2\gamma_a\gamma_c)\text{'} \quad (82)$$

where $\gamma_a = a/l$, $\gamma_b = 2h/a$, $\gamma_c = c/a$ are the dimensionless hollow cubic structures parameters, a is the side length, h is the wall thickness, and c is the contact section. The porosity of the unit cell is defined as

$$\Phi = 1-(1-\Phi_i)\{1-(1-\gamma_b)^3-3\gamma_c^2\}\gamma_a^3+3\gamma_a^2\gamma_c^2\}. \tag{83}$$

Gomaa [174] also developed a model for predicting of thermal conductivity of the porous media. To improve previous reported models, Gomaa [174] used variable porosity and multi-fluid-saturation. Normal cubic packing of spheres is used to represent the porous media and porosity is varied by flattening the spheres at their contacts. The final equation represent the thermal conductivity of rock as a function of solid and saturated-fluid thermal conductivities, fluid saturation, formation resistivity factor and Biot number ($Bi = 2R_p h/\lambda_S$, where $2R_p$ is the width of the unit cell, h is the heat transfer coefficient). This model based on the fundamental principles of heat transfer. Fluid distribution was determined as a function of saturation S_W. The following form of ETC equation was proposed by Gomaa [174]

$$\lambda = \phi(1-S_{crit})\lambda_{fC} + (1-\phi+\phi S_{WC})\lambda_{S-cr}, \tag{84}$$

where S_{crit} is the critical wetting phase saturation; λ_{S-cr} is the thermal conductivity of solid matrix-critical region; λ_{fC} is the thermal conductivity of the center fluids. If the pore spaces contain fluids $f_1, f_2, f_3, \ldots f_n$ and the wetting phase w with the saturations $S_1, S_2, S_3, \ldots$ and S_W with $S_1+S_2+S_3+\ldots+S_W=1$. To calculate of the ETC with Eq. (84) the quantities λ_{fC} and λ_{S-cr} should be known. If $S_W > S_{crit}$

$$\lambda_{fC} = \frac{S_1}{1-S_{crit}}\lambda_1 + \frac{S_2}{1-S_{crit}}\lambda_2 + \cdots \frac{S_n}{1-S_{crit}}\lambda_n + \frac{S_W - S_{crit}}{1-S_{crit}}\lambda_W, \tag{85}$$

and if $S_W < S_{crit}$

$$\lambda_{fC} = \frac{S_1}{1-S_W}\lambda_1 + \frac{S_2}{1-S_W}\lambda_2 + \cdots \frac{S_n}{1-S_W}\lambda_n, \tag{86}$$

where λ_1, λ_2, ... λ_n are thermal conductivities of fluids 1, 2, 3,....n, respectively. The wetting phase saturation geometry relationship for the microscopic model is

$$S_W = \left[\frac{6\pi}{8-(4/3)\pi}\right]\left[\frac{t^3}{3}+\frac{2}{3}H^3 - tH^2\right], \tag{87}$$

$t = \sqrt{1-(a/R_P)^2}$, $H = H_0/R_P$, R_P is the outer radius of the region II; a is the radius of grain contact; H_0 is the smallest z for the wetting phase –solid contact. Gomaa [174] represents a cross –section of the porous media fluid system. He is considering that the total heat flux crossing a unit area of this section is divided into two parts: (1) heat flux crossing the central portions of pore spaces; and (2) heat flux crossing the solid matrix contact region part.

Özbek [177] developed improved version of the Gomaa's model [174]. He developed the model to predict the thermal conductivity of multi-fluid-saturated rocks with variable porosity. This model used cubic packing of uniform diameter spheres but continuously variable porosity. Six parameters ϕ, λ_S, λ_W, λ_{nW} (thermal conductivity of non-wetting fluid), R_W (wetting fluid-solid contact resistivity), and R_{nW} (non-wetting fluid-solid contact resistivity) are needed for the predicting of the thermal conductivity of rocks as a function of saturation

$$\lambda = 0.059\exp(-3.31\phi) + 0.017S_W^{0.5} + 0.194\lambda_S^{0.783} + 0.02\exp(18.13\lambda_W) + 0.016\exp(86.36\lambda_{nW}) - 2\times10^{-4}R_W - 1.1\times10^{-4}R_{nW} - 0.07, \tag{88}$$

This equation represents experimental ETC values within 5 %.

Ghaffari [334] developed a two-dimensional heat transfer model for the same cubic pack of flattened spheres used in the models by Gomaa [174] and Özbek [177]. The temperature distribution was expressed in the form two differential equations, one for the rock solids portion and one for the fluid portion of the region. Ghaffari [334] used numerically technique to solution of the heat transfer equations with appropriate boundary and interface conditions. To apply the model, the following input quantities must be known: λ_S, λ_{Wf} (wetting fluid thermal conductivity), λ_{nWf} (nonwetting fluid thermal conductivity), S_W (water-saturation), and ϕ, r_2 are the structural parameter. For unconsolidated sands r_2 is close to zero. Cheng and Toksöz [381 and 382] and Zimmerman [335] successfully used the model of non-intersecting pores of a spheroidal shape to study mechanical, acoustical, and transport properties of rocks. A spheroid is a regenerate ellipsoid which has two axes of equal length. The parameter α which characterizes the shape of the spheroid is defined as ratio of the length of unequal axis to the length of one of the equal axes. At $\alpha \to \infty$ the spheroid can represent a needle-like tubular pore, a spherical pore at $\alpha \to 1$, or a thin "penny-shaped" crack at $\alpha \to 0$. Fricke [336] showed that for small porosities the ETC can be represented as linear function of porosity

$$\frac{\lambda_e}{\lambda_S} = 1 - \beta\phi, \tag{89}$$

where $\beta = \frac{1-r}{3}\left[\frac{4}{2+(r-1)M} + \frac{1}{1+(r-1)(1-M)}\right]$, $r = \lambda_f/\lambda_S$, and M is a factor that depends on the aspect ratio of the pore. For two regimes of oblate ($\alpha < 1$) and prolate ($\alpha > 1$) spheroids M defined as

$$M = \frac{(2\theta - \sin 2\theta)}{2\tan\theta \sin^2\theta}, \text{ where } \theta = \arccos\alpha, \tag{90}$$

and

$$M = \frac{1}{\sin^2\theta} - \frac{\cos^2\theta}{2\sin^3\theta}\ln\left[\frac{1+\sin\theta}{1-\sin\theta}\right], \text{ where } \theta = \arccos(1/\alpha). \tag{91}$$

The effect of pore shape on mechanical and thermal properties is small for prolate spheroid, while very sufficient for oblate pore. In limiting cases thin cracks:

$$\beta \to \frac{(1-r)(1+2r)}{3r} \ (\alpha \to 0), \text{ spherical pores;} \tag{92}$$

$$\beta \to \frac{3(1-r)}{2+r}, \text{ needle –like pores;} \tag{93}$$

and

$$\beta \to \frac{(1-r)(5+r)}{3(1+r)} \ (\alpha \to \infty). \tag{94}$$

For dispersions of spheroids, ETC can be estimated as (Fricke [336])

$$\frac{\lambda_e}{\lambda_S} = \frac{(1-\phi)(1-r) + r\beta\phi}{(1-\phi)(1-r) + \beta\phi}. \tag{95}$$

According this relation, the spherical pores cause the minimum changes in thermal conductivity, while crack-like pores have the maximum possible effect.

The ratio of the ETC of gas (air)- saturated rocks to the ETC liquid-saturated rocks is predicted as

$$\frac{\lambda_e}{\lambda_S} = \frac{(1-0.661\phi)(1+1.37\phi)}{(1-41.3\phi)(1-0.538\phi)}. \tag{96}$$

Prediction capability of the Eq. (95) is very accurate (within 3 %). The effect of pressure on the ETC of rocks can be taken into account through the crack porosity $\phi_C(P)$ as a function of pressure

$$\phi_C(P) = 0.0038\exp(-P/2) \tag{97}$$

in the equation

$$\lambda_e = 3.23\frac{1-0.661\phi_C(P)}{1+41.3\phi_C(P)}. \tag{98}$$

The prediction accuracy of the ETC by this equation is very high over the entire pressure range. This method of prediction required knowledge only of the high-pressure thermal conductivity (3.23 W·m^{-1}·K^{-1}) value only.

Boomsma and Poulikakos [337] developed geometrical ETC model for saturated porous metal foam. The model based on the idealized three-dimensional basic geometry of foam, the tetrakaidecahedron. The foam structure was represented with cylindrical ligaments which attach to cubic nodes at their centers. The final equation for the ETC is

$$\lambda = \frac{\sqrt{2}}{2}\frac{1}{(R_A + R_B + R_C + R_D)}, \tag{99}$$

$$R_A = \frac{4d}{[2e^2 + \pi d(1-e)]\lambda_S + [4 - 2e^2 - \pi d(1-e)]\lambda_f}, \tag{100}$$

$$R_B = \frac{(e-2d)^2}{(e-2d)e^2\lambda_S + [2e - 4d - (e-2d)e^2]\lambda_f}, \tag{101}$$

$$R_C = \frac{(\sqrt{2}-2e)^2}{2\pi d^2(1-2e\sqrt{2})\lambda_S + 2[\sqrt{2} - 2e - \pi d^2(1-2e\sqrt{2})]\lambda_f}, \tag{102}$$

$$R_D = \frac{2e}{e^2\lambda_S + (4-e^2)\lambda_f}, \tag{103}$$

where $e = r/L$, r is the break up the unit cell into four distinct vertical layers, L is the length between node centers in the tetrakaidecahedron. The value of e (about 0.339) is calibrated by comparison of the model calculations against the experimental data.

Hsu et al. [338] improved the accuracy of the Zehner and Schlunder [44] model at large solid-fluid conductivity ratios by incorporating conduction through a finite-area contact point. Hsu et al. [36 and 338] proposed algebraic expressions for the stagnant thermal conductivity model for some two-dimensional and three-dimensional spatially periodic media based on a lumped-parameter method and phase–symmetry model.

The geometries of the porous media include arrays of touching and non-touching in–line square and circular cylinders (two-dimensional), as well as touching and non-touching in-line cubes (three-dimensional). In this model the cell is divided into parallel layers (which may consists of fluid, solid, or composite layers) in the direction of the macroscopic temperature gradient. Because of the symmetry of thermal paths, the authors choose a unit cell, which is one quadrant of the cell. The unit cell can be considered to be consisting of three layers: a rectangular solid layer and two rectangular composite layers consisting of both solid and fluid phases. Hsu et al. [338] assumed that unit cell is a composite medium consisting of three layers. The layers represent parallel thermal resistances, therefore ETC of the unit cell (three parallel layers) is

$$\lambda = \gamma_a \gamma_c \lambda_S + \gamma_a (1-\gamma_k)\lambda_{Sf1} + (1-\gamma_a)\lambda_{Sf2}, \tag{104}$$

where λ_{Sf1} and λ_{Sf2} are the thermal conductivities of the composite layers, $\gamma_a = a/l$, $\gamma_c = c/a$ (touching parameter), a (length of one side of the solid square cylinder; diameter of the circular cylinder; or length of one side of a cube), c (width of a connecting plate; width of a square connecting column) and l (length of the unit cell) are the cell dimensions parameters. The values of λ_{Sf1} and λ_{Sf2} can be obtained based on a series-layer model as

$$\frac{1}{\lambda_{Sf1}} = \frac{(a/l)}{\lambda_S} + \frac{(l-a)/l}{\lambda_f}, \tag{105}$$

$$\frac{\lambda_{Sf1}}{\lambda_f} = \frac{1}{1+(k-1)\gamma_a}, \tag{106}$$

and

$$\frac{\lambda_{Sf2}}{\lambda_f} = \frac{1}{1+(k-1)\gamma_a\gamma_c}, \tag{107}$$

where $k = \lambda_f / \lambda_S$ is the fluid/solid thermal conductivity ratio. Substituting Eqs. (106) and (107) into Eq. (105) yields

$$\frac{\lambda}{\lambda_f} = \frac{\gamma_a\gamma_c}{\lambda} + \frac{\gamma_a(1-\gamma_c)}{1+(k-1)\gamma_a} + \frac{(1-\gamma_a)}{1+(k-1)\gamma_a\gamma_c}, \tag{108}$$

where γ_a and γ_c are related by equation

$$1-\phi = \gamma_a^2 + 2\gamma_c\gamma_a(1-\gamma_a). \tag{109}$$

Hsu et al. [36 and 338] showed that touching parameter γ_c is important when the solid/fluid thermal conductivity ratio, k, is high. The results of comparison various models reported by Nozad et al. [310] and Prasad et al. [383] and experimental data with the model by Hsu et al. [338] showed that the agreement is excellent if *c*/*a* is 0.01. For non-touching square cylinders *c*=0, γ_c=0, and $\gamma_a = (1-\phi)^{1/2}$, then the Eq. (94) becomes

$$\frac{\lambda}{\lambda_f} = \left(1-\sqrt{1-\phi}\right) + \frac{\sqrt{1-\phi}}{1+(k-1)\sqrt{1-\phi}} . \tag{110}$$

Described above technique was applied for the stagnant thermal conductivity of an array of in-line cubes with smaller connecting bars and circular cylinders (of diameter a) with connecting plates (of thickness *c* and height *h*). For example, the results for the stagnant thermal conductivity of an array of in-line cubes with smaller connecting bars are

$$\frac{\lambda}{\lambda_f} = \left(1-\gamma_a^2 - 2\gamma_a\gamma_c + 2\gamma_c\gamma_a^2\right) + \frac{\gamma_a^2\gamma_c^2}{k} + \frac{\gamma_a^2 - \gamma_a^2\gamma_c^2}{\left(1-\gamma_a+\gamma_a k\right)} + \frac{2\left(\gamma_a\gamma_c - \gamma_a^2\gamma_c\right)}{\left(1-\gamma_a\gamma_c + k\gamma_a\gamma_c\right)} . \tag{111}$$

For non-touching cubes, γ_c=0, and $\gamma_a = (1-\phi)^{1/3}$,

$$\frac{\lambda}{\lambda_f} = \left[1-(1-\phi)^{2/3}\right] + \frac{(1-\phi)^{2/3}}{1+(k-1)(1-\phi)^{1/3}} . \tag{112}$$

As was showed by Hsu et al. [338], these equations good represent the experimental ETC data and the results strongly depend on touching parameter, γ_c, values. For touching cubes model the best result was achieved with γ_c=0.13 while for touching square cylinder model the best result is γ_c=0.01 and 0.02. These models are also in good agreement with the values of ETC predicted by the Zehner and Schlunder model [44]. They obtained an analytical expression for the ETC of a packed bed by combining conduction and lumped parameter modeling techniques. Their axisymmetric unit cell consists of an eighth sphere in a cylindrical volume.

The inner cylinder $0 \le r \le 1$ contains both solid and fluid phases. The annular region $1 \le r \le R$, however, contains only fluid. The ETC is found by assuming unidirectional heat flow throughout the unit cell.

In addition, the inner cylinder and the outer annulus are assumed to be parallel thermal resistances. The model is

$$\lambda = \left(1-\frac{1}{R^2}\right)\lambda_f + \frac{1}{R^2}\lambda_{Sf}, \tag{113}$$

where λ_{Sf} is the ETC of the inner cylinder, $\frac{1}{R^2} = \sqrt{1-\phi}$. They chose a cylindrical unit cell that contains a spherical particle. They considered heat transfer by conduction through two parallel paths: conduction through the gas-filled voids in the outer cylinder and conduction through the solid and gas phase in the inner cylinder. Based on a semi-analytical approach, they obtained the following expression for the effective stagnant thermal conductivity of a packed bed

$$\frac{\lambda}{\lambda_f}=\left(1-\sqrt{1-\phi}\right)+\frac{2\sqrt{1-\phi}}{1-\sigma B}\left[\frac{(1-\sigma)B}{(1-\sigma B)^2}\ln\left(\frac{1}{\sigma B}\right)-\frac{1+B}{2}-\frac{B-1}{1-\sigma B}\right], \tag{114}$$

where shape factor B is $B=1.25\left(\frac{1-\phi}{\phi}\right)^{10/9}$. Upon B=0, B=1, and $B=\infty$, the values of r=0, r^2+z^2 =1, and r^2=1, which correspond to the z-axis, a sphere, and a cylinder, respectively. The first term in this equation represents heat conduction through the outer cylinder consisting of the fluid phase while second term represents heat conduction through the inner composite cylinder consisting of both solid and fluid phases. As was showed by Kaviany [71], this equation is accurate for moderate values of the solid/fluid thermal conductivity ratios (λ_S/λ_f) . Nozad et al. [310] obtained numerical solutions for the stagnant thermal conductivities of arrays of touching and non-touching square cylinders. His model is numerically instable at high ratio (λ_S/λ_f)>100, and perturbation method was used to overcome these difficulties. The numerical results for the touching square cylinders with the touching parameter c/a=0.02 are in excellent agreement with their experiment. Hsu et al. [36 and 338] note that γ_C is equivalent to the contact parameter of Nozad et al. [310]. Kaviayany [71] obtained numerical solution for the stagnant thermal conductivities of arrays of in-line and staggered non-touching circular cylinders in addition to those of in-line square cylinders. They found that their numerical results with c/a=0.002 agree best with experimental data than for c/a=0.01. Saez et al. [384] obtained numerical solution for the stagnant thermal conductivity of non-touching cubes. Their model at given porosity and thermal conductivity ratio (λ_S/λ_f) gives slightly higher values of stagnant thermal conductivity than those based on a two-dimensional model.

Assumed regular geometry of dispersed phase Pande et al. [40] developed an integrated theory for the ETC of all kind two-phase materials. They gave an expression for the prediction of ETC of a granular system by considering regular geometry of the dispersed phase

$$\lambda=\lambda_f\left[1+3.7396\left(\frac{\lambda_S-\lambda_f}{\lambda_S+2\lambda_f}\right)\phi^{2/3}\right]. \tag{115}$$

For two-phase system with a dispersion of one sphere one cube with porosity $\phi=\left(\frac{4\pi}{3}\right)\left(\frac{a^3}{R_0^3}\right)$, the ETC are:

$$\lambda=\lambda_f\left(1+3.844\phi^{2/3}\right),\ \text{when}\ (\lambda_S/\lambda_f)\to\infty, \tag{116}$$

$$\lambda=\lambda_f\left(1-1.545\phi^{2/3}\right),\ \text{when}\ (\lambda_S/\lambda_f)\to 0,\ \text{and} \tag{117}$$

$$\lambda = \lambda_f \left[1 + 2.307(\lambda_S - \lambda_f / \lambda_S + 2\lambda_f)\phi^{2/3}\right], \text{ when } (\lambda_S / \lambda_f) \approx 1, \tag{118}$$

where a_0 is the radius of spheres, R_0 is the distribution interval in a continuous phase with conductivity of λ_f. These relations were compared with experimental data and with prediction by Rayleigh's [373] and Meredith's [374] models. The predictions of these models are within 1.5 % to 4.1 %, while Maxwell's [332], Rayleigh's [373], and Meredith's [374] models are within 3.9 % to 19.6 %. Beniwal et al. [385] extended the model by Pade et al. [40] for statistically homogeneous and regular multiphase systems.

Adnani et al. [386] developed ETC model for binary mixtures of packed particle with stagnant gas at a high ratio λ_S / λ_g (up to 1300). A set of correlations for ETC of binary mixtures as a function of gas pressure and particle size is derived. The ETC was derived as a function of gas pressure and particle diameters (d_{1a} / d_{Sm}) by solving the conduction equation for assumed geometry. The final ETC correlations for various $(\lambda_S / \lambda_{std})$ are

$$\frac{\lambda}{\lambda_{std}} = A_1 + B_1\left(\frac{\lambda_S}{\lambda_{std}}\right) + C_1\left(\frac{\lambda_S}{\lambda_{std}}\right)^2, \text{ when } 1 < (\lambda_S / \lambda_{std}) < 20 \text{ and } \phi = 15\ \%, \tag{119}$$

$$\frac{\lambda}{\lambda_{std}} = D_1 + E_1 \log\left(\frac{\lambda_S}{\lambda_{std}}\right), \text{ when } 20 < (\lambda_S / \lambda_{std}) < 1300 \text{ and } \phi = 15\ \%, \tag{120}$$

$$\frac{\lambda}{\lambda_{std}} = A_2 + B_2\left(\frac{\lambda_S}{\lambda_{std}}\right) + C_2\left(\frac{\lambda_S}{\lambda_{std}}\right)^2, \text{ when } 1 < (\lambda_S / \lambda_{std}) < 20 \text{ and } \phi = 18\ \%, \tag{121}$$

$$\frac{\lambda}{\lambda_{std}} = D_2 + E_2 \log\left(\frac{\lambda_S}{\lambda_{std}}\right), \text{ when } 20 < (\lambda_S / \lambda_{std}) < 1300 \text{ and } \phi = 18\ \%, \tag{122}$$

where the coefficients A_i, B_i, C_i, D_i (i=1,2) are (ψ <0.125 and $1 < (\lambda_S / \lambda_{std}) < 1000$)

$$A_1 = 0.37 + 1.62\psi - 6.9\psi^2;\ A_2 = 0.44 + 1.97\psi - 27\psi^2 + 150\psi^3;$$

$$B_1 = 0.79 - 1.22\psi;\ B_2 = 0.7;\ C_1 = 0.013;\ C_2 = -0.013;$$

$$D_1 = -34.3 + 1050\psi - 1.25\times10^4\psi^2 + 4.9\times10^4\psi^3;$$

$$D_2 = -15.4 + 760\psi - 9710\psi^2 + 3.8\times10^4\psi^3;$$

$$E_1 = 31.4 - 667\psi + 7680\psi^2 - 3\times10^4\psi^3;$$

$$E_2 = 18.2 - 500\psi + 5920\psi^2 - 2.28\times10^4\psi^3;$$

$\psi = \frac{2-\alpha}{\alpha}\frac{\gamma}{\gamma-1}\frac{1}{\Pr_g}\frac{\eta_{std}}{P}\frac{T}{273}\frac{1}{d_{sm}}$, d_{sm} is the particle diameter as a characteristic length;

$\lambda_{std} = \lambda_g(1+\sigma)$; $\sigma = 4\frac{2-\alpha}{\alpha}\frac{\gamma}{\gamma-1}\frac{1}{\Pr_g}\frac{\eta_{std}}{P}\frac{T}{273}\frac{1}{d_{gap}}$, d_{gap} is the gap thickness; α is the thermal accommodation coefficient; $\gamma = C_P/C_V$; P gas pressure; Pr Prandtl number; η_{std} mean free path of gas molecules at 1 atm and 273 K. These correlations were derived by assuming zero contact area between particles. In practice, contact area is always larger than zero; therefore, actually values of ETC calculated with these relations would lower than measured values.

Yu and Cheng [387] developed two fractal models for the prediction of ETC of the bidispersed porous media. The model based on fractal theory and electrical analogy technique.

The prediction from the model was compared with the values calculated with lumped-parameter model and from experimental data. The results are good agreement with both lumped-parameter model and from experimental data. Bidispersed porous media is contains many clusters formed by particle agglomeration. There are two types of pores: macropores (on the order of 10^{-3} m) between the clusters and micropores (on the order of 10^{-5} m) between the particles inside the each cluster. When a bidispersed porous medium is saturated with a fluid the fluid fills both the micro- and macropores. The derived final models are: Model-1 (analytically derived)

$$\lambda = \frac{A_n}{A}\lambda_n + \left(1-\frac{A_n}{A}\right)\lambda_{mc}, \tag{123}$$

where A is the total area of a representative cross section and A_n is an equivalent area of a cross section having the same porosity as the nontouching particles or clusters, with $0 \le A_n/A \le 1$, λ_n is the thermal conductivity of the nontouched particles or clusters with fluid, λ_{mc} is the thermal conductivity of the mixed chains of particles or clusters with fluid. Hsu et al. [338] previously reported the expression for λ_n (see above Eq. 104). In order to determine the second term in Eq. (123) the authors assumed that the contact resistance between particles is only along the heat flow direction and the cube particle (or cluster) model is considered. The dimensionless ETC of the parallel chains in a unit cell is

$$\lambda_{mc}^* = \frac{\lambda_{mc}}{\lambda_f} = \frac{k_{\max}^2}{A}\left(\frac{k_{\max}}{L_0}\right)^{D_T-1}\frac{D_f}{1+D_T-D_f}\left\{\left[\frac{\gamma_{a1}(1-\phi)^{-1/2}}{\beta\sqrt{1-\phi}+2\left(1-\sqrt{1-\phi}\right)}\right]+\frac{1-\gamma_{a1}}{\gamma_{c1}^2(1-\phi)(\beta-1)/\gamma_{a1}^2+1}\right\}^{-1}, \tag{124}$$

where $\frac{k_{max}^2}{A}$ is the ratio of the maximum area of cross section of a mixed chain to the representative area A; $\frac{k_{max}}{L_0}$ is the ratio of the maximum diameter of cross section of a mixed chain to the representative length L_0 along the heat flow direction; and the porosity in a cluster is $\phi = 1-(1-3\gamma_{c1}^3)\gamma_{a1}^3 - 3\gamma_{c1}^2\gamma_{a1}^2$.

Combining Eqs. (122), (123) and (124) we have the fractal effective thermal conductivity expression for a cluster. The ETC for bidispersed porous media can be derived by similar method

$$\lambda_{mc}^* = \frac{\lambda_{mc}}{\lambda_f} = \frac{k_{max}^2}{A}\left(\frac{k_{max}}{L_0}\right)^{D_T-1} \frac{D_f}{1+D_T-D_f}\left\{\left[\frac{\gamma_{a1}(1-\phi)^{-1/2}}{\beta\sqrt{1-\phi}+2\left(1-\sqrt{1-\phi}\right)}\right]+\frac{1-\gamma_{a1}}{\gamma_{c1}^2(1-\phi)(\beta-1)/\gamma_{a1}^2+1}\right\}^{-1}, \quad (125)$$

$$\lambda = \frac{A_n}{A}\left[\left(1-\sqrt{1-\phi}\right)+\frac{\sqrt{1-\phi}}{1+\left(1/\lambda_c^*-1\right)\sqrt{1-\phi}}\right]+\left(1-\frac{A_n}{A}\right)\frac{k_{max}^2}{A}\left(\frac{k_{max}}{L_0}\right)^{D_T-1}\frac{D_f}{1+D_T-D_f}\times$$

$$\times\left\{\left[\frac{\gamma_a/\gamma_a'}{\lambda_c^*\gamma_a'+2\left(1-\gamma_a'\right)}+\frac{\left(1-\gamma_a\right)}{\gamma_c^2\gamma_a'^2\left(\lambda_c^*-1\right)/\gamma_a^2+1}\right]\right\}^{-1}, \quad (126)$$

where $\gamma_a = l_c / a_c$ and $\gamma_c = c_c / l_c$ are the geometric length scale ratio and contact length scale ration at the macroscale for clusters in a bidispersed medium, respectively; $\gamma_a' = a_c / k = \sqrt{(1-\phi)(1-\phi_c)}$; D_T is the tortuosity fractal dimensions ($1 < D_T < 2$), D_T =1 represent a straight capillary path or straight chain of particles (or clusters), at D_T =2 we have tortuous line so irregular that it fills an entire plane. Higher value of D_T corresponds to a highly tortuous capillary (or chain). The average porosity of a bidispesed medium is given by $\phi = \left[\left(k^2 - a_c^2\right)+a_c^2\phi_c\right]/k^2$, where k is the size of a capillary or a chain in the medium.

Bayuk et al. [235] developed method to calculate of the ETC of oil-, water-, and gas-saturated sandstones. The method based on the selfconsistent approach. The developed model represent of ETC as a function of porosity ϕ, aspect ratio α (ratio of minimum and maximum axes of ellipsoid, internal rock structure, shape of pores), the thermal conductivities of rock matrix, λ_S, and saturated fluid, λ_f. According to the model, each inclusion (mineral grain or crack) assumed to be of elliptical shape (ellipsoid) is embedded in a medium having thermal conductivity λ_0 chosen arbitrarily. For inclusions, differing in shape and orientation the ETC can be represented as

$$\lambda = \left\{ \sum_{j=1}^{N} \phi_j \lambda^{(j)} \int P\left(F^{(j)};\theta,\varphi,\psi\right)\left[\lambda^{(0)}\left(I - F^{(j)}\right) + \lambda^{(j)}F^{(j)}\right]^{-1} dF^{(j)} d\theta d\varphi d\psi \right\} \times \times \left\{ \sum_{j=1}^{N} \phi_j \lambda^{(j)} \int P\left(F^{(j)};\theta,\varphi,\psi\right)\left[\lambda^{(0)}\left(I - F^{(j)}\right) + \lambda^{(j)}F^{(j)}\right]^{-1} dF^{(j)} d\theta d\varphi d\psi \right\}^{-1}, \quad (127)$$

where θ, φ, ψ are the Eulerian angles specifying the rotation of inclusions in space, ϕ_j is the volume concentration of the *i*-th components; $P\left(F^{(j)};\theta,\varphi,\psi\right)$ is the distribution function over shape and orientation of the *j*-th component, *N* is the number of components. This relation for ETC was applied for dry, water-, and gas-saturated rocks. The model parameters were calculated using the procedure of non-linear optimization (fitting procedure). Three main parameters (inclusion shapes, orientation, and spatial distribution) considerably affect the ETC.

Gruescu et al. [49] developed estimating techniques for the evaluating ETC in partially saturated porous material and study of the influence of microstructures (morphology and mineralogy) on the ETC of triphastic porous materials. The numerical simulations are compared with experimental data on seven samples. The developed model is

$$\lambda = \lambda_S + 3\lambda_S \left(\sum_{i=2}^{r} \varphi_i \frac{\lambda_i - \lambda_S}{\lambda_i + 2\lambda_S} \right) \Big/ \left(1 - \sum_{i=2}^{r} \varphi_i \frac{\lambda_i - \lambda_S}{\lambda_i + 2\lambda_S} \right), \quad (128)$$

with $\varphi_1 = \frac{\varphi_{quartz}}{1-\varphi_P}$, $\varphi_2 = \frac{\varphi_{calcite}}{1-\varphi_P}$, where λ_i is the thermal conductivity of constituents (quartz, calcite); φ is the volume fraction of the inhomogeneous space; $\varphi_P = \varphi_g - \varphi_l$; φ_l is the volume fraction of the liquid phase; and φ_g is the volume fraction of the gaseous phase; r=2 porous phases (liquid and gas). The shape of minerals and clay matrix being supposed spherical and their distribution is isotopic.

Based on the microscopic temperature Lee and Yang [388] evaluated the ETC in principal direction of a nonhomogeneous anisotropic porous medium. They considered a periodically wedge-shaped solid in a two-dimensional enclosure filled with a fluid. The correlation equation proposed by Lee and Yang [388] is

$$\lambda = 1 + (\sigma - 1)(1 - \varepsilon) f(\varepsilon, \sigma, \alpha, \beta), \quad (129)$$

$$f(\varepsilon,\sigma,\alpha,\beta) = \left(1 + m\varepsilon^n\right)^{-1},$$
$$m = \left(2/\alpha + 3/\alpha^2\right)\beta(1 + 6\beta)(\ln\sigma)^5\left[2.522 \times 10^{-3} - 8.957 \times 10^{-4}(\ln\sigma) + 9.714(\ln\sigma)^2\right],$$
$$n = \left(20 - 29/\alpha + 15/\alpha^2\right)\left(40 + 3\beta - 13\beta^2\right)\left(0.0095 + 0.001666\ln\sigma\right)$$

where $\sigma = \lambda_S / \lambda_f$, ε is the fraction of fluid, $\alpha = a/p$ is the aspect ratio (a and p are the height and width of the wedges), $\beta = b/p$ dimensionless quantity (b is the width of the wedge base). This equation applicable for the range $0 \le \varepsilon \le 1$, $1 \le \sigma \le 1000$, $0.2 \le \beta \le 1$ and $\alpha \ge 1$. The maximum uncertainty of the correlation (129) is less 5 %. Special case $f(\varepsilon, \sigma, \alpha, \beta) = 1$ corresponds to the classic mixing rule Eq. (50).

Cheng and Torquato [331] consider the effect of the interfacial resistance on the ETC of periodic arrays of spheres with imperfect interfaces. Thermal resistance at the interface can dramatically decrease of ETC of the composite (see for example, Lipton and Vernescu [389]). This interfacial resistance is measure d by a jump in the temperature across the interface. They consider a composite material consisting of periodic arrays of equal-sized spherical inclusions of radius a and conductivity λ_2 embedded in a uniform matrix of conductivity λ_1, in which there exists a thermal resistance at the interface between the two phases. The strength of this resistance is characterized by a dimensionless parameter R. The final result for the ETC calculation is

$$\lambda = \lambda_1 (1 - 4\pi B_{10} / E_0) \text{ for the simple cubic lattice case,} \tag{130}$$

$$\lambda = \lambda_1 (1 - 8\pi B_{10} / E_0) \text{ for the body-centered cubic lattice case,} \tag{131}$$

and

$$\lambda = \lambda_1 (1 - 16\pi B_{10} / E_0) \text{ for the face-centered cubic lattice case.} \tag{132}$$

Here, the coefficient B_{10} incorporates the effect of the imperfect interface. The exact solution for ETC is

$$\lambda = \lambda_1 \left(1 - \frac{3\phi}{D} \right), \tag{133}$$

where

$$D = -\beta_1^{-1} + \phi + c_1 \beta_3 \phi^{10/3} + c_2 \beta_5 \phi^{14/3} + c_3 \beta_3^2 \phi^{17/3} + c_4 \beta_7 \phi^6 + c_5 \beta_3 \beta_5 \phi^7 + c_6 \beta_9 \phi^{22/3} + O(\phi^{25/3})$$

with β_i is defined by

$$\beta_i = \frac{\alpha - 1 - iR}{\alpha + (i+1)/i + (i+1)R}, \quad \alpha = \lambda_2 / \lambda_1 .$$

The numerical constants c_i for three cubic lattice arrays are tabulated. Equation (133) enables one to estimate ETC for composite materials with imperfect interfaces, with the interfacial resistance incorporated in the constants β_i. At the critical value $R_C = \alpha - 1$, the ETC of composite material equals exactly to the matrix conductivity. When $R > R_C$ the spherical inclusions reduce ETC of the composite material below that of the matrix. Equation (133) is useful for porosities below 30 %. Torquato and Rintoul [390] derived following approximate equation for ETC of a suspension (random or periodic) with dimensionless R

$$\lambda = \lambda_1 \frac{1 + 2\phi\beta_1 - 2(1-\phi)\varsigma_2\beta_1^2}{1 - \phi\beta_1 - 2(1-\phi)\varsigma_2\beta_1^2}, \tag{134}$$

where ς_2 is the aforementioned three-point microstructural parameter (tabulated as a function of porosity ϕ). Miloh and Benveniste [391] studied the ETC of composite media with ellipsoidal inhomogeneity and highly conducting interface. Thermal contact resistance between two media may be due to roughness or acoustic mismatch at their common boundaries or may also often arise due to presence of a thin interphase layer of low conductivity. The final equation for the ETC derived by Miloh and Benveniste [391] is

$$\lambda = \lambda_1 + c_2 \frac{3\lambda_1(\lambda_2 - \lambda_1) + 6\lambda_1(\alpha / a)}{(1 - c_2)[\lambda_2 + 2\lambda_1 + 2(\alpha / a)] + 3c_2\lambda_1}, \tag{135}$$

where a is the radius of spherical particles; λ_1 is the thermal conductivity of matrix; λ_2 is the thermal conductivity of particles; α is the interface parameter. Brucker and Majdalani [64] developed model for ETC of porous block using computer technologies. They found explicit solution of the heat transfer equation for various physical configurations that constitute the building blocks of complex heat sinks. Singh et al. [253] developed model to predict the ETC of the multi-phase porous by assuming an effective continuum medium. They considered the cylindrical geometry of multi-phase media having a line source and sensor along the z axis. There is a regular periodic dispersion of spherical particles with different conductivity values at an interval R_0 in a continuous medium. Their ETC model is

$$\lambda = \lambda_0 \left[1 + \sum_{i=1}^{\gamma} 3.844 \left(\frac{\lambda_i - \lambda_0}{\lambda_i + 2\lambda_0} \right) \psi_i \psi^{-1/3} \right], \tag{136}$$

where $\psi = \frac{4}{3}\pi \frac{a^3}{R_0^3}$ is the total volume fraction (porosity) of dispersed phases for the simple cubic lattice (there is one sphere per cube). This equation is valid only for dilute dispersions, λ_0 is the thermal conductivity of continuous medium; λ_i is the thermal conductivity of spheres of phases; a is the radius of spheres. In the case of loose systems, $\lambda_S / \lambda_{gas} \to \infty$. In such systems neighboring interactions are not small. This consideration yields the following expressions for the ETC of loose two-phase systems. When $\lambda_g \rightarrow \lambda_0$, $\lambda_S \rightarrow \lambda_i$ and $\lambda_g / \lambda_S \rightarrow 0$

$$\lambda = \lambda_g \left[1 + 3.844\psi_S^{2/3}\right], \text{ for } 0.5 > \psi_S > 0. \tag{137}$$

For the reverse case when $\lambda_S \rightarrow \lambda_0$, $\lambda_g \rightarrow \lambda_i$ and $\lambda_S / \lambda_g \rightarrow \infty$

$$\lambda = \lambda_S \left[1 - 1.616\psi_g^{2/3}\right], \text{ for } 0.5 > \psi_g > 0. \tag{138}$$

This model does not correspond to a real loose two-phase system. For a real loose two-phase system by making small dispersions ξ of solid or gas phase in effective continuous medium ETC equation becomes

$$\lambda = \lambda_{ecm} \left[1 + 3.844\xi_S^{2/3}\right], \text{ for } \xi_S > 0, \tag{139}$$

$$\lambda = \lambda_{ecm} \left[1 - 1.616\xi_S^{2/3}\right], \text{ for } \xi_g > 0, \tag{140}$$

where ξ_S and ξ_g are the volume fractions of solid and gas phases in ECM, λ_{ecm} is the ETC of ECM. This approach has been applied for the prediction of ETC of moist porous materials by Singh et al. [254]. The prediction capability of the model has been tested for different type of soils saturated with different liquids. The final expression for ETC of three –phase system (solid-water-gas) is

$$\lambda = \lambda_a \left(1 + 3.844 \frac{\lambda_W - \lambda_a}{\lambda_W + 2\lambda_a} \psi_{ma}^{2/3}\right), \tag{141}$$

when the dispersion of the water in the air is small $0 < \psi_{ma}^{2/3} < 0.4$. When $\psi_{ma}^{2/3}$ lies between 0.4 and 1.0, the ETC of moist soil air becomes

$$\lambda = \lambda_W \left[1 + 3.844 \frac{\lambda_a - \lambda_W}{\lambda_a + 2\lambda_W} \left(1 - \psi_{ma}^{2/3}\right)\right], \tag{142}$$

where λ_a and λ_W are the thermal conductivity of air and water, respectively.

Singh et al. [392] developed a geometrical model for estimation of the ETC by using the resistor approach of two-phase systems with spherical inclusions. The models for spherical and cubic particles are

$$\lambda = \frac{\lambda_f\left[\lambda_f + 0.806F^{2/3}(\lambda_S - \lambda_f)\right]}{\lambda_f + F^{2/3}\left[0.806(\lambda_S - \lambda_f)(1 - 1.2407F^{1/3})\right]}, \tag{143}$$

$$\lambda = \frac{\lambda_f\left[\lambda_f + F^{2/3}(\lambda_S - \lambda_f)\right]}{\lambda_f + F^{2/3}(\lambda_S - \lambda_f)(1 - F^{1/3})}, \tag{144}$$

where $F = 1 - \exp\left[-0.29\phi_S^2 \ln(\lambda_S / \lambda_f)\right]$ is the porosity correction.

Verma et al. [393] proposed following equations to predict the ETC of a multiphase system (inclined slabs)

$$\lambda = \left(\lambda_{II}^2 + \lambda_{\perp}^2\right)^{1/2}, \tag{145}$$

where $\lambda_{II} = [\phi\lambda_1 + (1-\phi)\lambda_2]\cos\theta$ and $\lambda_{\perp} = \left[\frac{\phi}{\lambda_1} + \frac{(1-\phi)}{\lambda_2}\right]\sin\theta$. When $\theta = 90^o$, $\lambda_{\perp} = \left[\frac{\phi}{\lambda_1} + \frac{(1-\phi)}{\lambda_2}\right]$ and $\lambda_{II} = 0$ we obtain the expression for parallel resistors Eq. (51). If $\theta = 0^o$, $\lambda_{II} = [\phi\lambda_1 + (1-\phi)\lambda_2]$ and $\lambda_{\perp} = 0$ we obtain the expression for perpendicular resistors Eq. (50). The angle θ (structural parameter) depends on geometry of the particles. The authors defined the angle θ in term of sphericity. Angularities of the grains in a porous system greatly alter its thermophysical properties. For example, ETC of the spherical particles radically differs from the non-spherical particles. The angle θ can be define as

$$\tan\theta = B_0 F\phi^2\left(\psi\frac{\lambda_2}{\lambda_1}\right)^{1/2}, \tag{146}$$

where ψ is the sphericity of the particles; $F = \frac{3-\phi}{2\phi}$ is the formation factor, B_0 is the constant of proportionality whose values depends upon the mode of packing of the particles and the nature of the system. For two-phase system B_0 has been evaluated numerically. For two-phase granular systems it has been found to be 1.5, while for two-phase suspension, emulsions, and solid-solid mixtures are 1.7, 0.65, and 1.05, respectively.

Berryman [394] derived theoretical expression for ETC of fluid-saturated porous materials. The derived expression for ETC depends on two formation factor-one associated with the pore space and other one with the solid frame. This model accurately predict minimal and maximal bounds of the ETC fluid-saturated porous materials

$$\min(S_1, S_2) \le \lambda \le \max(S_1, S_2), \text{ where } S_2 = \lambda_1 + \frac{\lambda_2 - \lambda_1}{F_2} \text{ and } S_1 = \lambda_2 + \frac{\lambda_1 - \lambda_2}{F_1}, \tag{147}$$

$\lim_{\lambda_1 \to \infty} \frac{\lambda}{\lambda_1} = \frac{1}{F_1}$ and $\lim_{\lambda_2 \to \infty} \frac{\lambda}{\lambda_2} = \frac{1}{F_2}$ are the formation factors, when one of the constituents is much more strongly conducting than the other. Zeng et al. [261] developed model for ETC of intersecting square rod, cylindrical rod, and spherical structures for aerogel. The results are

$$\lambda = \left(\frac{d}{D+d}\right)^2 \lambda_S + 2\left[\left(\frac{D}{D+d}\lambda_S\right)^{-1} + \left(\frac{d}{D+d}\lambda_g\right)^{1}\right]^{-1} + \left(\frac{D}{D+d}\right)^2 \lambda_g, \tag{148}$$

for cubic array of intersecting square rods and

$$\lambda = \frac{\pi d^2 \lambda_S}{4D^2} + \frac{\lambda_g d(D-d)(\pi - B_2)}{D^2 B_1} + \left(1 - \frac{d}{D}\right)^2 \lambda_g, \tag{149}$$

for cubic array of intersecting cylindrical rods, where *d* is the characteristic dimension of the structure (a few nanometers); and *D* is the pore size (about 15-20 nanometers). Both geometric parameters *d* and *D* are determined from the porosity ϕ and the specific surface area S_S as

$$\phi = 1 - d^2(d+3D)/(d+D)^3\ ,\ S_S = 12dD(d+D)^{-3} / \rho_{por} \text{ square rods}$$

and

$$\phi = 1 - \left[(3/4)\pi d^2 D - \sqrt{2}d^3\right]/D^3\ ,\ S_S = 3\pi dD - 6\sqrt{2}d^2/\left(D^3 \rho_{por}\right)$$

for cylindrical rods.

Schulz [395] studied the reliability of general relationship for the ETC of two-phase materials with very high porosity (up to 96 %). The following equations are recommended

$$\lambda = \lambda_S (1-\phi)^X\ \text{, where } X = \frac{1-\cos^2\alpha}{1-F} + \frac{\cos^2\alpha}{2F} \text{ for closed porosity,} \tag{150}$$

$$(1-\phi)(\lambda_S - \lambda)\left[\frac{2(1-\cos^2\alpha)}{\lambda_S + \lambda} + \frac{\cos^2\alpha}{\lambda}\right] = \phi(2-\cos^2\alpha) \text{ for open porosity,} \tag{151}$$

and at $\cos^2\alpha = 1/3$, $F = 1/3$,

$$\lambda = \lambda_S \left(\frac{4-9\phi}{10}\right) + \left[\left(\frac{4-9\phi}{10}\right)^2 + \frac{1-\phi}{5}\right]^{1/3}, \tag{152}$$

where F and α are the shape factor and orientation factors for the pores, respectively. These equations are taking into account conduction by gas across the pores. Shape factor depend on the axial ratio a/b of the spheroid (a/b=1 sphere; $a/b \to \infty$ cylinder; and a/b=0 lamella). These equations were tested against literature data for the thermal conductivity data of porous metals, graphite foam, ZrO_2. These equations were derived by applying the continuum theory to two-phase system (solid and pore). This theory is taking into account the effect of pore shape and their orientation.

Krupiczka [238] developed two models for predicting of ETC of single fluid-saturated porous media. Normal cubic packing of spheres and horizontal packing of cylinders are used to represent the structures.

The author determined of temperature distribution within the single unit cell by solving the Laplace equation, and then determined the conductivity. Krupiczka [238] recommended equation for the fluid-saturated porous media is (for models composed of cylinders and spheres)

$$\frac{\lambda}{\lambda_f} = \frac{\sigma}{\sigma - 1}\left[\frac{2\sigma}{\sqrt{2\sigma - 1}} arctg(2\sigma - 1) - \frac{\pi}{2}\right], \tag{153}$$

for the model made up of cylinders $(\phi = 0.215)$, and

$$\frac{\lambda}{\lambda_f} = \frac{\pi}{2}\left(\frac{\sigma}{\sigma - 1}\right)^2\left[\ln\sigma - 1 + \frac{1}{\sigma}\right] + 1 - \frac{\pi}{4}, \tag{154}$$

for the model made up of spheres $(\phi = 0.476)$. The correlation by simpler functions is

$$\frac{\lambda}{\lambda_f} = \left(\frac{\lambda_S}{\lambda_f}\right)^{A + B\log(\lambda_S/\lambda_f)}, \tag{155}$$

where $A = 0.280 - 0.757\log\phi$ and $B = -0.057$, λ is the ETC of fluid-saturated porous media; λ_f is the thermal conductivity of fluid; λ_S is the thermal conductivity of solid; ϕ is the fractional porosity. This relation is valid for porous media with (λ_S/λ_f) >50 and at porosities between 0.215 and 0.476. Equation (155) was tested for 165 data points. About 76 % of the calculated values of ETC agreed within 30 % of experimental values.

Kunii and Smith [236] developed a lumped parameter model for heat transfer in packed beds of unconsolidated materials. They consider a unit cell comprised of two spherical particles. Overall heat flow is assumed to be unidirectional and they also assumed that heat transfer through the unit cell can be represented by two mechanisms in parallel: (1) conduction and radiation through the void space; and (2) heat transfer through the solid phase (conduction through the contact surface, fluid layer in the vicinity of the contact point, radiation between particle surface, and the solid). Kunii and Smith [236] also developed model for predicting of thermal conductivity of single fluid-saturated porous media (for unconsolidated and consolidated materials such as sandstone). This model based on normal cubic and rhombohedral packing. Final derived equation presents the thermal conductivity as a function of thermal conductivity of the constituting elements and porosity. This equation is given as

$$\frac{\lambda}{\lambda_f} = \left(\frac{\phi + \beta(1-\phi)}{\varepsilon + (2/3)(\lambda_f / \lambda_S)} \right), \tag{156}$$

$$\varepsilon = \varepsilon_2 + (\phi - 0.259)(\varepsilon_1 - \varepsilon_2)/0.217 \text{ for } \phi_1 \geq \phi \geq \varepsilon_2,$$
$$\varepsilon = \varepsilon_1 \text{ for } \phi > \varepsilon_1; \text{ and } \varepsilon = \varepsilon_2 \text{ for } \phi < \varepsilon_2,$$

where β is the packing factor, 1 for normal cubic (loose packing) packing and 0.895 for rhombohedral (close packing) packing; ε is the ratio of effective thickness of the stagnant fluid near grain contacts to the diameter of a grain evaluated as a function of (λ_S / λ_f) and packing factor; ε_1 and ε_2 correspond to cubic and tetrahedral packing of uniform spheres. The authors provided the equation to calculate the values of ε for normal cubic and rhombohedral packing. This relation correlates reported data within 20 %. This model assumes no radiation and no heat transfer through contact surfaces. Keese [223] has examined of the correlations (155) and (156).

Biancheria [46] obtained a general relation for ETC of ellipsoid pore model. de Vries [237] applied the work of Eucken [368] and Burger [396] to the thermal conductivity of soil. He (see also [336]) suggested the following relation for a two-phase fluid-solid medium

$$\lambda = [\phi\lambda_f + (1-\phi)F\lambda_S]/[\phi + (1-\phi)F], \tag{157}$$

where $F = (1/3)\sum_{i=1}^{3}\left[1 + (\lambda_S / \lambda_f - 1)g_i\right]^{-1}$, g_i is the relative dimensions of a particle in i-th direction, $\sum_{i=1}^{3} g_i = 1$. The Maxwell's relation Eq. (52) is a special case of the above equation. If $g_1 = g_2 = g_3$ (a sphere) de Vries' Eq. (157) reduces to that of Maxwell Eq. (52).

Yagi and Kunii [307] derived following theoretical relation for ETC packed beds

$$\frac{\lambda}{\lambda_f} = \delta\left(\frac{\lambda_S}{\lambda_f}\right) + \frac{(1-\phi-\delta)\beta}{\eta\left(\frac{\lambda_f}{\lambda_S}\right) + \frac{w\lambda_f}{\lambda_f + wD_P h_{rs}}} + \phi\beta\frac{D_P h_{rv}}{\lambda_f}, \tag{158}$$

where $\beta = l_P / D_P$ (length between grain centers /average packing diameter); $w = l_V / D_P$; l_V is the effective thickness of fluid film adjacent to the contact surface of two solid particles; h_{rs} is the solid to solid radiation heat transfer coefficient; h_{rv} is the void to void radiation heat transfer coefficient; $\eta = l_S / D_P$; l_S is the effective length of solid relating to conduction; δ is the solid phase contact surface parameters. For fine gas-saturated particles this relation reduced to simple equation

$$\frac{\lambda}{\lambda_f} = \frac{(1-\phi)\beta}{(\lambda_f / \lambda_S) + w}. \tag{159}$$

Kimura [397] suggested a relation for ETC based on the resistor model equation

$$\frac{\lambda}{\lambda_f} = \frac{(1-g)}{(1-d)(\lambda_f / \lambda_S) + d\lambda_f / (\lambda_f + dh_r d_P)} + g\left(1 + \frac{h_r d_P}{\lambda_S}\right), \tag{160}$$

where d is the total thickness of fluid layer per unit length of bed (varies with ϕ and ratio λ_f / λ_S); d_P is the average packing diameter; $g = \left(1-\phi^{1.3}\right)$; h_r is the radiant heat transfer coefficient. Kimura [397] determined the value of g from a plot of $g / \phi^{1.5}$ versus λ_f / λ_S.

Schumann and Voss [398] have estimated ETC of the unit cell by dividing it into vertical slices of infinitesimal width and assuming unidirectional heat flow through the slice. Consequently, they computed the thermal resistance of the slice, integrate the result in the horizontal direction, and apply pre-determined boundary conditions. Schumann and Voss [398] derived the following theoretical equation for ETC of granulated materials

$$\lambda = \lambda_f \tau^3 + \left(1-\tau^3\right)\theta, \tag{161}$$

where $\theta = \frac{\lambda_S \lambda_f}{\lambda_f + \rho(\lambda_f + \lambda_S)}\left\{1 + \frac{\rho(1+\rho)(\lambda_f - \lambda_S)}{\lambda_f + \rho(\lambda_f - \lambda_S)} \ln\left[\frac{\lambda_f (1+\rho)}{\rho \lambda_S}\right]\right\}$, $\tau = \rho(\rho+1)\ln\left(1+\frac{1}{\rho}\right) - \rho$, ρ is the bulk density. This relation assumes only point contacts between grains, therefore calculated values of ETC expect to be less than measured. This was confirmed by Wilhelm [399]. He showed that Eq. (161) does not account for contribution through particle contact points. Consequently, they propose an empirical correction factor, but it has limited applicability (Krupiczka [238]). His measurements are differs from Eq. (161) by 18.5 %. Wilhelm [399] suggested a factor $\Delta = 10^{m+n(\lambda_S / \phi)-5}$, where m=0.859 and n=3.12 be subtracted from the Eq. (146). After Wilhelm [399] corrections Eq. (161) agrees with the measured values of ETC within 8.5 %.

The statistical model developed by Huang [172] is

$$\frac{\lambda}{\lambda_S} = A + B\left(\frac{\lambda_f}{\lambda_S}\right), \quad A = (1-\phi)\exp[-n/(1-\phi)], \tag{162}$$

where $B=\phi\exp(-n/\phi)+\frac{H^2}{\phi[1-\exp(-n/\phi)]}$, $H(\phi)=1-\phi\exp(-n/\phi)-(1-\phi)\exp[-n/(1-\phi)]$, n is the geometric pore factor. This model at n=1 reduced to ETC for dry sandstones (consolidated porous materials). At n=0 the model approaches a simple arithmetic mean (for pure solids). For unconsolidated porous materials n >1. For gas-filled porous rocks at high pressures ETC equation is

$$\frac{\lambda}{\lambda_S}=\alpha+\beta\left[\frac{\lambda_g}{(1+1/SP)\lambda_S}\right], \tag{163}$$

where $\alpha=(1-\phi)\exp[-n/(1-\phi)]$, $\beta=\phi\exp(-n/\phi)+\frac{H^2}{\phi[1-\exp(-n/\phi)]}$, $S=2r/l_0P_0$, l_0 is the mean free path at the reference pressure P_0. This relation allows calculating the ETC of gas – saturated rocks at any given pressure P. The effect of β on ETC is much less than the effect of α at low porosity rocks. For high porosity rocks the effect of β on ETC is much greater than the effect of α. This model has only one empirical parameter. Huang [172] model yields the best results at assuming geometric pore factor is equal to porosity.

Walsh and Decker [169] defined the maximum and minimum possible values of ETC by using the mathematically analogous problem in magnetic permeability. For low porosity systems $(\phi << 1)$ the maximum and minimum values of ETC can be expressed as

$$\frac{(\lambda_S-\lambda)}{\lambda_S}=\frac{3\phi(1-\sigma)}{(2+\sigma+\phi)}\ \text{(minimum)}, \tag{164}$$

and

$$\frac{(\lambda_S-\lambda)}{\lambda_S}=\frac{\phi(1+2\sigma)(1-\sigma)}{(3\sigma+\phi-\phi\sigma)}\ \text{(maximum)}, \tag{165}$$

where $\sigma=\lambda_f/\lambda_S$. These relations are applicable for rocks, where the conductivity of the matrix is usually greater than that of the pore fluid $(\sigma<1)$. For low-porosity rocks fully saturated with water the ETC is very close to the thermal conductivity of matrix λ_S.The effect of pore fluid on ETC should be given by Eq. (165) if instead of the total porosity, only the porosity ϕ_C due to cracks used in calculation. Therefore the equation for estimating the actual effect of pore fluid is given by the expression

$$\frac{(\lambda_S-\lambda)}{\lambda_S}=\frac{\phi_C}{(3\sigma+\phi_C)}, \tag{166}$$

where the values of ϕ_C can be found from volumetric strain $-\Delta V/V_0$ as a function of pressure. The "crack porosity" is the intercept on the $\Delta V/V_0$ axis of the linear portion of the pressure –strain curve extrapolated to zero pressure. The authors derived the expression for ETC of randomly distributed inclusions of arbitrary spheroidal configuration as

$$\frac{\lambda_S-\lambda}{\lambda-\lambda_f}=R\left(\frac{\phi_C}{3\sigma}\right), \tag{167}$$

where $R=\frac{\int[\alpha/(\sigma+\pi\alpha/2)](dnc^3/d\alpha)d\alpha}{\int[\alpha/\sigma](dnc^3/d\alpha)d\alpha}$, α is the aspect ratios; c is the radii; n is the number of cracks at each α. All necessary information about the distribution of crack configuration required for evaluating the integrals for R can be estimated from measurements of the elastic compressibility as a function of pressure. For example, the values of $(dnc^3/d\alpha)$ is proportional to slope of a plot of $(\beta_e-\beta)$ version P. Using this information R can be calculated by numerical integration.

Deissler and Eian [308] developed two unidirectional conduction models to correlate ETC measurements of magnesium oxide powder saturated with gases (air, argon, helium). They first considered a square array of infinite cylinders. ETC is obtained by applying a procedure similar to that of Schumann and Voss [398]. The result is

$$\frac{\lambda}{\lambda_f}=\frac{\sigma}{\sigma-1}\left[\frac{2\sigma}{\sqrt{2\sigma-1}}\arctan(2\sigma-1)-\frac{\pi}{2}\right], \tag{168}$$

where $\sigma=\lambda_S/\lambda_f$. This relation was determined at assuming that the shape of the solid-fluid interface is fixed ($\phi=0.214$). Their second model consists of a cubic array of spheres. For this model ETC is given by

$$\frac{\lambda}{\lambda_f}=\frac{\pi}{2}\left(\frac{\sigma}{\sigma-1}\right)^2\left(\ln\sigma-1+\frac{1}{\sigma}\right)+1-\frac{\pi}{4}. \tag{169}$$

This model also assumes a fixed solid-fluid interface; thus it applies when $\phi=0.475$. Similar approaches are employed by several authors [343,400, and 401] to derive unidirectional heat flow models for ETC of soil, snow, and powder, respectively. Deissler and Boegli [309] extend of the Deissler and Eian [308] model by considering a three – dimensional cubic array of spheres. The unit cell consists of an eighth-sphere located in a cylinder. The unit cell is axisymmetric and it resembles. The porosity also is the same $\phi=$ 0.475 because the geometry of the solid-fluid interface is fixed. The thermal conductivity equation is numerically solved to obtain a curve of (λ/λ_f) versus σ. Crane and Vachon [29] and Jaguaribe and Beasley [402] employed a resistor network model to predict ETC of porous media. Jaguaribe and Beasley [402] assumed porous media consisting of a bundle of infinite cylinders.

Hadley [34] develops a semi–empirical correlation to predict ETC of two- and three-phase porous media. He considered the volume-average temperature of a system comprised of n_P phases. This temperature $\langle T \rangle$ is the sum of the superficial phase temperatures. Then the gradient of $\langle T \rangle$ is taken and the spatial averaging theorem is applied to the result. He is next considering ETC of a dilute of spheres, which is valid for $\phi \to 1$. The final expression for ETC derived by Hadley [34] is

$$\frac{\lambda}{\lambda_f} = (1-\alpha_P)\frac{\phi f_0 + \sigma(1-\phi f_0)}{1-\phi(1-f_0)+\sigma\phi(1-f_0)} + \alpha_P \frac{2\sigma^2(1-\phi)+(1+2\phi)\sigma}{(2+\phi)\sigma+1-\phi}, \tag{170}$$

where α_P is a parameter representing the degree of consolidation. To determine the empirical parameters f_0 and α_P, Hadley [34] considered limiting cases for σ. The consolidation parameter is related to ETC of an evacuated porous medium ($\sigma \to \infty$). That is

$$\frac{\lambda}{\lambda_f} = 2\alpha_P \frac{(1-\phi)}{2+\phi}. \tag{171}$$

This equation together with experimental data were used by Hadley [34] to determine the values of α_P as a function of $(1-\phi)$. The values of f_0 were determined from intermediate σ e.g., water-saturated media.

It was found that f_0 is varying from 0.8 to 0.9. Equation (171) reproduced reported data for ETC within 18 %. Equation (171) can be rewritten as

$$\frac{\lambda}{\lambda_f} = \frac{\left[(1-\phi)F + \frac{\lambda_S}{\lambda_f}\{1-(1-\phi)F\}\right]}{\left[1-(1-\phi)(1-F)+\frac{\lambda_S}{\lambda_f}(1-\phi)(1-F)\right]}, \tag{172}$$

where F is between 0 and 1. Verma et al. [255] defined the parameter F as

$$F = \exp\left[-\psi\left(\frac{\lambda_f}{\lambda_S}\right)^{1/3}\right], \tag{173}$$

where ψ is the sphericality of the particles. The values of F for granular systems are 0.82 and 0.75 for emulsion –like systems. For the two-phase system with spherical inclusions they found the relation

$$\lambda = \frac{\lambda_f\left[2.598\phi^{1/3}(\lambda_S - \lambda_f) + 3.224\phi^{-1/3}\lambda_f\right]}{\left[\left(1 - 1.2407\phi^{1/3}\right)\left\{2.598 5\phi^{1/3}(\lambda_S - \lambda_f) + 3.224\phi^{-1/3}\lambda_f\right\} + \lambda_f\right]}. \tag{174}$$

Misra et al. [403] improved this relation by replacing ϕ by porosity correction term $F_P = \exp\left[-C_2(1-\phi)^{2/3}\right]$ as

$$\lambda = \frac{\lambda_f\left[2.598F_P^{1/3}(\lambda_S - \lambda_f) + 3.224F_P^{-1/3}\lambda_f\right]}{\left[\left(1 - 1.2407F_P^{1/3}\right)\left\{2.5985F_P^{1/3}(\lambda_S - \lambda_f) + 3.224F_P^{-1/3}\lambda_f\right\} + \lambda_f\right]}. \tag{175}$$

where $C_2 = 2.736\exp\left[-0.004(\lambda_S / \lambda_f)\right]$.

Dul'nev et al. [329] developed the model to predict ETC porous materials

$$\frac{\lambda}{\lambda_S} = (h/L)^2 + (\lambda_f/\lambda_S)(1-h/L)^2 - 2(\lambda_f/\lambda_S)(h/L)(1-h/L)\left[(1-h/L)(1-\lambda_f/\lambda_S)\right]^{-1}$$

$$(h/L)^3 - (3/2)(h/L)^2 + (1-\phi)/\phi = 0. \tag{176}$$

He is also proposed ETC model (Dul'nev et al. [341]) as

$$\frac{\lambda}{\lambda_S} = c^2 + \nu(1-c)^2 + \frac{2\nu(1-c)c}{1-c+\nu c},\ \nu = \lambda_f/\lambda_S,\ c = l_1/L,\ \phi = 2c^3 - 3c^2 + 1, \tag{177}$$

l_1 -geometrical (or structural) parameter, L-length of the cell, and method for calculation of the ETC moistened porous materials on the bases the three-component structure. This model predicts ETC of porous materials within 20 %.

Cheng and Vachon [405] developed a model for randomly distributed particles in a continuous phase. The model is

$$\frac{1}{\lambda} = 2\int_0^{\frac{B}{2}}\left[\frac{dx}{\lambda_C + B(\lambda_d - \lambda_C) - Cx^2(\lambda_d - \lambda_C)}\right] + \frac{1-B}{\lambda_C}, \tag{178}$$

where $B = \left(\frac{3\phi}{2}\right)^{1/2}$, $C = 4\left(\frac{2\phi}{3}\right)^{1/2}$, and $\phi = \phi_d$, x is the dimension of dispersed phase along the x-axis; λ_C is the thermal conductivity of the continuous phase; λ_d is the thermal conductivity of dispersed phase; and ϕ_d is the volume fraction of the dispersed phase. This model is valid for porosities in the range $\phi \le 0.667$. Kazatchenko et al. [481] developed unified pore-structure model for thermal conductivity of carbonate rocks. They studied the effect of pore-space microstructure on the physical properties of double-porosity carbonate rocks. They found low sensitivity of thermal conductivity to the secondary pores for various shapes.

Effective medium theory (EMT), unlike the Maxwell-Eucken model (see below Eq. 179), does not have any continues and dispersed phases. According EMT, the effective thermal conductivity of a two-phase system can be estimated using the following equation [546]

$$\lambda = \frac{1}{4}\left[\lambda_f(3\upsilon_f - 1) + \lambda_S(3\upsilon_S - 1) + \sqrt{\left[\lambda_f(3\upsilon_f - 1) + \lambda_S(3\upsilon_S - 1)\right]^2 + 8\lambda_f\lambda_S}\right], \quad (179)$$

where υ_f and υ_S are the volumes contained by pores/fluids and solid matrix, respectively. EMT supposes a heterogeneous material in which the two material components (solid and fluid phases) are distributed randomly, with neither phase necessarily continuous or dispersed [547].

5.2.4. Empirical Models

An alternative approach to characterizing ETC of rocks attempts to find a correlation between conductivity and some of well-known physical properties. The empirical models based on the regression analysis of the measured ETC data as a function of other easily measured physical properties as porosity, permeability, temperature, pressure *etc*. The shortcomings of the empirical models are they may be applicable only to the particular suite of rocks being investigated. For empirical model it is not clear how adjustable fitted parameters are related to the structure of the porous sample. Therefore, this type empirical models only practical interest.

Asaad [166] proposed following expression for the ETC of consolidated rock-fluid system

$$\lambda = \lambda_S\left(\frac{\lambda_f}{\lambda_S}\right)^m, \quad (180)$$

where $m = c\phi$, $c \approx 1.0$ (for unconsolidated rocks). If c=1.0, this equation is identical to the geometrical mean Eq. (49). The values of *c* are close to unity for unconsolidated samples, but a considerably deviation from unity is observed for the consolidated samples. Somerton [72] used this correlation to calculate the ETC of fluid saturated rocks. For consolidate sandstones and limestones the correlation factor *c* is higher.

The parameter *c* is structural characteristics of the rock and related to the Archie cementation factor *m*. This model was successfully modified by Novikov [554] to apply for the prediction of the thermal conductivity of fluid (water, air, kerosene, ethylene glycol, heavy oil, solutions) saturated sandstones.

Anand et al. [170] developed following correlation to calculate the thermal conductivity of dry rocks

$$\lambda_D = 0.34\rho - 3.2\phi + 0.53K^{0.1} + 0.013F - 0.031, \quad (181)$$

where ρ is the bulk density (g·cm^{-3}), K is the absolute permeability (*md*), F is the formation resistivity factor. The authors concluded that porosity is the most important parameter in thermal conductivity of dry rocks while density and permeability have equal effect and formation resistivity factor is the least important. This relation is not included rock-solid conductivity because knowledge of the mineral composition is usually lacking. The deviations between measured and calculated values of ETC are within 10-15 %. The authors also developed correlation equation for thermal conductivity of fluid-saturated rocks

$$\frac{\lambda}{\lambda_D} = 1 + 0.03\left[\left(\frac{\lambda_f}{\lambda_a}\right) - 1\right]^{0.33} + 4.57\left[\frac{\phi}{1-\phi}\frac{\lambda_f}{\lambda_D}\right]^{0.482m}\left(\frac{\rho_{sat}}{\rho_D}\right)^{-4.3}, \tag{182}$$

where λ_a is the thermal conductivity of air, ρ_{sat} and ρ_D are the densities of fluid-saturated and dry rock, respectively, and m is the Archie's cementation factor ($F = 1/\phi^m$). This relation is predicting thermal conductivity fluid-saturated rocks within 10-15 %. The parameter m in Eq. (182) relating to the porosity.

On the basis of experimental data analysis Somerton et al. [2] concluded that porosity, water saturation, and conductivities of solid and liquid phases are sufficient to develop the correlation equation for thermal conductivity of unconsolidated sands. They included porosity and water saturation as variables into correlation equation

$$\lambda = 0.735 - 1.3\phi + S_W^{1/2}, \tag{183}$$

where S_W is the water saturation. The authors also developed equation for thermal conductivity of sands for which the solid conductivity λ_S is known

$$\lambda = 0.735 - 1.3\phi + 0.363\lambda_S S_W^{1/2}. \tag{184}$$

This equation is excellent reproduced the thermal conductivity of oil-saturated sands. However, this model is not applicable for air-saturated and brine –saturated sands. Practical important range of porosity is 0 to 0.4.

Wyble [408] proposed empirical equation to describe the effect of pressure on the properties (conductivity, permittivity, porosity) of rocks

$$\frac{W_P}{W_0} = B + (1-B)10^{-\alpha P}, \tag{185}$$

where W_P is the value of the property under consideration at pressure P; W_0 is the value of the property at zero pressure; B is the fraction to which the curve is asymptotic; and α is the coefficient of exponential decreases. Thikhomirov [170] developed a correlation equation to predict the temperature dependence of ETC of dry and fluid saturated rocks. The proposed correlation is

$$\lambda_{293} = 6.24\exp(-1.63S_W), \tag{186}$$

where S_w is the water content. This correlation describes the measured values of ETC water-saturated rocks within 6-12 %. ETC of the water-saturated rock can be expressed by using ETC of a dry sample as

$$\lg \lambda_{S293} = \lg \lambda_{293} + 1.18 S_W \lg \lambda_{W293}, \tag{187}$$

where λ_{293} is the ETC of dry rock at the same temperature, λ_{W293} is the thermal conductivity of pure water at the same temperature. ETC of water-saturated rocks at any temperature can be predicted by relation

$$\lambda = T^{-0.55} 26.31 \exp(0.61\sigma + 0.6 S_W). \tag{188}$$

This correlation agrees with measured values of ETC water-saturated rocks within 16 %. Another relation was proposed by Tikhomirov [170] by using ETC value for the rock at 20 °C

$$\lambda = 0.047 \lambda_{20}^{4.98} T^{(0.17 \log T - 1.61\lambda_{20} + 0.12)}, \tag{189}$$

where λ in cal·s^{-1}·cm^{-1}·C^{-1}; λ_{20} is the thermal conductivity at 20 °C; T is the temperature in K. This relation predicts the measured values of ETC with maximum deviation within 5 % to 8.5 %. Although Tikhomirov [170] suggested using of the above equation for dry rocks only, it was also applied by Anand [171] to results for the brine saturated samples. The maximum differences between measured and predicted values of ETC are within 3 %. This relation was applied also for samples saturated with organic liquids. In this case the agreement is poor. He is developed also regression analysis equations for relation between ETC and saturation as

$$\lambda_{20} = 6.24 \exp(-1.63 w) \text{ and } \lambda_{20} = 1.3 \exp(0.58\rho + 0.4 w), \tag{190}$$

where λ_{20} is the ETC of saturated rock at 20 °C; w is the fractional saturation by weight; ρ is the rock density. Anand et al. [171] showed that Tikhomirov's equation does not fit most of experimental data satisfactory. They provide modified expression of the same general form as Tikhmirov's [170] equation

$$\lambda_T = \lambda_{68^0} - 0.71(T/1000 - 0.528)(\lambda_{68^0} - 0.80)\left[\lambda_{68^0}(T/1000)^{-0.55\lambda_{68^0}} + 0.74\right], \tag{191}$$

where λ_T is the thermal conductivity at temperature T; λ_{68^0} is the thermal conductivity at temperature 68 °F. This relation was applied for fluid-saturated rocks.

Sugawara and Yoshizawa [224] proposed following empirical generalized equation for the fluid saturated porous materials

$$\lambda = (1 - A)\lambda_S + A\lambda_f, \tag{192}$$

where $A=\frac{2^n}{2^n-1}\left[1-\frac{1}{(1+\phi)^n}\right]$ some factor accounting for the pore geometry; n>0 ($\approx$0.5) is the empirical exponent determined by mode of packing, pore size, pore shape, and emissivity inside of the pore.

The parameter A becomes equal to porosity if the pores are considered cylindrical and parallel with direction of heat flow. For more complicated structure A depends on porosity and cementation as well. This equation was applied for porous rocks (sandstones with different porosities) as a typical example of porous material. The value of n for sandstones is 2.5. Zierfuss [315] used this relation to describe ETC of some carbonate rocks and clay sandstones.

The model proposed by Woodside and Messmer [82] is

$$\lambda=\alpha\left[\frac{\lambda_f\lambda_S}{\lambda_S(1-\gamma)+\lambda_f\gamma}\right]+\beta\lambda_S+\delta\lambda_f, \tag{193}$$

where $\beta+\alpha+\gamma=1$ and $\beta+\alpha\gamma=(1-\phi)$. Here α, β, and γ are parameters for cubic formation and $\delta=\frac{1}{F}=\phi-0.03$. Chaudhary and Bhandari [411] took a geometric mean of $\lambda_{\max}$ and λ_{min} given by Eqs. (50) and (51) as

$$\lambda=\lambda_{\max}^n\lambda_{\min}^{1-n},$$

where weighting factor

$$n=\frac{k(1-\log\phi)}{\log\left[\phi(1-\phi)\left(\frac{\lambda_S}{\lambda_f}\right)\right]} \tag{194}$$

In Eq. (194) k is found by the best fit method applied to the experimental data. The authors also found that for rocks with porosity within 19.5 to 32.6 % the values of n lies between 0.8 and 0.9. The extreme values of n (n=1 and 0) are not obtained since in both cases the relation Eq. (178) diverges. This is different kind of weighted geometric mean model. Veerendra and Chaudhary [562] considered different cases for the relation between the fluid λ_f and matrix λ_S thermal conductivities. In the case $\lambda_f>\lambda_S$, the lower and higher values of thermal conductivity are: $\lambda_L=\lambda_S\exp(\alpha\phi)$ and $\lambda_H=\lambda_f\exp[-\alpha(1-\phi)]$, where $\alpha=(1-\lambda_S/\lambda_f)$. The effective thermal conductivity for two-phase mixture with ϕ between 0.2 and 0.8 is $\lambda=(1-\phi)\lambda_L+\phi\lambda_H$. In the case $\lambda_S>\lambda_f$, $\lambda_H=\lambda_S\exp(\beta\phi)$ and $\lambda_L=\lambda_f\exp[-\beta(1-\phi)]$, where $\beta=(\lambda_f/\lambda_S-1)$, and the effective thermal conductivity is $\lambda=(1-\phi)\lambda_H+\phi\lambda_L$. For high ratio λ_S/λ_f proposed a corrective term for the effective thermal conductivity $\lambda=(1-\phi)\lambda_H+\phi\lambda_L\pm\phi(1-\phi)\sqrt{\lambda_L\lambda_H}$.

Jamalova [301] suggested that the relation between ETC and saturation is linear for porosities less than 50 %. She also found semi-empirical relation for ETC

$$\lambda = \lambda_W \frac{3 - 2v(1-\sigma)}{3\sigma + v(1-\sigma)}, \tag{195}$$

where $\sigma = \lambda_W / \lambda_S$, $v = w/[w-(1-w)\rho_W/\rho_S]$is the quantity which defines the weight content of water, w is the fractional saturation on weight bases, ρ_W and ρ_S are the density of water and solid matrix, respectively. Bogomolov [412] proposed the model for thermal conductivity which taking into account that solid spheres are packed into a tetrahedral packing. The model is

$$\lambda = 3\pi\lambda_f \ln\left(\frac{0.43 + 0.31\phi}{\phi - 0.26}\right). \tag{196}$$

Kaufman [413] proposed the following equation for ETC

$$\lambda = \lambda_{dry}\left[1 + \frac{wf(w)}{100}\right], \tag{197}$$

where $f(w) = 1.15 - 6.05\rho + 14.3w$ (for inorganic materials of complex structure, ρ is the density of rock, and w is the fractional saturation on weight basis) percentage increase of ETC of the material for every percent increase of saturation. Zierfuss and Viliet [227] include the electrical formation resistivity factor F to the correlation equation for ETC as

$$\log F\lambda = A + B\phi + C\phi^2 + D\phi^3, \tag{198}$$

where A, B, C, and D are the empirical constants which depend on the nature of the saturating fluid and rock solids. Litovsky and Shapiro [414] reviewed experimental and theoretical modeling results for thermal conductivity of ceramic and refractory materials (Al_2O_3, SiO_2, MgO, Cr_2O_3, SiC, Y_2O_3, and their composites) with porosity less than 30 %. Keese [223] proposed empirical relations to calculate ETC of rock-saturated samples as a function of fractional porosity (ϕ), air permeability (K), thermal conductivity of saturated fluid (λ_f), thermal conductivity of solid phase (λ_S), formation resistivity factor (F), grain size distribution parameter (D_{90}/D_{10}), and median grain diameter (D_{50}). For example,

$$\lambda = 0.00243 - 0.00558\phi + 0.42\times10^{-5}K + 4.293\lambda_f + 0.773\times10^{-4}F \tag{199}$$

for quartz sands and

$$\lambda = 0.0034 - 0.00672\phi + 0.107\times10^{-3}K + 2.233\lambda_f - 0.00354D_{50} \tag{200}$$

for extracted oil sand.

These relations reproduced experimental ETC data within 7.9 to 20.6 %. Eucken [368] first modified Maxwell's equation (Maxwell-Eucken model) and then applied the analogy

$$\lambda = \lambda_S \frac{1+2\phi\varepsilon}{1-\phi\varepsilon},\ \varepsilon = \frac{1-Q}{2Q+1},\ Q = \lambda_S / \lambda_f . \tag{201}$$

This equation was derived for isometric pores (spherical or cubical) with equal dimensions in all three axes. At $(\lambda_S / \lambda_f) \to \infty$ Eucken's model reduced to Buntebarth and Schopper [302]. Burger [396] extended Maxwell's work to ellipsoidal as well as spherical particles. Dachanov and Djakanov [415] gave the following empirical relation for prediction of ETC rocks from density only

$$\lambda = \frac{\rho^{3.1}}{4680}, \tag{202}$$

where ρ is the density of the sandstone in g·cm^{-3}, and λ is the ETC in cal·s^{-1}·cm^{-1}·C^{-1}. The ETC predicted by this equation very low compared to the actual values. Anand [171] found another relation like this which based on the literature measurements

$$\lambda = \frac{\rho^{3.42}}{2910}. \tag{203}$$

The model developed by Budiansky [416] is

$$\frac{(1-\phi)}{2+\lambda_S/\lambda} + \frac{\phi}{2+\lambda_f/\lambda} = \frac{1}{3}. \tag{204}$$

Robertson and Peck [417] model is

$$\lambda^{1/2} = (1-\phi)\lambda_S^{1/2} + \phi\lambda_f^{1/2}. \tag{205}$$

This model is not useful for liquid saturated rocks.

Hashin and Shtrikman theory [363] predict upper and lower bounds of the ETC as

$$\lambda_{\min} = \lambda_f + \frac{1-\phi}{[1/(\lambda_S-\lambda_f)]+\phi/3\lambda_f} \text{ and } \lambda_{\max} = \lambda_S + \frac{\phi}{[1/(\lambda_f-\lambda_S)]+(1-\phi)/3\lambda_S}. \tag{206}$$

ETC of rocks should generally fall in between these bounds. Some researchers (for example [102]) prefer to use the average of the upper and lower bounds or the square-root average

$$\sqrt{\lambda} = \phi\sqrt{\lambda_f} + (1-\phi)\sqrt{\lambda_S}. \tag{207}$$

Popov et al. [203] also reported the results of the study of relationship between ETC of rocks and other petrophysical properties. Pribnow and Umsonst [192] and Pribnow et al. [193] developed a layered model to predict the thermal conductivity of rocks from the mineral content. The estimated values of the thermal conductivity are strongly controlled by the quartz, amorphous, and or crystalline content. ETC of parallel and perpendicular to the layers provides the upper λ_U and lower λ_L limits and can be calculated by the arithmetic and the harmonic mean $\lambda_U = \sum_{i=1}^{n}(\phi_i \lambda_i)$ and $\lambda_L = \sum_{i=1}^{n}\left(\frac{\phi_i}{\lambda_i}\right)$, where $\sum_{i=1}^{n}\phi_i = 1$, λ_i is the thermal conductivity of the i-th considered mineral. The geometric mean of λ_U and λ_L represent a statistical spatial distribution of parallel and series arrangement $\lambda_G = \sqrt{\lambda_U \lambda_L}$.

Decker [421] proposed an empirical relationships between ETC and porosity and bulk density in three major families of sedimentary rocks. The relationships were derived from detailed characterization of the mineralogy, petrography, and petrophysical properties of 80 carefully selected rock samples. The model is valid for porosities between 1% and 40 %. The relation is of the form

$$\ln\lambda = \ln\lambda_m \,\text{int ercept} + \text{slope}\phi \text{ and } ln\,\lambda = ln\,\lambda_m + \rho, \tag{208}$$

where λ_m is the thermal conductivity of the constituent minerals.

Adler [423] proposed equation for ETC of two component (rock-fluid) system for intermediate porosities

$$\frac{\lambda}{\lambda_S} = \frac{(2\sigma - 1) - 3\phi(\sigma - 1) + \{[(2\sigma - 1) - 3\phi(\sigma - 1)]^2 + 8\sigma\}^{1/2}}{4\sigma}, \tag{209}$$

where $\sigma = \lambda_S / \lambda_f$. At porosities above 0.7, the differences between measured and predicted with Eq. (209) are large. Emirov et al. [274,275,277, and 278] developed empirical relation to calculate the combined effect of pressure and temperature on the ETC of fluid-saturated rocks. The relation is

$$\lambda(P,T) = CT^{n(P)}, \tag{210}$$

where exponent $n(P)$ is the function of pressure. The values of exponent n for each measured isobar were determined using the fitting procedure. For gas saturated sandstones the values of n varied from 0.057 to 0.428, while for water-saturated sandstones n is varied from 0.171 to 0.294 in the pressure range from 0.1 to 250 MPa. The exponent n is monotonically decreasing with pressure.

5.2.5. Well-Log Correlation Models

Correlation of thermal conductivity with physical properties obtained from geophysical well logging data was previously discussed in the works [496-508]. In several studies [330,406,418-420] the method of calculation of ETC from wire-line logs were reported. According to this method one or more logging measurements or some derived properties used directly to calculate of ETC via empirical relationship. This method reviewed in [418]. Other way to calculate ETC from logging measurements is the major mineral or rock components are identified and the volumetric fractions of these components are derived from regular wireline data. This composition together with component thermal conductivities is then used to calculate the ETC assuming an appropriate mixing law (see for example [54, 246, and 360]).

There are three ways in which well-logs can be used to infer estimates for in –situ ETC: (1) Empirical relationships between the ETC and parameters derived from well logs, such as porosity, bulk density, sonic (*p*-wave) velocity, and seismic travel times. This approach is not limited to well logs, if petrophysical parameters are known from laboratory measurements. A useful discussion of different approaches provided by Blackwell [248] (see also [509-511]); (2) extension of the mixing –model approach to the borehole scale: the volume fractions of the different mineral (or fluid) phases are either taken directly from induced γ -ray spectroscopy logs (Williams and Anderson [424]) or determined from a joint analysis of other logs such as γ -ray, sonic traveling, γ -density, and neutron porosity (Demongodin et al. [364]). Then an appropriate mixing model is applied; (3) Williams and Anderson [424] derived a phonon conduction model for ETC which utilizes temperature, acoustic velocity, and bulk density measurements from well-logs. The accuracy of the model claimed by the authors is 15 %. The predicting ETC from well log data was reported by Goss et al. [406]. They measured ETC and several other properties for a suite of Imperial Valley rock samples. Then developed an empirical correlation from the data and applied this relationship to well-log data for the borehole section from which the core samples had been taken. The correlation equation was expressed as

$$\lambda = 1.33 - 0.026\phi - 0.38V_P, \tag{211}$$

where V_P is the compressional (*P*) wave velocity. The predicting capability of the correlation is within 20-25 %. Popov et al. [203] found close correlation between ETC and porosity. About 900 sedimentary rock samples from different oil-gas fields were studied to establish the correlation between parameters $\lambda_{sat} / \lambda_{dry}$ and rock porosity and permeability. They measured ETC for 272 dry and water-saturated cores from four zones in the depth intervals of 1654-1763, 2067-2116; 2191-2219; and 2440-2457 *m*. A close correlation was found between the ratio $\lambda_{sat} / \lambda_{dry}$ and with gamma-gamma density log and with $\lambda_{sat} / \lambda_{dry}$ and gamma ray. The found correlations allowed predicting the ETC of rocks in situ conditions. Dea [425] proposed following correlation equation to estimate ETC from wellbore-log data

$$\lambda = 1.27 - 2.25\phi + 0.39\lambda_S S_W^{1/2}, \tag{212}$$

where S_W is the saturation of wetting –phase fluid. If the quartz content can be estimated, the approximate values of the thermal conductivity of sandstones can be obtained from following relation

$$\lambda_S = 7.7Q + 2.85(1 - Q), \tag{213}$$

where Q is the fraction quartz content.

Goss et al. [405] and Houbolt and Wells [407] demonstrated a strong correlation between ETC and sonic velocity. Vacquier et al. [330] obtained correlations between ETC and a set of geophysical log parameters (gamma ray, neutron porosity index, sonic travel time, and bulk density) calibrated by rock samples from gas wells in the Paris basin

$$\lambda = a + b\rho_b + c\Delta t + d(1-\phi_N)^2 + eV_{SH}, \tag{214}$$

where logging parameters are ρ_b bulk density (g/cm^3), Δt interval travel time (s/m), $(1-\phi_N)^2$ is the solidity and V_{SH} is the "volume of shale" (from GR measurements), *a*, *b*, *c*, and *d* are calculated by multiple regression from experimental data. Most of these correlations produce adequate ETC predictions for individual lithologies. It is important to develop further some global correlations. Brigaud et al. [124] described a procedure for predicting in-situ ETC profiles in wells by using lithologic information, geophysical well logs, and temperature data. In-situ ETC is expressed as a function of lithology, porosity, and temperature. This procedure was applied for three wells in the Uinta basin of northeastern Utah. 565 ETC data and lithology descriptions of the corresponding cuttings, along with sonic and neutron logs were available. The measurements were used to test the validity of the ETC prediction.

Hartman et al. [256] derived the relationship between ETC of rocks and other petrophysical properties. The derived equation for ETC can be used to predict rock ETC from logging data. A regression analysis of ETC, bulk density, and sonic velocity yield ETC with an average accuracy of better than 0.2 Wm^{-1}K^{-1}. The derived equation has the form

$$\lambda = a_0 + a_1V_P + a_2\rho + a_3\phi, \tag{215}$$

where V_P and ρ are the sonic velocity and density, respectively. The experimental bulk densities (measured by γ-ray absorption due to Compton scattering) for saturated and dry rocks were described with equations $\rho_{bS} = 161.5\mu_S$ and $\rho_{bD} = 162.1\mu_D$ where μ_S and μ_D are the absorption coefficient for saturated and dry rocks, respectively. Sonic velocity measurements were fitted to two exponential equations as a function of pressure

$$V(P) = 2360 + 787\exp(0.356P) + 35.7P \text{ and } V(P) = 2760 + 1180\exp(0.243P). \tag{216}$$

Nosal [422] derived and tested a relationship between ETC of sedimentary rocks and parameters usually available from geophysical logs. Data for 52 samples of three major families of sedimentary rocks are used: quartz sandstone; limestones, and dolomites. The proposed model is

$$\lambda = F\lambda_S + (1-F)\lambda_S^{1-g}\lambda_f^g, \tag{217}$$

where $g = \phi/(1-F)$, $F = \lambda_D/\lambda_S$, λ_D is the thermal conductivity of dry rock. The exponent *g* is related to the tortuosity $\phi/(1-F)$, which expresses the increase in heat-flow path length due to the distribution of porosity. This model was applied to water and oil saturated rock samples.

Wen-Long Cheng et al. [565] developed the method of estimation of geological formation thermal conductivity by using approximation based on well-log temperature data, which can be easily obtained through conventional well log. As well known, most thermal conductivity values of samples were obtained under room temperature and atmospheric conditions [566 and 567], which were quite different from the in situ conditions. Seto and Bharatha [567] and Somerton et al. [2] tried to minimize the measurement deviation by improving the simulated conditions, whereas the in situ conditions still could not be realized easily in laboratory. An alternative solution to the sampling difficulties in laboratory is to measure the thermal conductivity by means of in situ techniques, including methods related to direct measurements or well-log correlations [565]. In the works [214 and 568] the direct thermal conductivity measurement techniques have been developed. It is highly desirable to estimate the thermal conductivity from logging data such as density logging, sonic logging, electric resistivity logging. Another technique for estimating thermal conductivity is the temperature –gradient logs methods [566 and 567], which is based on the Fourier law of heat conduction. According the method proposed by Cheng et al. [565], the wellbore temperature is a function of thermal conductivity λ_e, heat flow rate $\frac{dQ}{dz}$, and the ratio of formation heat capacity and wellbore heat capacity, ω [565]

$$T_W = T_{ei} + \frac{1}{2\pi r_{to} U_{to}} \frac{dQ}{dz} + \frac{8\omega^2}{2\pi^3 \lambda_e} \frac{dQ}{dz} \int_0^\infty \frac{1-\exp\left(-\frac{\alpha_e \tau}{r_h^2} u^2\right)}{u^3 \Delta(u,\omega)} du. \qquad (218)$$

The thermal conductivity λ_e was calculated by solving this equation [565]. The relative uncertainty in thermal conductivity calculation was within 0.6 %.

Chan and Jeffrey [221] measured ETC of granite in the situ thermal conditions at the deep of 340 m. The results were compared with laboratory experiments. The difference is within 1 to 5 %. The results of measurements were analyzed by using the two models (Ohm's law model for group of thermal resistance in series element arrangement and parallel element arrangement of the thermal resistance).

There are many other empirical correlation equations concerning the relationship between ETC and porosity, for example

$$\lambda = \lambda_a \phi^{1/3} + \lambda_S \left(1-\phi^{2/3}\right) \text{ by Ribaud [409],} \qquad (219)$$

$$\lambda = \lambda_S \frac{\phi^{2/3} + Q\left(1-\phi^{2/3}\right)}{\phi^{2/3} - \phi + Q\left(1-\phi^{2/3}+\phi\right)}, \text{ by Russell [410],} \qquad (220)$$

where $Q = \lambda_S / \lambda_a$.Three different methods (numerical, predictive analytical, and the bounds obtained by variation methods) for calculation of the ETC of composite porous materials were developed by Staicu et al. [51]. The numerical calculations used the actual microstructure of the porous media. Rzhevsky and Novik [426] model gave too high estimates at low thermal conductivity.

Conclusions: Unfortunately, most developed models have their disadvantages: some overestimate while others underestimate systematically the true bulk thermal conductivity. Most of them are valid only for a specific range of volume ratios (porosities), and yield completely unreasonable results outside this range. Parallel and series models are easy to understand, but have the disadvantage of being rather special case, applicable mostly to bedded sediments. They lead to the well known arithmetic and harmonic means, respectively, and define upper and lower limits for all other models. The models where pores assumed as spherooidal inclusions in a homogeneous and isotropic material are capable of explaining the complete experimental ETC data. However, additional information on the spheroid's aspect ratio or orientation, respectively, is required by these models. Most predictive models are work to within 10-15 % accuracy (for typical conductivity ratios, i.e., $\lambda_S / \lambda_f < 10$). For large ratios some break down more than others, and the geometrical mean is one of them.

6. Theoretical Backgrounds of the Temperature Dependency of ETC of Solids at Constant *P*. Application for Dry and Fluid-Saturated Sandstones

Above (see sect. 5.2) we detailed discussed porosity models for ETC of rocks and their accuracy and predictive capability. In this section we will discuss of the temperature dependence models for ETC rocks. Studies of the temperature and pressure dependence of the ETC of rocks showed that temperature is dominating. At low temperatures when heat transfer occurs by pore conduction, the radiation is insignificant. The changes of composition and structure of the minerals and possibly the phase state of the fluid are very important. The heat trapped in the rocks would cause the thermal expansion of the pore fluid and increase the fluid pressure.

Thermal-expansion of rocks is relatively small in magnitude; however, thermal-expansion behavior may have significant effects on the structure of rocks. Because of the multi-mineral composition of rocks, heating causes micro-fracturing due the differential thermal expansion of mineral grains. Differences in thermal-expansion characteristics of various minerals in the assemblage of mineral grains can cause structural damage upon heating the rock.

Even a given mineral the coefficient of thermal expansion may be different in different crystallographic directions. These differences in thermal expansion results in stress concentrations at grain contact points, when a rock heated, leading to the possibility of fracture of individual mineral grains and disaggregating of the rock. Therefore, the effects of heating rocks are altering their properties and temperature dependence of the ETC of rocks is complicated.

At normal temperatures for uniform crystals thermal conductivity has been found to be inversely proportional to absolute temperatures ($\lambda \approx T^{-1}$). The experimental results for ETC as a function of temperature were fitted to linear-Eucken law [439] (see also [115, 122, and 205])

$$\lambda^{-1} = A + BT, \tag{221}$$

where *A* (the thermal resistance at zero *T*=0) is related to the scattering of phonons by impurities and imperfections, *B* (rate of increase in the thermal resistance) is related to phonon-phonon scattering and approximately proportional to an inverse power of sound velocity (Ziman [74]) and caused by the increase of atomic oscillations. Measurements of the temperature dependence of ETC on water-saturated rocks produce a better approximation to the intrinsic behavior of the rock than does making these measurements on dry samples. The knowledge of this behavior is crucial for the conversion of laboratory measurements to in-situ conditions where thermally-induced cracks are likely to be nonexistent or filled with water. For ETC, the influence of thermally-induced pore space is small but significant and becomes relevant for temperature estimates to large depths.

For amorphous materials thermal conductivity is found to increase with increasing temperatures. Eucken [368] proposed the following equation for thermal conductivity of mixtures of amorphous materials and mixed crystals

$$\lambda^{-1} = AT + B + \frac{C}{T}, \tag{222}$$

where *A*, *B*, *C* are constants which depend upon the relative influence of the crystals or amorphous substances. The relation Eq. (222) was successfully used by Dos Santos [556] to describe the experimental thermal conductivity of porous ceramic materials (mixture of crystalline and glassy phases).

Thermal conductivity can be divided into lattice conductivity and radiative conductivity. Lattice conductivity (or phonon conductivity) is produced by the diffusion of thermal vibration in a crystalline lattice, while radiative conductivity is produced by infrared electromagnetic waves.

At temperatures below 500 °C radiative conductivity is negligible in comparison to lattice conductivity. Theoretically, lattice conductivity tends to be inversely proportional to temperature ($\lambda_{lat} \propto T^{-1}$ for three-phonon mode), while radiative conductivity tends to be proportional to the cube of the temperature ($\lambda_{rad} \propto T^3$) (Clauser [440]). Some authors suggested that variation of the lattice conductivity with temperature is $\lambda_{lat} \propto T^{-5/4}$, for four-phonon mode approach [441, 179, and 442]. The relationship $\lambda_{lat} \propto T^{-1}$ is valid only for structurally perfect isotropic single crystals. However, many rocks are composed of mixtures of highly disordered crystals of different compositions, therefore, ETC of rocks tends to decrease more slowly than $\lambda_{lat} \propto T^{-1}$ and may tend to actually increase, in some case, with increasing temperature (Somerton [72]). For example, the thermal conductivity of feldspar aggregates increase with temperature (Birch and Clark [79 and 164]), and the thermal conductivity of glasses and vitreous materials also increase with increasing the temperature [72, 251, and 279].

The variation $\lambda_{lat} \propto T^{-1}$ results from a first-order perturbation treatment, which is valid for the weak interaction. With temperature increasing the phonon mean free path approaching to lower limit l_0, which can be comparable to the linear dimensions of the unit cell of the crystal lattice (Ziman [74], Spitzer [75]). As was mentioned above, since the mineral components of a rock are imperfect, the phonon free path is reduced which results in a high thermal resistivity than that a perfect crystals. There are additional structural effects which reduce the phonon free path, i.e., grain boundaries, microcracks and the volume porosity as well as the shape and orientation of crystals and their fragments. All these structural effects are assumed to be independent of T for a first approximation and they cause a constant thermal resistivity.

As was shown by Leibfried and Schlömann [448], Lee and Kingery [449], and Clark [220] that at high temperatures (above the Debye temperature, $T > \Theta$) the thermal conductive (lattice or phonon conductivity) of many electrically nonconducting (dielectrics) solids, including the majority of geophysical interesting materials, is given by Eq. (221). At high temperatures the phonon mean free path reaches a lower limit equal to the lattice dimensions (Ziman [74], Spitzer [75]). When lower limit is reached, the temperature dependence of the λ disappears and Eq. (221) can be replaced by

$$\lambda_{\lim} = kV\left(\frac{\rho}{\langle M\rangle}\right)^{2/3}, \tag{223}$$

where k is the Boltzmann constant, V is the sound velocity, ρ is the density, and $\langle M\rangle$ is the mean atomic weight (Spitzer [75]). A theory for the lattice (phonon) conductivity gives following relation for the lattice conductivity

$$\lambda_{lat} = \frac{a}{3\gamma_{Th}^2}\frac{1}{T}K_T^{3/2}\rho^{1/2}, \tag{224}$$

where a is the lattice parameter, ρ is the density, K_T is the isothermal bulk modulus, and γ_{Th} is the Grüneisen parameter. This relation is valid at high temperatures ($T/\Theta >> 1$). The lattice conductivity decrease approximately linearly $\lambda_{lat} \sim T^{1}$, but increases rapidly with density ($\lambda_{lat} \sim \rho^{1/2}$). At high temperatures λ_{rad} increase very rapidly with temperature and can be dominate λ_{lat}. Roufosse and Klemens [450] reformulated of th7e theory of thermal resistivity of dielectric crystals at ordinary and high temperatures in terms of anharmonic three-phonon interactions. They found that thermal conductivity varies inversely with temperature (see also [571])

$$\lambda = \frac{3}{4\pi^3\sqrt{2}}\frac{Mv^2}{\gamma_{Th}^2}\frac{\omega_D}{a}\frac{1}{T} \text{ or } \lambda = \frac{3^{1/3}}{2^{1/6}\pi^{4/3}\gamma_{Th}^2}\frac{K^3}{(2\pi)^3\hbar^3}Ma\Theta^2\frac{\Theta}{T}, \tag{225}$$

even if the phonon mean free path is shorter than the cell dimensions, because the major contribution to the anharmonic interaction comes from the highest harmonic of the fundamental reciprocal-lattice vectors. In Eq. (223) ω_D is the Debye frequency, M is the mass of a unit cell, and a is the dimension of the unit cell (a^3 is the volume of the unit cell). The same relation was reported by Leibfried and Schlömann [448], for the thermal conductivity. But the thermal conductivity given by Roufosse and Klemens [450] and Leibfried and Schlömann [448] is in the ration 1.0 to 6.8. Roufosse and Klemens [76] studied the lattice thermal conductivity of minerals at high temperatures. They showed that lattice thermal conductivity of minerals should not depart markedly from T^{-1} law even when the average phonon mean free path is comparable to lattice spacing.

As was shown by Lee and Deming [443], usually the rocks with thermal conductivity less than 2 $W{\cdot}m^{-1}{\cdot}K^{-1}$ at room temperature tend to increase with T, while that thermal conductivity higher than about 2 $W{\cdot}m^{-1}{\cdot}K^{-1}$ tend to decrease with temperature. Some have shown an increase of ETC with increasing temperature [171,224,444, and 445]. Anand [171] results of ETC measurements for six sandstones showed that in general, ETC decreases linearly with temperature increasing. The decrease is much more pronounced for liquid-saturated samples than for dry samples. Most of the recent publications, however, show decrease in thermal conductivity values with temperature increasing. All of the measurements by Somerton [230] and Khan [228], for example, showed a decrease in ETC with temperature increasing. Pribnow et al. [100] also found that for marine sedimentary rocks the ETC increases with temperature. This behavior is due to positive temperature coefficient for seawater (λ increases with T up to about 420 K, see below Figure 33) and a negative coefficient for the rock matrix (λ_S decreases with T). Experimental results by Roy et al. [102] show that temperature has a significant effect on ETC of rocks matrix components in the range of 0-200 °C. The authors found that for rocks with a large thermal conductivity at room temperature the decrease in thermal conductivity with temperature is more pronounced than for rocks with a low thermal conductivity at room temperature. Similar trends for thermal diffusivity of sandstone and limestone was found by several authors [249,409, and 446,]. Birch and Clark [164] reported appreciable variation in the ETC of igneous and sedimentary rocks between the temperatures 273 K and 673 K.

6.1. Empirical and Semiempirical Models

The temperature dependence for most porous rocks in the upper part of the crust at temperatures up to 600 ˚C can be expressed by the simple relation (Čermak and Rybach [11]) (see above Eq. 221). This relation has been used by Seipold [198] to represent of the thermal conductivity results for granites and amphibolites. Buntebarth [186 and 447] studied the effect of temperature on the ETC of rocks samples. Buntebarth [186] determined A and B from measurements on 113 samples of metamorphic rocks from KTB borehole in the temperature range from 50 ˚C to 200 ˚C. The various widely used empirical and semiempirical predictive models for the temperature dependence of ETC for porous rocks are presented below. Kukkonen et al. [190] analyzed various correlation equations for temperature and pressure dependence of ETC of the porous rocks. Several types of functions have been used by Seipold and Huenges [201], Seipold [162], Clauser and Huenges [348] for the fitting of measured ETC data for porous rocks. The most common used functions is

$$\lambda(T) = A + B/(350 + T). \tag{226}$$

Different approaches to calculate $\lambda(T)$ were developed by Zoth and Hänel [451]. They derived seven different parameters *A* and *B* for various porous rock types. Seipold [162 and 163] used different expressions for the temperature dependence of thermal conductivity

$$\lambda(T) = A + 1/B(T - 532), \tag{227}$$

where *B* is the coefficient which depends on rock type and *T*.

At higher temperatures Seipold [162 and 163] included an additional cubic term CT^3 to take into account the contributions of electromagnetic radiation. Vosteen and Schellschmidt [111] measured a set of magmatic, metamorphic, and sedimentary rocks samples from different depth levels of the Eastern Alpine crust to obtain an estimate of the temperature dependence of ETC. They found that temperature dependence of ETC of the crystalline (magmatic and metamorphic) rocks is different to that of sedimentary rocks. Vosteen and Schellschmidt [111] detailed compared the accuracy of the various models for temperature dependence of the thermal conductivity of porous rocks. They studied the effect of temperature on thermal conductivity for different type of rocks (crystalline and sedimentary). They also proposed two general equations for thermal conductivity of rocks. A general equation for the temperature dependence of λ for Eastern Alpine rocks can be written as

$$\lambda(T) = \frac{\lambda(0)}{0.99 + T(a - b/\lambda(0))}, \tag{228}$$

where $\lambda(0)$ is the thermal conductivity of rock at 0°C. For sedimentary rocks $\lambda(0)$ is

$$\lambda(0) = 0.54\lambda(25) + 0.5\sqrt{1.16[\lambda(25)]^2 - 0.39\lambda(25)},\ a=0.0034,\ b=0.0039. \tag{229}$$

The equation for crystalline rocks is valid in the temperature range from 0°C to 500°C, while for sedimentary rocks from 0°C to 300°C. Vosteen and Schellschmidt [111] found that for rocks with a large λ_{amb} at ambient conditions the decrease in thermal conductivity with *T* is more pronounced than for rocks with a low λ_{amb}. Equations (228) and (229) yield nearly the same results as the equations of Zoth and Haenel [451]. The relation has been widely used by various authors (see, for example [111,256,348,569, and 570]).

Abid [526] based on the measured thermal conductivity data proposed empirical relation for the effective thermal conductivity of fluid saturated sandstone

$$\lambda(T) = \frac{(1.08\lambda_{25} - 0.15)^2}{\lambda_{25}(1.09 + 0.0033T) - 0.009T - 0.149} - 0.0048T + 0.065, \tag{230}$$

where λ_{25} is the thermal conductivity of sandstone at room temperature (25 °C). In the case of ice saturation λ_{25} should be replaced into λ_0 (at zero temperature). This relation predicts the measured values of ETC fluid saturated sandstone within 3.4 % to 9.0 %.

Sass et al. [122] model was used by Hartmann et al. [256] for fluid- saturated rocks. The prediction accuracy of the equation by Sass et al. [122] is within 5-7 %. The models proposed by Chapman et al. [452] and Sass et al. [122] have less uncertainty when applied to rocks with relative high ETC (> 3 W·m^{-1}·K^{-1} at 20-25 °C), while the model proposed by Sekiguchi [80] and Somerton [72] tend to have relatively low uncertainty over entire range of rocks ETC (1.4 to 8.0 W·m^{-1}·K^{-1} at 20-25 °C). The model by Funnell et al. [453] is good for the rocks with low ETC (<3 W·m^{-1}·K^{-1} at 20-25 °C). This model tends to systematically overestimate in situ thermal conductivity, while the temperature model by Chapman et al. [452] tends to systematically underestimate in situ thermal conductivity.

The temperature behavior of ETC depends also on rock-type. Because of, as was mentioned above, the multi-mineral composition of rocks, heating causes micro-fracturing due the differential thermal expansion of mineral grains. As thermal expansions increase with temperature, but differently for all minerals, "thermal cracking" by differential expansion may create contact resistances between mineral grains, thus contributing to the observed decrease of conductivity with temperature. Closed cracks act as scattering centers for the heat carrying phonons (Seipold [162], Seipold and Huenges [201]) (interrupting of the crystal lattice scatters the phonons). This effect is probably not as severe in fluid-saturated rocks as it is in dry rocks.

Rocks are very complicated and consist of oxides and minerals with various chemical elements. Therefore, the temperature dependences of ETC of rocks strongly depend on the conductivity of its skeleton solid mineral components (rock-forming minerals and cementation substance) from which it is formed. For example, the decrease of ETC with T is quite different, depending on the feldspar content. Rocks with high feldspar content show an increase in ETC with temperature of some plagioclase feldspars which compensate the decrease in ETC with increase in temperature observed for other minerals and rocks. The rocks that are rich in feldspar show a decrease in ETC with T about 10 %, while for rocks that are poor in feldspar decrease more than 40 % (Clauser and Huenges [348]). ETC of some rocks with high feldspar content is relative independent of temperature or even increase with temperature (Čermak and Rybach [11]). The pyroxene-granulite which contents 34 % plagioclase shows increase of the ETC with T (Abdulagatov et al. [297]). ETC of rocks are decrease with temperature as porosity and amorphous phase in rocks increase. ETC of crystalline rocks decreases almost linearly at temperatures up to 700 °C (Abdulagatov et al. [297]). Thermal conductivity of rocks with a high amorphous phase (for example, volcanic or magnetite containing rocks) linearly increases with temperature. For amorphous materials (for example, fused quartz, silica glasses, and vitreous materials) the thermal conductivity increases with temperature as $\lambda \propto T^{n}$ (Ziman [74]), where n>0. The increase of the thermal conductivity of fused quartz with temperature was studied by Ratcliffe [251]; Horai and Susaki [109]; Abdulagatov et al. [294] (see also Figure 7). Therefore, temperature dependence of the ETC in rocks materials depends whether crystalline or amorphous structure dominate. More complicated behavior of ETC is found for rocks with mixture of amorphous and crystalline structure. The thermal conductivity for mixed crystals varies more slowly than $\lambda \propto T^{-n}$, where 0<n<1. Pyroxene-granulite is a mixed crystalline and amorphous structure. Since ETC for pyroxene-granulite increase with temperature; therefore, in this sample the amorphous structure dominates crystalline components. Therefore, precise thermal conductivity measurements can provide information about order of crystallization of the rocks. Horai and Susaki [109]) found that for andesite ETC is far less temperature dependence. Thus, the changes of thermal conductivity with temperature depends on various factors (opacity, composition, shape, size, and orientation of crystals and their fragments) (Drury and Jessop [361], Buntebarth [186 and 447]). Most sandstone contains a predominance of quartz. Quartz is an anisotropic mineral. Its thermal conductivity varies from 6.5 to 11.3 $W{\cdot}m^{-1}{\cdot}K^{-1}$ along different optical axes. The average value is 7.7 $W{\cdot}m^{-1}{\cdot}K^{-1}$ (Horai [107]). Jõeleht et al. [454]) derived the value of 6.7-7.1 $W{\cdot}m^{-1}{\cdot}K^{-1}$ for quartz thermal conductivity by fitting measured ETC to the different models (Wiener model, based on the resistor concept, 6.7 $W{\cdot}m^{-1}{\cdot}K^{-1}$, layered model 7.1 $W{\cdot}m^{-1}{\cdot}K^{-1}$, geometric mean model 6.5 $W{\cdot}m^{-1}{\cdot}K^{-1}$). This low value of the quartz derived by Jõeleht et al. [454] can be explained due to the sample was not pure quartz but may contain 15 % other silicates (feldspars, clay minerals). An addition of 10-15 % clay minerals (2 $W{\cdot}m^{-1}{\cdot}K^{-1}$) to quartz is decrease the matrix thermal conductivity from 7.7 to 6.5 $W{\cdot}m^{-1}{\cdot}K^{-1}$. Since quartz has significantly larger thermal conductivity than other minerals (7.7 $W{\cdot}m^{-1}{\cdot}K^{-1}$) the quartz content is the dominant factor in the thermal conductivity of the rocks. The quartz bearing samples are more sensitive to temperature variations. High content of quartz results in a high value of conductivity and a rapid decrease of λ with increasing temperature. Rocks having low content of quartz exhibit much smaller temperature dependence of ETC. McKenna et al. [365] used quartz content and porosity as a independent parameters to correlate experimental ETC data as $\lambda = 4.08 - 0.111\phi + 0.026Q$. This equation is applicable for sandstone. Özbek [177] considered the various methods of estimating solids thermal conductivity for a mineral assemblage. Quartz, having the highest conductivity of common rock-forming minerals, dominates thermal behavior of the assemblage. The ETC for higher quartz content solid materials can be estimated by the Eq. (213), Özbek [177], where Q is the fraction quartz content. The technique of determination ETC of rocks from the mineralogical composition was developed in the works [192, 193, and 424]. Since the mineral components of rocks are imperfect, the matrix thermal conductivity can be estimated from experimental $\lambda - P$ curves at each measured constant temperature using linear extrapolating procedure high pressure data to the zero pressure (see Abdulagatov et al. [292 and 293]). The axis intercept is equal to the values of $\lambda_S(T)$. The experiments demonstrate the great variation in the effects of T on the ETC of rock samples. Whereas the ETC of sandstone as a rock type poor in quartz and mica are less dependent on variation in PT-conditions, exhibits the quartz –and mica-bearing gneisses a high sensitivity to PT –conditions. There is a relation between the amount of ETC and its temperature sensitivity. The higher ETC is the higher its sensitivity to temperature variations. This effect may be result from the thermal properties of quartz where its proportion contributes to the amount of the thermal conductivity of rocks. The thermal conductivity of a sandstone decreases with increasing temperature. This decrease is less than of the quartz matrix since the saturating fluid (gas, oil, and water) increases in conductivity with temperature. The saturation modifies the ETC of rocks slightly. For 100 % saturations the ETC vary less than 20 %.

Figures 15 to 17 show the temperature dependence of ETC at fixed pressure (0.1 MPa) and the values predicted by various models. As these figures demonstrate the models [336,426, and 453] shows good agreement (within 1-2 %) with our ETC data for gas-saturated sandstone, while some empirical predictive models [27, 80, 170, and 452] are systematically lower at high temperatures by (4 % to 11 %). These models do not take into account the mineralogical composition of porous rocks, for example, the quartz content which considerably affects the slope (rate) of ETC changes. Excellent agreement within 1.2 and 2.7 % was found between the Funnell et al. [453] and geometric mean models (see Figure 16) and our experimental data, respectively. Figure 17 demonstrate the temperature dependence of the predicted ETC of water-saturated sandstone from various porosity models with the temperature dependence of $\lambda_S(T)$, $\lambda_f(T)$, and $\phi(T)$. As one can see, the agreement is good enough (maximum AAD is about 9 %). Excellent agreement within 0.2 % was found with Zimmerman's [335] model, while the values of ETC predicted with Hsu et al. [36] model systematically lower than the present measurements by 0.5 %. Kunii and Smith [236] prediction deviate from our ETC data by 1.3 % (prediction values systematically are lower). The deviations about 3.5 % was observed for the values of ETC predicted with Hadley's [34] model.

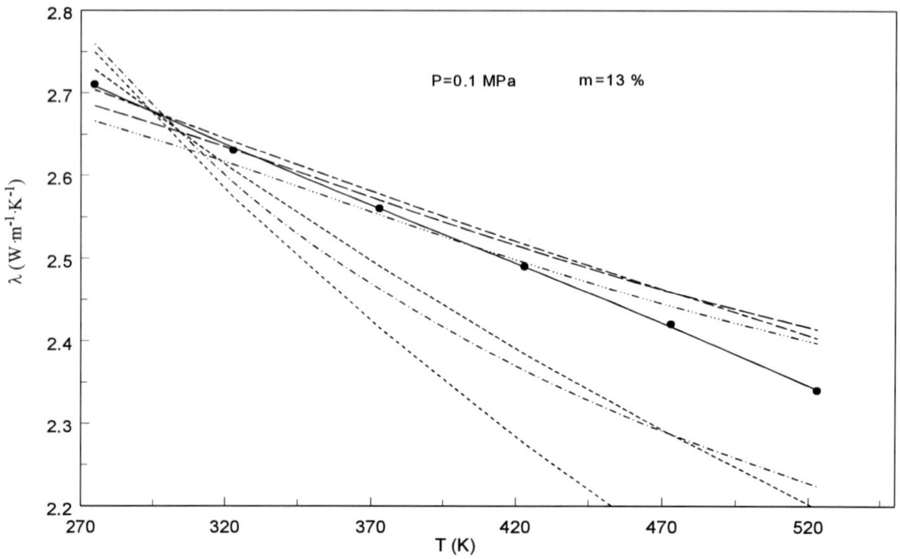

Figure 15. Measured and predicted values of ETC of gas-saturated sandstone as a function of temperature at atmospheric pressure. (——), this work Eq. (228); (– – –), Tikhomirov [170]; (– · – · –), Sekiguchi [80]; (—— – ——), Funnell et al. [453]; (—————), Rzhevsky and Novik [426]; (– ··· – ··· –), Fricke [336]; (······), Sass et al. [27].

As was mentioned above, the thermal conductivity for mixed crystals varies more slowly than $\lambda \propto T^{-n}$, where 0<*n*<1. Our measured ETC data for fluid-saturated sandstones as a function of temperature along the constant pressure were fitted to the power law, $\lambda = AT^{-n}$. We found that for gas-saturated sandstone the values of exponent *n* varying within 0.3 to 0.4 (0<*n*<1) depending on pressure, therefore, the sandstone under study is mixed crystals (see also section 1.2 and [74]). Eucken [368] model, Eq. (222), and the linear-Eucken law, Eq. (221), [439] also were fitted to the present ETC data. These models with the parameters (*A*=0.000269, *B*=0.2707, and *C*=5.6102, Eq. 222) and (*a*=0.30532 and *b*=0.0002302) are represent our experimental results for ETC porous sandstone within 0.1 % and 0.2 %, respectively. For water-saturated sandstones the values of exponent *n* varying within 0.17 to 0.19 depending on pressure (for oil-saturated sandstones the values of *n* are changed from 0.20 to 0.22). For example, for water-saturated sandstones *A*=10.584, 10.301, 11.148, and 12.057 and *n*=0.18, 0.17, 0.18, 0.19 at constant pressures of *P*=0.1 MPa, 50, 100, and 250 MPa, respectively. Eucken's [368] model also was fitted to our ETC data for water- and oil-saturated sandstones. These models with parameters (*A*=0.0005796 and λ_0=4.446 W·m^{-1}·K^{-1}) at 0.1 MPa and *A*=0.0006265 and λ_0=4.845 W·m^{-1}·K^{-1} at 250 MPa) are excellent (within 0.1

%) represented our experimental results for ETC water-saturated sandstone as a function of temperature at fixed pressure. For oil-saturated sandstones we have A=10.191 and n=0.2041 at P=0.1 MPa and A=12.946 and n=0.2225 at P=250 MPa. This is means that oil-saturated sandstone under study is mixed crystals. The Eucken's [368] model parameters for oil-saturated sandstones are A=0.0006527 and λ_0 =3.796 W·m^{-1}·K^{-1} at 0.1 MPa and A=0.0007247 and λ_0 =4.416 W·m^{-1}·K^{-1} at 250 MPa.

It is reasonable to assume that the porosity dependence models discussed above (section 5.3) still holds at any temperature T, where $\lambda_S(T)$, $\lambda_f(T)$, and $\phi(T)$ are the conductivity of solid, fluid, and porosity considered as a function of temperature. We used some theoretical models for the porosity dependence to estimate the temperature effect on ETC. Figures 15 to 17 shows the temperature dependence of ETC of gas-, water-, and oil-saturated sandstones at fixed pressure together with the values predicted by various empirical models and with the porosity models where $\lambda_S(T)$, $\lambda_f(T)$, and $\phi(T)$ are a function of temperature. As this figure demonstrates, the models by Funnell et al. [453], Asaad [166], and Anand [171 and 222] shows a good agreement (deviations within 1.6 %) with the present data, while some empirical predictive models ([27,170, and 452]) are systematically lower than measured values at high temperatures by (4 to 6 %).

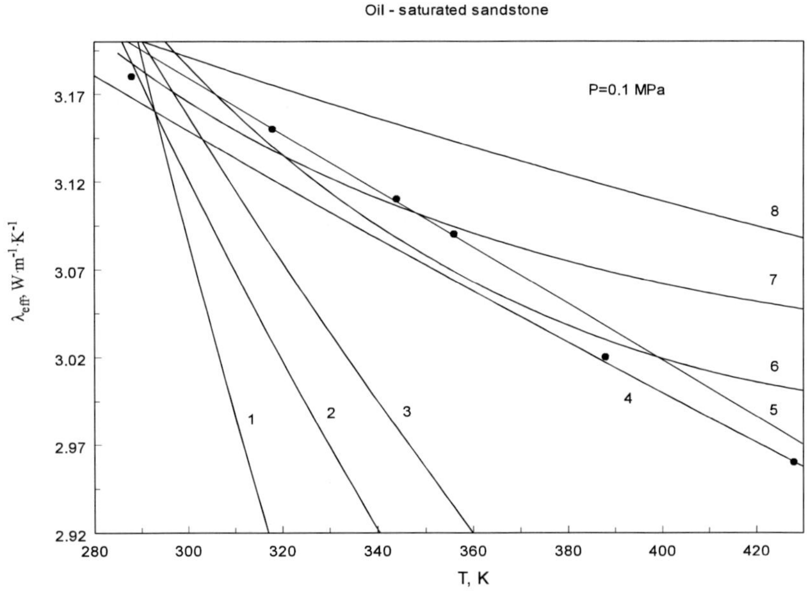

Figure 16. Measured [292] and predicted values of ETC of the oil-saturated sandstone as a function of temperature at atmospheric pressure. 1- Chapman et al. [452]; 2- Tikhomirov [170]; 3- Sass et al. [27]; 4- Funnell et al. [453]; 5- Eq. (81); 6-Anand [171,222]; 7-Asaad [166]; 8 - Zimmerman [335].

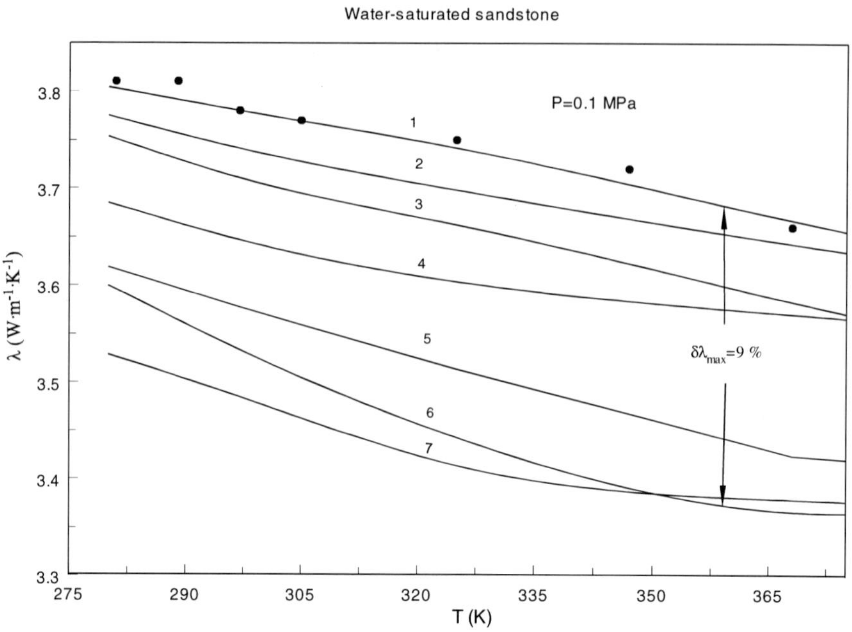

Figure 17. Measured and predicted values of ETC of the water-saturated sandstone as a function of temperature at atmospheric pressure (porosity is considered as a function of temperature). 1- Zimmerman [335]; 2- Hsu et al. [36]; 3- Kunii and Smith [236]; 4- Hadley [34]; 5- Zehner and Schlünder [44]; 6- Keller et al. [322]; 7- Buntebarth and Rueff [185].

These models no taken into account the mineralogical composition of the porous rocks, for example, quartz content which considerable affected on the slope (rate) of the ETC changes. The same figure includes also the prediction of the theoretical models where the porosity and matrix thermal conductivity are considered as a function of temperature (Zimmerman [335]).

6.2. Temperature Coefficient of the ETC of Fluid- Saturated Sandstones

In order to estimate the effect of temperature on the ETC of dry, water-, and oil-saturated sandstones, the temperature, $\beta_T = \frac{1}{\lambda}\left(\frac{\partial \lambda}{\partial T}\right)_P$, coefficients of the ETC, were calculated using the our experimental ETC data for sandstones. The derived values of β_T are shown in Figure 18-20 as a function of temperature for various constant pressures. The ETC of water- and oil-saturated sandstones affected up to 19 % and 12 %, respectively, by the temperature at constant P and porosity in the pressure range from 0.1 to 250 MPa and at temperatures from 288 K to 520 K. As one can see the values of β_T changed within (0.42-0.60)$\times 10^{-3}$ K^{-1}, in the temperature range from 288 K to 520 K and at pressures up to 250 MPa. The values of β_T almost linearly decrease with temperature increasing.

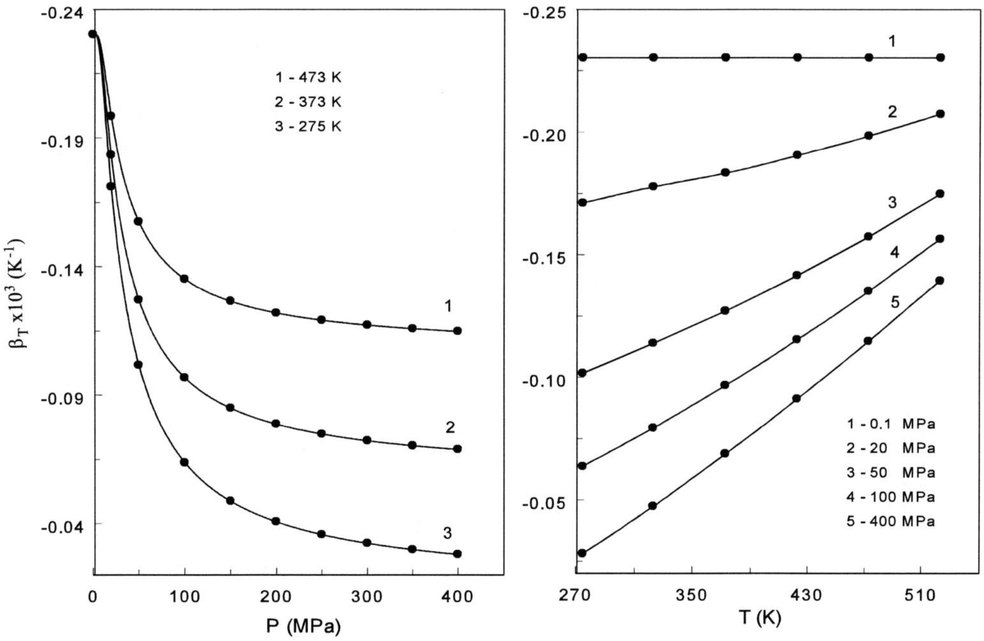

Figure 18. Derived values of temperature coefficient β_T of ETC for gas-saturated sandstone as a function of pressure P (left) and temperature T (right) at selected isotherms and isobars. Symbols are derived from experimental ETC data [159]. Solid lines are calculated from Eq. (232).

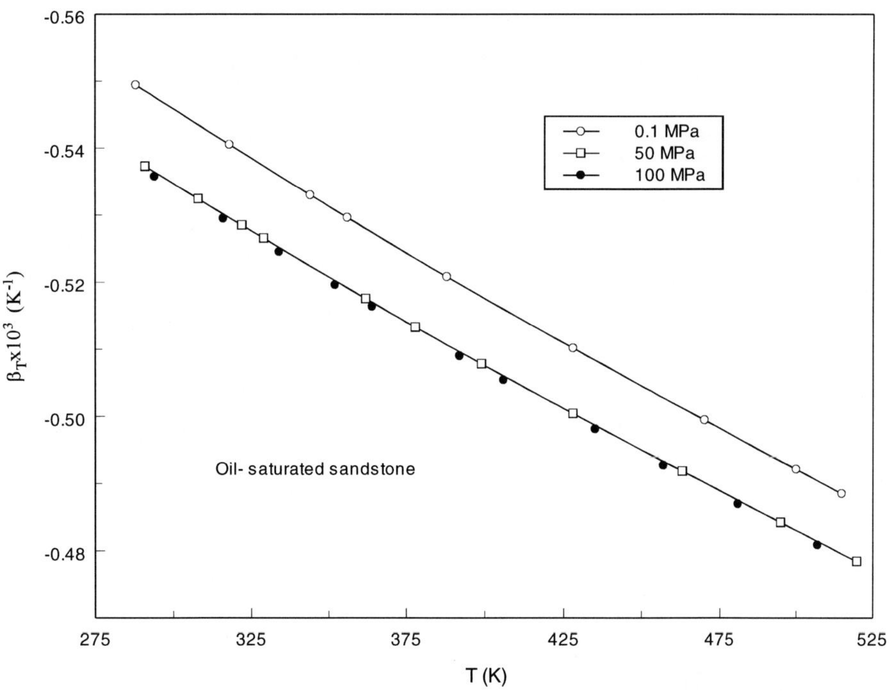

Figure 19. Derived values of temperature coefficients of ETC for oil-saturated sandstone as a function of temperature T at selected isobars. Symbols are derived from experimental ETC data [292]. Solid lines are calculated from Eq. (232).

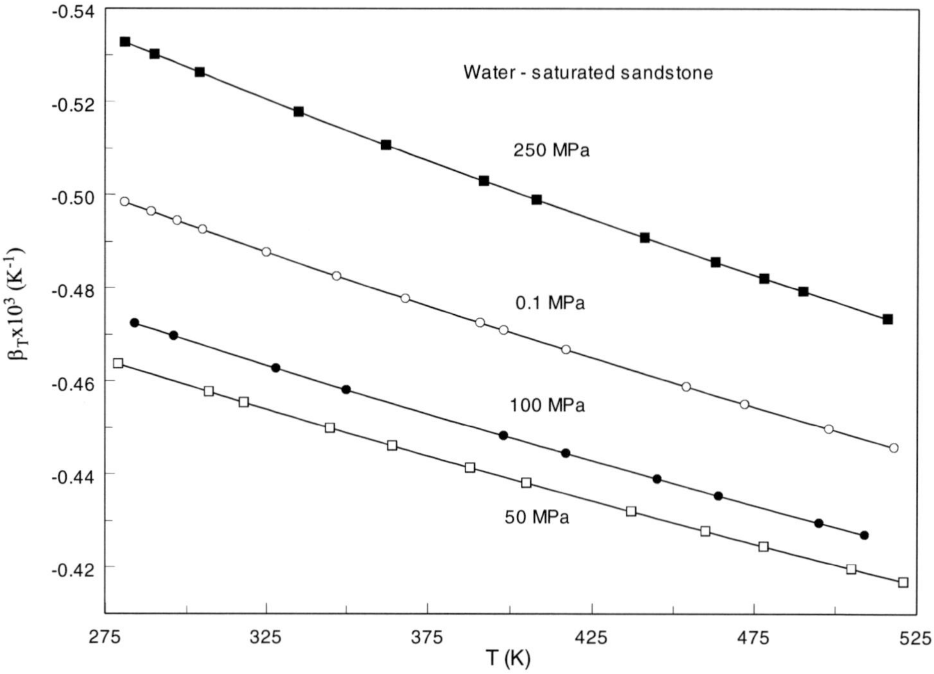

Figure 20. Derived values of temperature coefficients of ETC for water-saturated sandstone as a function of temperature T at selected isobars. Symbols are derived from the ETC data [314]. Solid lines are calculated from Eq. (232).

7. Pressure Coefficient of ETC Fluid-Saturated Sandstones

Beck et al. [553] and Goncharov et al. [558] simplified Eq. (11) into a linear pressure dependence of lattice thermal conductivity $\lambda(P,T_0)=\lambda(P_0,T_0)\left(1+\frac{K_0'}{K_0}P\right)$. This model was used by Beck et al. [553] to represent measured high pressure thermal conductivity measurements for MgO, NaCl, and KCl.

Tang and Dong [557] reported first-principle calculation of harmonic and anharmonic lattice dynamics of MgO and their pressure dependences. The implications for lattice thermal conductivity at high pressure are discussed based on a simple kinetic transport theory. They estimated the pressure dependence of lattice thermal conductivity in MgO based on a simple kinetic transport model $\lambda=\frac{1}{3}\sum_i c_i v_i^2 \tau_i$, where c_i, v_i, and τ_i are the mode heat capacity, group velocity, and lifetime (relaxation time) of the i-th phonon mode, respectively. The pressure dependence of thermal conductivity can be written as

$$\frac{\lambda(P)}{\lambda(0)}=\left[\frac{\lambda(P)}{\tau(P)}\frac{\tau(0)}{\lambda(0)}\right]\frac{\tau(P)}{\tau(0)}. \tag{231}$$

As was demonstrate in [557] $\frac{\lambda(P)}{\tau(P)}$ is the linear function of pressure and the slope $\frac{d\ln(\lambda/\tau)}{dP}$ increases with the increase of T. At ambient pressure is about 1.3 % and 1.7 % GPa^{-1} at 300 K and 2000 K, respectively. Tang and Dong [557] considered two models for the pressure effect on the lattice thermal conductivity. In the first model, the pressure dependence of the lattice thermal conductivity is only contributed by the increase of phonon frequencies at higher pressure, *i.e.*, $\frac{d\ln(\lambda)}{dP}=\frac{d\ln(\lambda/\tau)}{dP}$. In model two the 3-th order anharmonicity induced phonon-phonon scattering rate as proportional to the square of a relative anharmonicity parameter, $a_r=[A_{st}(0)(1+0.0153P)]/[(172.7+4.2P)]$. According to the model two $\frac{d\ln(\tau)}{dP}$ is about 1.8 % GPa^{-1} at ambient pressure. Overal pressure dependence of the lattice thermal conductivity is $\frac{d\ln(\lambda)}{dP}=\frac{d\ln(\lambda/\tau)}{dP}+\frac{d\ln(\tau)}{dP}$. At temperatures of 300 K the model two is predicted pressure coefficient of the thermal conductivity is about 1-2 % GPa^{-1} (lower than experiment) because the model underestimates the pressure coefficient of phonon lifetime $\frac{d\ln(\tau)}{dP}$.

Anderson and Bäckström [552] reported a method for measuring the pressure dependence of the thermal conductivity of hard material and single crystals. Two parallel metal strips are evaporated onto a flat surface of the specimen, one being used as heater, the other as a resistance thermometer. The method was tested using MgO crystal and glass as specimens at high –pressures (up to 1 GPa). The uncertainty of the measurements is 3 %.

Single-crystal thermal conductivity studies at high pressures are reported by Osako et al. [455]. They reported thermal conductivity of garnet and olivine at pressures up to 8.3 GPa and 1100 K by using a pulse heating method. Previously this method was developed by Dzhavadov [456]. Walsh and Decker [169] studied the effect of pressure and saturated fluid on the ETC rocks materials using contact method. Effect of pressure on ETC is smaller than the effect of temperature. The pressure effect on ETC strongly depends on the rocks nature, from mineralogical composition, porosity, and density. ETC of rocks increases with pressure. However, the effect of pressure on ETC of porous materials is different for distinct pressure range. The fractures and microcracks developed after stress release, when the sample was brought to the surface, begin to quickly close (some pores closing completely and others become narrow) again with increasing pressure leading to a decrease in internal thermal resistances of the sample (increasing mechanical contacts between the grains, improving the contact between the grains), see Figure 21, Shimozura [470]. This improves the thermal contact between mineral grains (reduces thermal contact resistance), increases the overall density of the rock or decreases the porosity. At a crossover pressure, between (50 and 100) MPa, depending on initial porosity of the rock, this process comes to an end. At high pressures when all of the cracks are assumed to be closed, a further pressure increase does not affect thermal conductivity significantly and the ERTC changes with pressure almost linearly, $\lambda(P)=\lambda_S(T)(1+\beta_P P)$. If the pressure continues to increase, there is reduction of the rock's intrinsic porosity which is not artificially created by stress release. Therefore, a rapid increase of ETC was noted at low pressures, below 100 MPa (see Figure 8). Some authors have missed this low pressure range where the rapid pressure dependence of the ETC of porous materials is observing. The same behavior of ETC in the low pressure range was found for most rocks (see for example [109,185,190,201,348,433,457-459]). For some rocks the increasing of the ETC at pressures up to 50 MPa was reached up to 100 % (see, for example, Volarovhich [433]). The linear behavior of pressure dependence of the ETC of rocks at high pressures is typical as reported by many other authors [109,455,190,348,460-463]. For rocks with small porosity the increase of the ETC is about 1.3 % per 10 MPa (Lebedev et al. [458], Lebedev [457 and 459]). Edmondson [464] has found the increases in ETC of sandstones to be (8 to 12) % per 7 MPa in the stress range of (6 to 25) MPa. Woodside and Messmer [82 and 161] reported almost the same results of 11.5 % per 7 MPa in the stress range of between (0 and 7) MPa and 2.5 % per 7 MPa in the range of from (14 to 28) MPa.

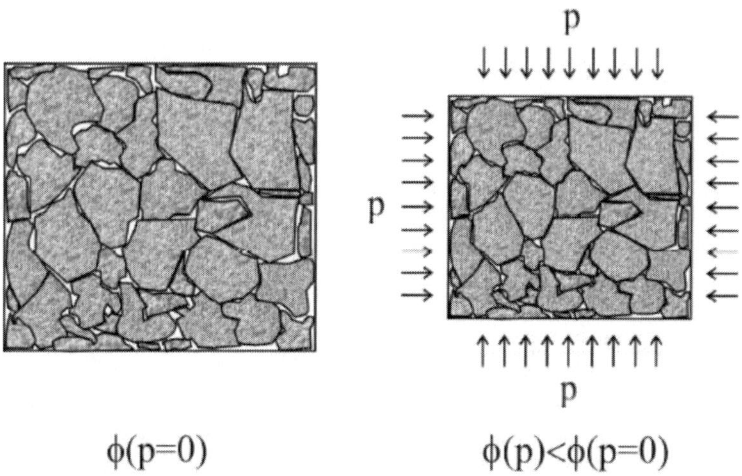

Figure 21. Schematic diagram of the configurations of microstructure of the porous rocks before (left, P=0) and after (right, under pressure P) stress applied.

Osako et al. [551] measured the pressure effect on the thermal conductivity of serpentine up to 8.5 GPa and at temperatures up to 800 K. The pressure coefficient of the thermal conductivity of antigorite is significantly smaller than those of olivine. Measurements were made by means of the pulse-heating method [455] in the Kawai-type apparatus. The measured values of the thermal conductivity as a function of pressure were fitted to different type function; first is a linear to 6 GPa $\lambda = 2.4 + 0.07P$ and other one a square function of pressure to 8.5 GPa $\lambda = 2.21 + 0.142P - 0.01P^2$. Our results for ETC [159, 291, 292, and 297] of sandstones increase by about (0.1 to 0.27) % per 10 MPa in the high pressure range and (1.3 to 2.0 %) per 10 MPa at low pressures (below 100 MPa).

In order to estimate the effect of pressure on the ETC of dry, water-, and oil-saturated sandstones we calculated the pressure, $\beta_P = \frac{1}{\lambda}\left(\frac{\partial \lambda}{\partial P}\right)_T$, coefficients of the ETC using our experimental ETC data. The derived values of β_P for gas-, water-, and oil-saturated sandstones are presented in Figures 22-24. As one can see from these figures, the pressure coefficient of ETC for fluid saturated sandstones is changes within (0.045 to 3.684)×10^{-3} MPa^{-1}. The pressure coefficient, β_P, slightly changes with temperature at high pressures, while the pressure dependence of β_P sharply changing at pressures below 100 MPa where

ETC exhibit rapid changes. In high pressure range (above 100 MPa) the values of β_P slightly decreases with pressure increasing. The pressure dependence of the ETC of rocks reported by some authors [190,455, and 462] (for thermal diffusivity) is very small (0.03 to 0.04) GPa^{-1}. Horai and Susaki [109] found that in the pressure range from (2 to 12) kbar the rate of pressure change of the ETC, $(d\lambda/dP)$, of rocks is >0.09 $W{\cdot}m^{-1}{\cdot}K^{-1}{\cdot}kbar^{-1}$. The highest rate of 0.33 $W{\cdot}m^{-1}{\cdot}K^{-1}{\cdot}kbar^{-1}$ was found for crystalline quartz. The values of pressure coefficient for amphibolites and granites reported by Seipold [199] are (0.08 to 0.25) GPa^{-1} are close to our results of (0.1 to 0.18) GPa^{-1}. The data reported by Seipold et al. [200] shows the values of pressure coefficients within (0.016 to 0.16) GPa^{-1}. Measurements of olivine by Beck et al. [463] shows a pressure dependence in λ of (5 to 6) GPa^{-1}.

The pressure dependence of thermal diffusivity for forsterite reported by Fujisawa et al. [465] shows a higher value than reported by other authors. The calculation by Hofmeister [77] shows the values of $(d\ln\lambda/dP)$ for some minerals vary within (0.021 to 0.28) GPa^{-1}. For example, theoretically predicted pressure dependence of λ by Hofmeister [77] for olivine is 4 % GPa^{-1}. Ljubimova et al. [181] reported the values of pressure coefficient for some rocks are within (0.02 to 0.70) GPa^{-1}. The pressure dependence $(d\ln\lambda/dP)$ is within $\approx$0.04 GPa^{-1} for olivine and 0.03 GPa^{-1} for granite [455]. Therefore, the literature values of the pressure coefficient of ETC for various rock types cover a wide range depending on the nature of rocks and their mineralogical and structural characteristics. Most thermophysical, electrical, and acoustical properties of porous materials show typical pressure dependence [169, 185, 427-432, and 466].

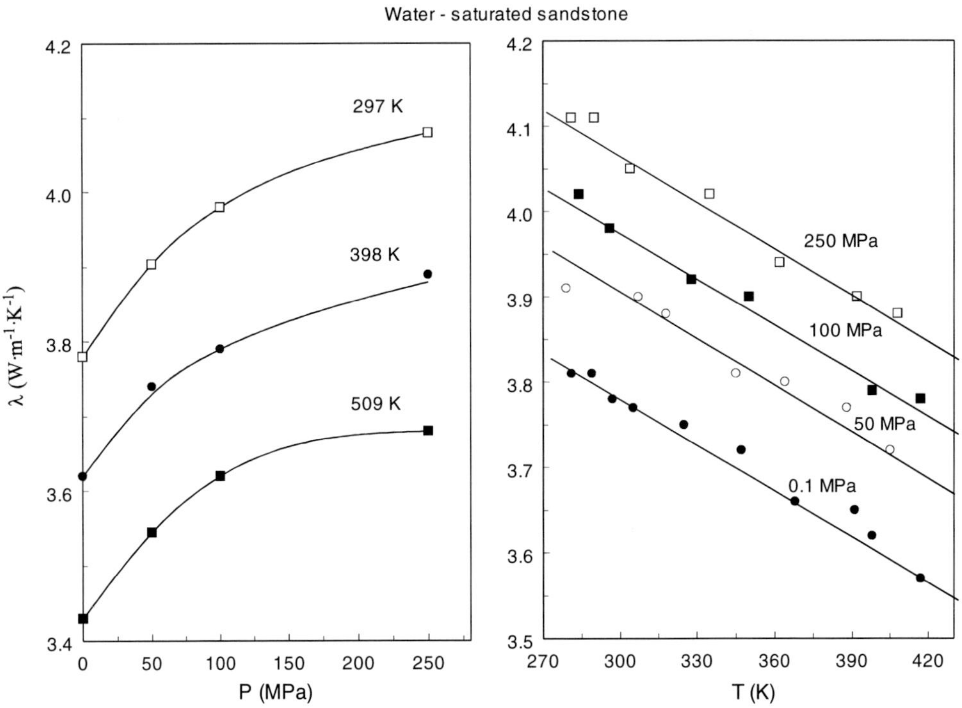

Figure 22. (Continued on next page.)

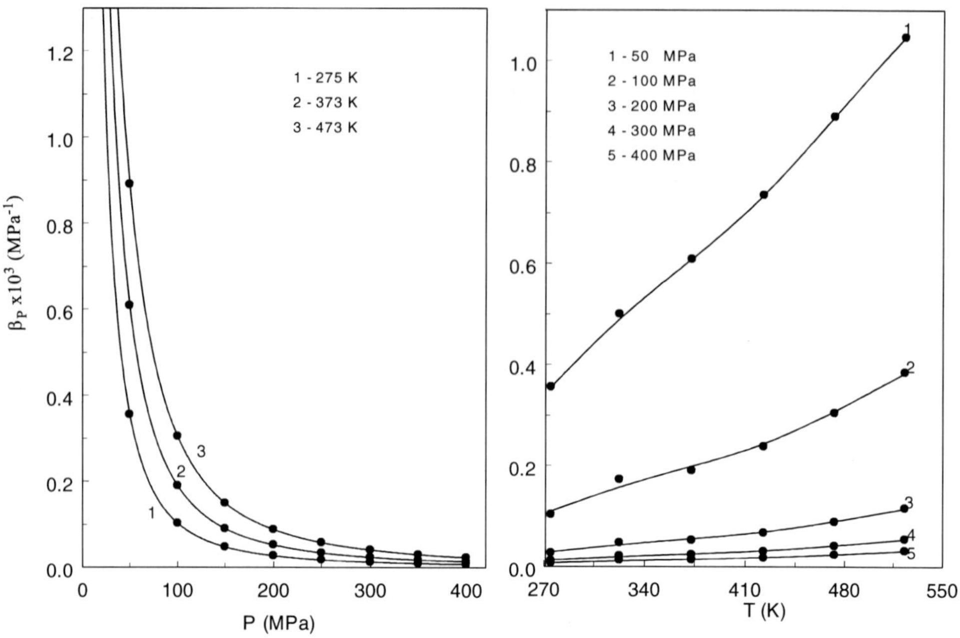

Figure 22. Derived values of pressure coefficient β_P of ETC for gas-saturated sandstone as a function of pressure *P* (left) and temperature *T* (right) at selected isotherms and isobars. Symbols are derived from experimental ETC data (Abdulagatova et al. [159]); Solid lines are calculated from Eq. (232).

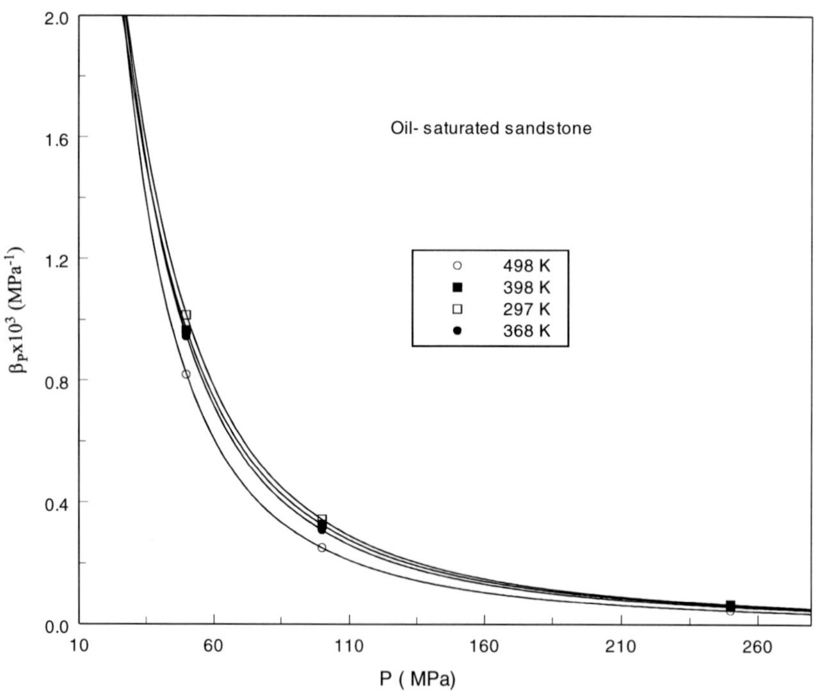

Figure 23. Derived values of pressure coefficients of ETC for oil-saturated sandstone as a function of pressure *P* at selected isotherms. Symbols are derived from experimental ETC data [292]. Solid lines are calculated from Eq. (232).

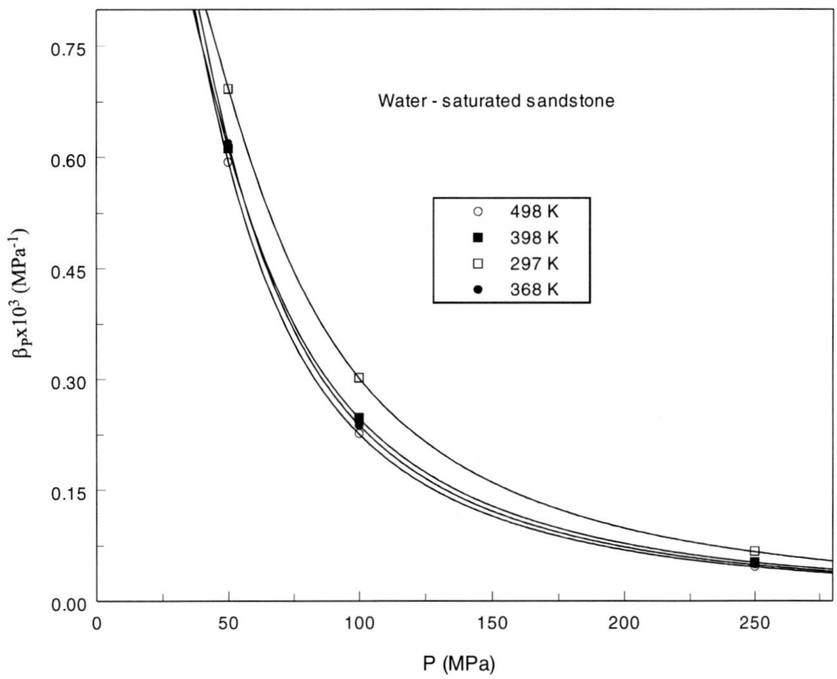

Figure 24. Derived values of pressure coefficients of ETC for water-saturated sandstone as a function of pressure P at selected isotherms. Symbols are derived from the ETC data [314]. Solid lines are calculated from Eq. (232).

Brace [427] shows that in the pressure range up to 50 MPa, very high volume compressibility was found due to the closure of great airfilled cracks. At pressures higher than 50 MPa a smaller constant compressibility was reached. Similar results were observed by Wang et al. [187] for granite samples after thermal stress cracking. The compressibility K_T above 100 MPa varies slightly from linear behavior with pressure [169,428-432]. At pressures about 400 MPa the porosity of the samples changed (exponentially) by about (1 to 2) %. Therefore, the ETC of porous materials is changes due to porosity changes by about (0.04 to 0.07) $W \cdot m^{-1} \cdot K^{-1}$.

Pressure coefficient derived from contact methods also unreliable because deformation can change the geometry and cracking reduces thermal contact [89]. The values of β_P vary by factor 3 between various reported ETC data [89] (Ross et al. [96], Bäckström [467], Brydsten et al. [468], Alm and Bäckström [469]). The ETC of water- and oil-saturated sandstone is relatively strong affected up to 8 % and 14 %, respectively, by pressure at constant T and porosity in the pressure range from 0.1 to 250 MPa and at temperatures from 288 K to 520 K.

To describe of the pressure dependence of ETC of rocks we have slightly modified the Zimmerman's model [335] by including one more parameter, namely the crossover pressure P_0 and $\lambda_0(T)$ as a function of temperature. The derived equation is

$$\lambda(T,P) = \lambda_{Sf} \exp(-P_0/P) + \lambda_0(P = 0.1, T), \tag{232}$$

where the parameters λ_{Sf} and λ_0 have simple physical meanings, namely, at $P \to \infty$ (at extremely high pressure limit), $\lambda(T, P=\infty) = \lambda_{Sf}(T) + \lambda_0(P=0.1, T)$, therefore, $\lambda_{Sf}(T)$ can be estimated as $\lambda_{Sf}(T) = \lambda(T, P=\infty) - \lambda_0(P=0.1, T)$ (see Figure 25) and at $P = 0$ (the hypothetical zero-pressure, low pressure limit, usually atmospheric pressure), $\lambda(T, P=0) = \lambda_0(P=0, T)$. All measured values of ETC for water- and oil-saturated sandstone were fitted to the Eq. (232). The derived values of $\lambda_{Sf}(T)$ together with the values of ETC at atmospheric pressure as a function of temperature were represented as

$$\lambda_{Sf}(T) = \frac{1}{A + BT} \text{ and } \lambda_0(P = 0.1, T) = \frac{1}{C + DT}, \tag{233}$$

where A=1.5889, B=1.3638×10^{-3}, C=0.26347, D=0.17197×10^{-3}, and P_0=32 MPa for oil-saturated and A=2.4824, B=1.5279×10^{-3}, C=0.2249, D=0.1304×10^{-3}, and P_0=50 MPa for water-saturated sandstone. Therefore, Eq. (232) together with Eq. (233) accurately describes the combined effect of temperature and pressure on the ETC of water- and oil-saturated sandstones. The uncertainty (AAD, average absolute deviation) of the representation is within 0.13 to 0.56 % over the entire range of pressure. As one can see, ETC of water- and oil–saturated rock at any T and P can be predicted just by knowing the two limited values of the ETC at high pressures, namely, $\lambda(T, P \to \infty)$, and the values of ETC at atmospheric pressure, $\lambda(T, P = 0.1)$, as a function of temperature.

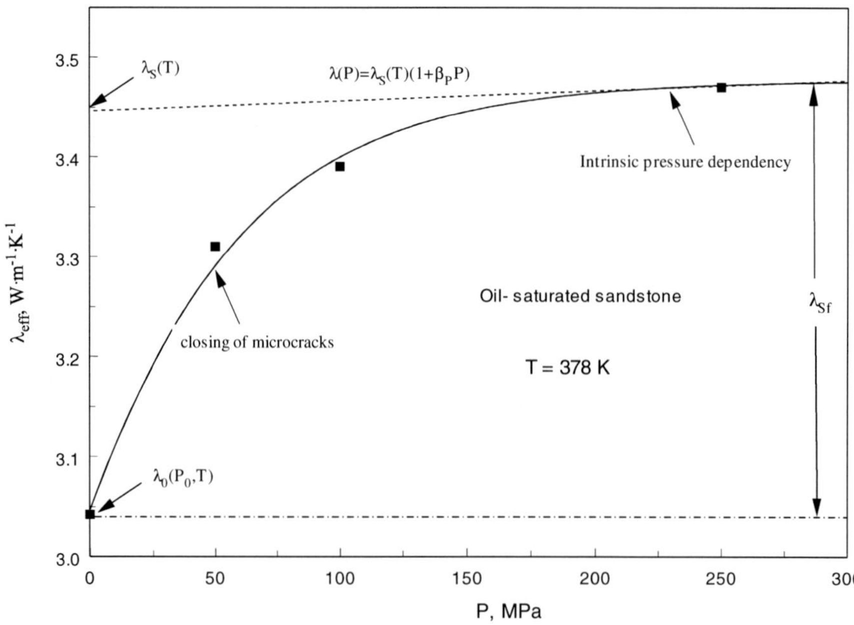

Figure 25. Experimental ETC [292] λ_{eff} of oil-saturated sandstone as a function of pressure P along the selected isotherm of 378 K together with values calculated from Eq. (232) (solid lines). Dashed line is linear extrapolation to the zero pressure.

Only one adjustable parameter P_0 (crossover pressure) can be estimated using the experimental data. This makes it much easier to predict the ETC of water- and oil-saturated rocks at any temperatures and pressures by using the minimal experimental information. This is physically very clear because crossover pressure P_0 strongly depend on the rocks characteristics and varies from (10 to 150) MPa. The value of P_0 is almost constant for given rock. The values of ϕ_0 are slightly changes with temperature due to thermal expansion. Figure 25 shows the comparison of the calculated with Eq. (232) and experimental values of ETC of oil-saturated sandstone.

All porosity dependence models discussed above (see sect. 5.3) usually are applicable at fixed *T* and *P*. The temperature and pressure dependences of ETC in these models can be taken into account through the porosity $\phi(P,T)$ and solid matrix thermal conductivity $\lambda_S(P,T)$ as a function of *T* and *P*. Obviously that porosity can be changed by stress and temperature. For example, the effect of pressure on the ETC of porous rocks in the Fricke [336] model can be taken into account through the porosity $\phi(P)$ as a function of pressure (the effect of pressure on the λ_S is small)

$$\frac{\lambda}{\lambda_S} = \frac{[1-\phi(P)]}{[1-\phi(P)]+\beta\phi(P)}. \tag{234}$$

The effect of pressure on the porosity can be estimated from the measured ETC as

$$\phi(P) = \frac{(\lambda/\lambda_S)-1}{(\lambda/\lambda_S)-(\lambda/\lambda_S)\beta-1}. \tag{235}$$

The derived values of $\phi(P)$ as a function of pressure for various temperatures are shown in Figures 26-28. As one can see from these figures, at low pressure range (below crossover pressure 100 MPa) the rapid changes in porosity (like isothermal compressibility, was observed), while at high pressures when all pores are closed, the porosity is almost constant. The derived pressure dependence of porosity can be fit with an exponential function of pressure [335].

$$\phi(P) = \phi_\infty + \phi_1 \exp(-P/P_0) \tag{236}$$

where ϕ_∞ is the limited values of the effective porosity at high pressures ($P \to \infty$) and $\phi_0 = \phi_\infty + \phi_1$ is the effective porosity at zero pressure (initial porosity of sandstone before applying the pressure); $P_0 = 50.81$ MPa. Equation (234) together with Eq. (236) is reproduce our measured ETC data for fluid saturated sandstones with high accuracy (AAD=1.0 %) over the entire pressure range. This method of prediction is required knowledge only high-pressure ($P \to \infty$) thermal conductivity values. Figures 29-31 demonstrate pressure dependence of the ETC predicted from various theoretical models where the porosity was assumed as a function of pressure. The pressure dependence of the solid matrix thermal conductivity $\lambda_S(P,T)$ is small and do not much affected on the ETC of sandstone. However, very important temperature dependence of $\lambda_S(P,T)$. For example, (see below Figure 39), the values of ETC of sandstone predicted from Anand's [171] model at low porosities with various solid matrix thermal conductivity $\lambda_S(P,T)$ are significantly differ (within 30 %). Therefore, accurate determination of the solid matrix conductivity is very important to develop predictive models.

The pressure dependence of ETC of oil-saturated sandstone was also estimated using the model developed by Zimmerman [335]. The experimental values of porosity as a function of pressure for Zimmerman [335] model can be estimated as

$$\phi_{\exp} = \frac{(1-r)-(1-r)(\lambda/\lambda_S)}{(1-r)-r\beta-(1-r)(\lambda/\lambda_S)+\beta(\lambda/\lambda_S)}, \tag{237}$$

using the measured values of ETC and fluid and solid thermal conductivities. The values of parameters (ϕ_∞ and ϕ_1) for the Eq. (236) were determined using fitting procedure for each measured isotherms.

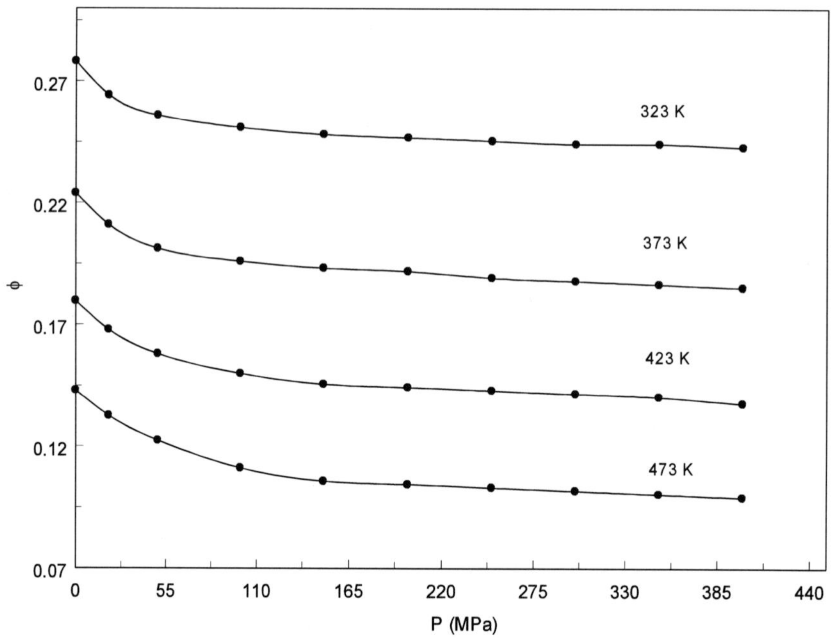

Figure 26. Derived values of porosity of gas-saturated sandstone as a function of pressure along the selected isotherms. The symbols are derived from experimental ETC data [159]; Solid lines are calculated from Eq. (236).

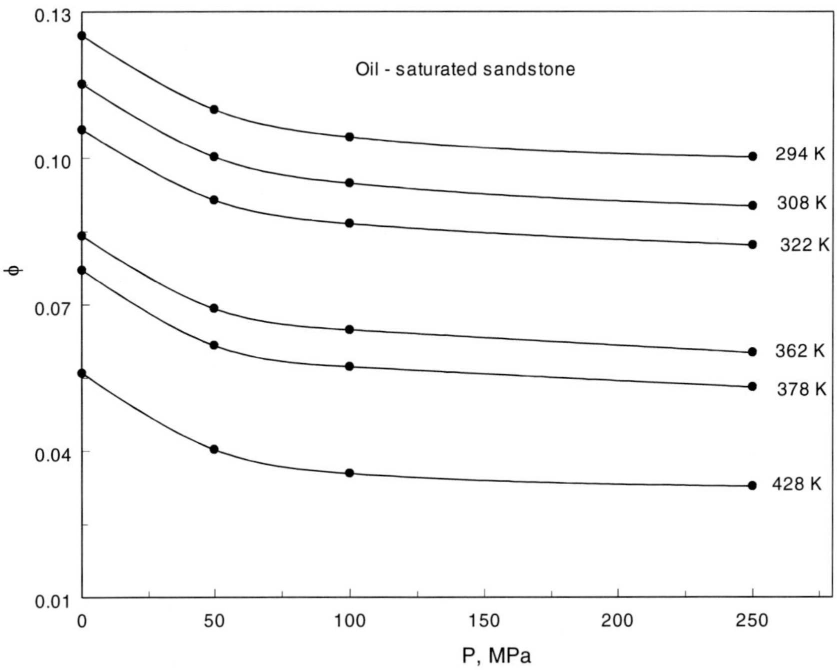

Figure 27. Derived values of porosity of the oil-saturated sandstone as a function of pressure along the selected isotherms. The symbols are derived from the experimental ETC data [292]; Solid lines are calculated from Eq. (236).

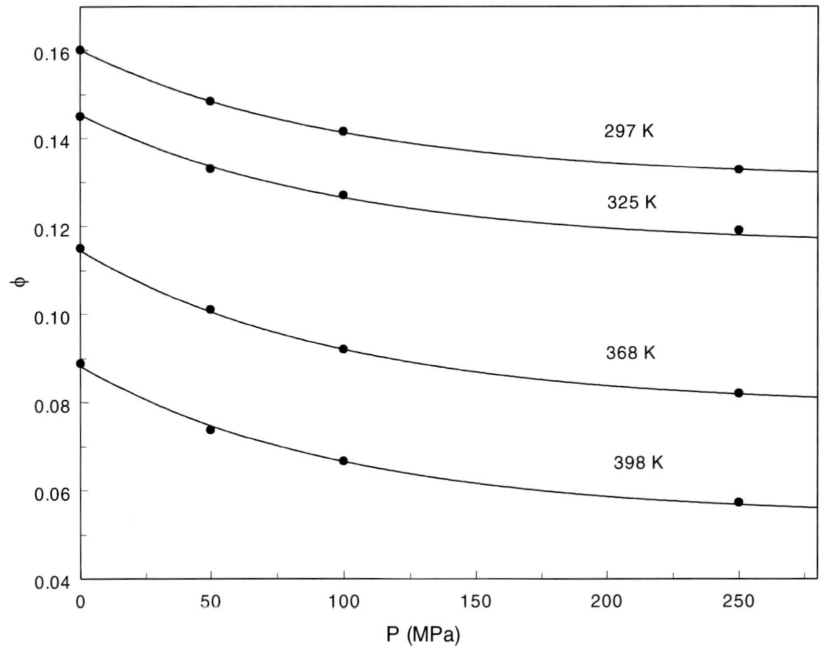

Figure 28. Derived values of effective porosity of the water-saturated sandstone as a function of pressure along the selected isotherms. The symbols are derived from the present experimental ETC data [314]; Solid lines are calculated from Eq. (236).

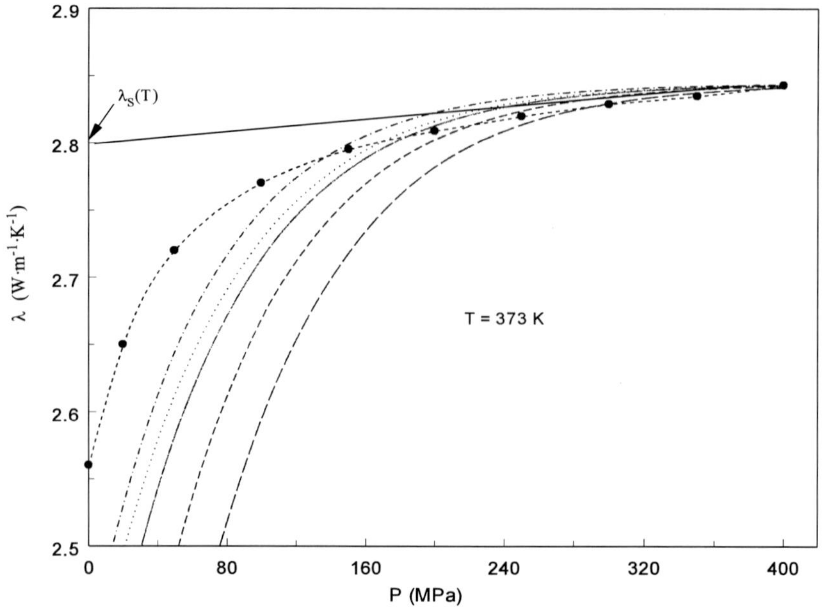

Figure 29. ETC of gas-saturated y sandstone as a function of pressure predicted from various models together with experimental value [159] for selected temperature of 373 K with variable porosity $\phi(P)$. (—— ——), Buntebarth and Schopper [302]; (— — —), Huang [172] (Eq. 162); (– ··· – ··· –), Maxwell [332]; (········), Sugawara and Yoshizawa [224]; (– · – · –), Russell [410]; (– – –); This work Eq. (232).

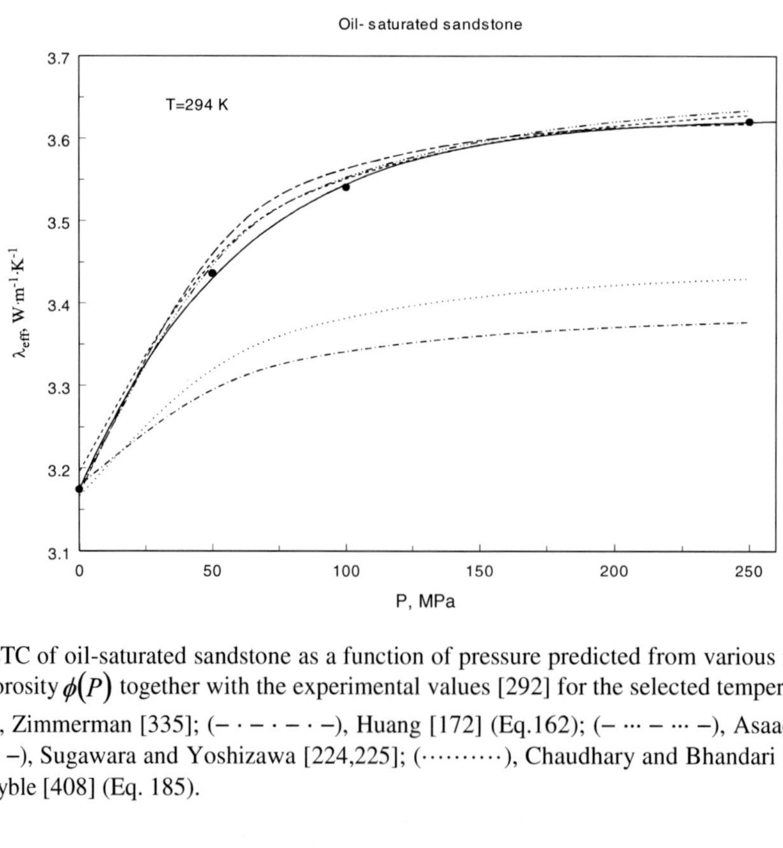

Figure 30. ETC of oil-saturated sandstone as a function of pressure predicted from various models with a variable porosity $\phi(P)$ together with the experimental values [292] for the selected temperature of 294 K. (———), Zimmerman [335]; (– · – · – · –), Huang [172] (Eq.162); (– ··· – ··· –), Asaad [166] (Eq. 180); (– – – –), Sugawara and Yoshizawa [224,225]; (··········), Chaudhary and Bhandari [411]; (—– ——–—), Wyble [408] (Eq. 185).

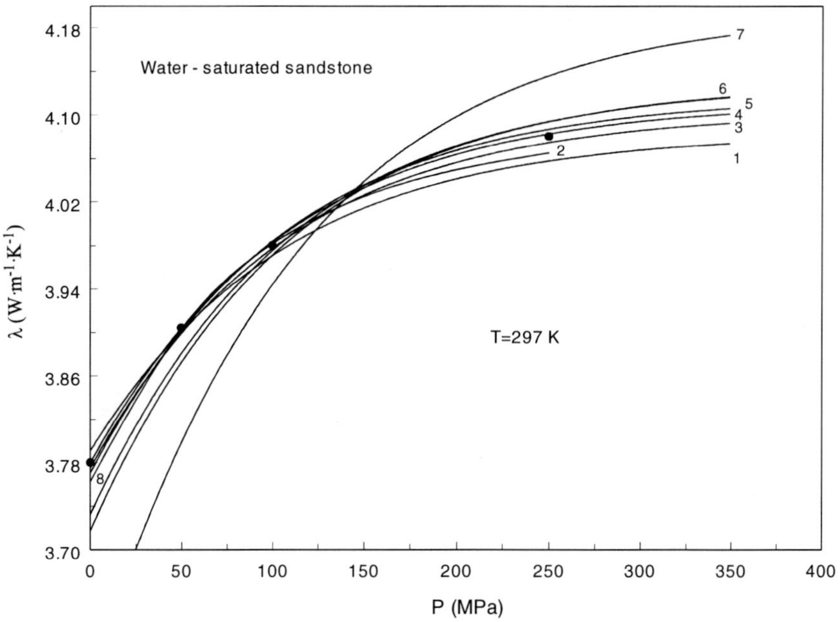

Figure 31. ETC of water-saturated sandstone as a function of pressure predicted from the various models with a variable porosity $\phi(P)$ together with the present experimental values for the selected temperature of 297 K. 1- Zehner and Schlünder [44]; 2- Wyble [408]; 3- Zimmerman [335]; 4- Keller et al. [322]; 5- Hadley [34]; 6- Sugawara and Yoshizawa [224]; 7- Krupiczka [238]; 8- Hsu et al. [36, 338].

The values of parameter P_0 is very slightly changing with temperature and can be considered as constant P_0=55 MPa. The parameter ϕ_1 is also very weak function of temperature ($\phi_1 \approx$ 0.018), while ϕ_∞ essentially changing with temperature increasing due to thermal expansion. The comparison between the calculated with Eq. (236) and our experimental ETC data for oil-saturated sandstone is shown in Figure 30.

The agreement between the measured and calculated values of ETC is excellent (average absolute deviation, AAD, between 0.1 and 0.6 %) over the entire range of pressure. As Figure 30 shows the low pressure anomaly of ETC strongly depend on the pressure dependence of porosity $\phi(P)$.

Therefore, the rapid changes of ETC at low pressure range are result of rapid decreases of the porosity with pressure increasing. This is means that sharp changes in porosity are result of large rate of the compressibility changes at low pressures. Figure 30 also shows ETC as a function of pressure calculated with various reported models where the porosity was considered as a function of pressure (Eq. 236). As this figure demonstrates, the prediction of ETC of sandstone sample by the models of Sugawara and Yoshizawa [225], Asaad [166], Wyble [408], and Zimmerman [335] is good enough.

7.1. Application ETC Models for Fluid-Saturated Sandstones

In order to apply the models (see above sect. 5.3) for prediction of the ETC of fluid-saturated rocks the accurate thermal conductivity data of saturated fluid $\lambda_f(T)$ (air, oil, water, *etc.*) and solid matrix $\lambda_S(T)$ are require.

7.1.1. Thermal Conductivity of Air

To calculate the thermal conductivity of air the reference (NIST) equation can be recommended [471]. The uncertainty for the dilute gas (air) thermal conductivity is 2 % with increasing uncertainties near the triple point. For the non-dilute gas, the uncertainty is 2 % for temperatures greater than 170 K. The uncertainty is 3 % at temperatures less than the critical point and 5 % in the critical region, except for states very near the critical point.

7.1.2. Thermal Conductivity of Oil

In order to calculate the ETC of sandstones saturated with castor oil at high temperatures and high pressures we used the data reported by Aliev and Magaramova [472] and Averkin et al. [473] at high temperatures (up to 573 K) and at pressures up to 250 MPa. They used the absolute steady state technique to accurately measurements of the thermal conductivity of castor oil (ПЭС-5).

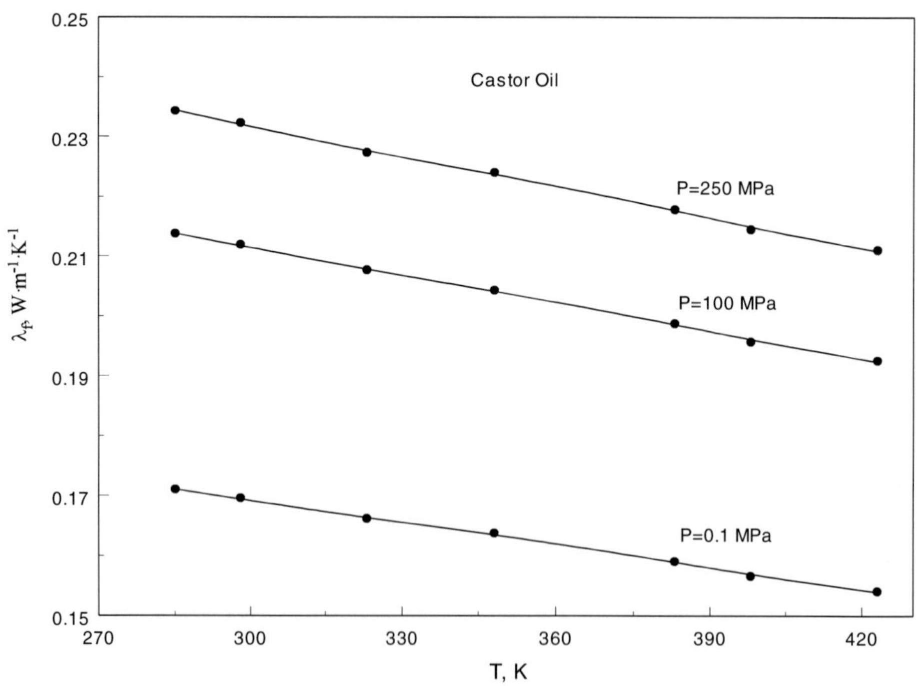

Figure 32. Temperature dependence of the thermal conductivity of castor oil along various selected isobars. Symbols are the experimental data reported by Averkin et al. [473]. Solid lines are the calculated from Eq. (238).

Averkin et al. [473] also studied the effect of pressure on thermal conductivity of castor oil. Neizer [474] reported the effect of temperature on thermal conductivity of castor oil in the range of temperature from 273 K to 423 K. The measured thermal conductivity data for castor oil in the temperature range from 293 K to 573 K were described with the correlation equation

$$\lambda_{oil} = 0.20666\text{-}0.0001247T\ (\mathrm{W{\cdot}m^{-1}{\cdot}K^{-1}}). \tag{238}$$

Figure 32 shows the thermal conductivity of castor oil as a function of temperature along the various constant pressures. As this figure shows, thermal conductivity of castor oil almost linearly decreases with temperature increasing like ETC of oil-saturated sandstone (see above Figure 9c).

7.1.3. Thermal Conductivity of Water

To calculate the thermal conductivity of the pure water at high temperatures and high pressures we used the internationally accepted (International Association for the Properties of Water and Steam, IAPWS standard) reference data reported by Kestin et al. [475]. The thermal conductivity of water at atmospheric pressure as a function of temperature is [476]

$$\lambda_w(T) = -0.76825 + 2.24957(T/298) - 0.874095(T/298)^2. \tag{239}$$

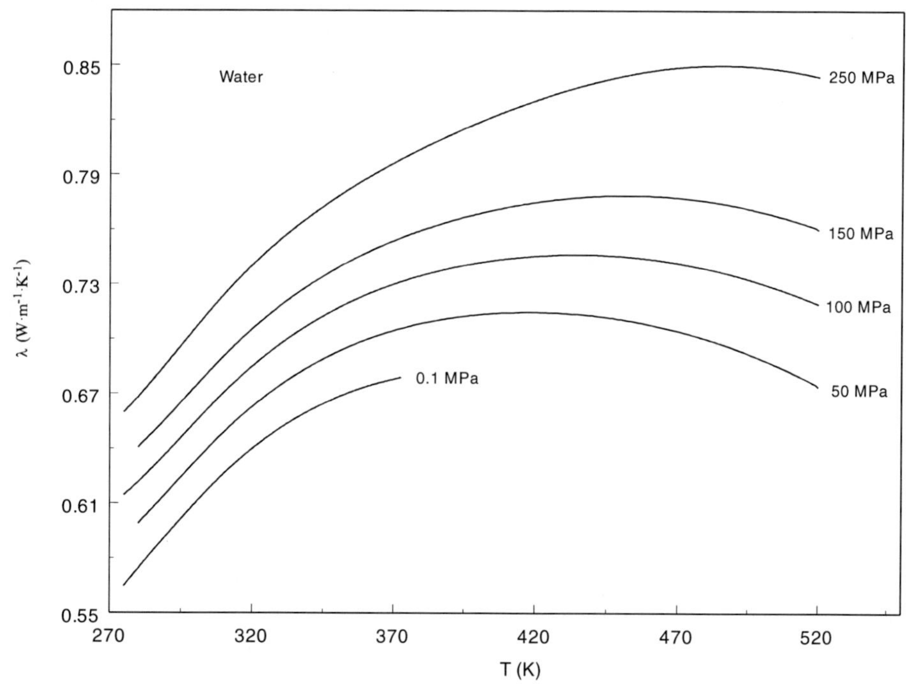

Figure 33. Temperature dependence of the thermal conductivity of water along various selected isobars calculated with IAPWS formulation Kestin et al. [475].

The maximum deviation of the experimental data employed, from this Eq. (239) is 1.1 %. This equation is valid at temperatures from 274 K to 360 K. In order to calculate of the thermal conductivity of pure water at high temperatures (from 251.17 to 1275.00 K) and at high pressures (up to 1000 MPa) the IAPWS formulation [475] can be recommended. Figure 33 shows the thermal conductivity of water as a function of temperature along the various constant pressures calculated with IAPWS formulation [475]. As this figure shows, thermal conductivity of water monotonically increase with temperature increasing up to 420-470 K (depending on pressure), goes through the maximum and then decrease at high temperatures. This behavior of the thermal conductivity of water considerable changes the ETC of water-saturated porous rocks.

7.1.4. Thermal Conductivity of Matrix

In order to apply the model discussed in the sect. 5.3 we also need to know the actual values of the conductivities of the matrix and matrix forming minerals. The conductivity non-clay minerals are quite well known (Horai [107]). This for clay-minerals is poorly defined.

The change of thermal conductivity with temperature depends on various factors (opacity, composition, shape, size, and orientation of crystals and their fragments) (Drury and Jessop [361]; Buntebarth [186 and 447]). The temperature dependence of the ETC of rocks strongly depends on the quartz content (see above, [177, 365, and 454]). In this work the matrix thermal conductivity was estimated from the experimental $\lambda - P$ curves for the dry sample at each measured constant temperature using linear extrapolating procedure (Abdulagatov et al. [293]) high pressure data to the zero pressure. The derived values of $\lambda_S(T)$ for the sandstones sample were fitted to the following correlation equation

$$\lambda_S(P) = \lambda_S(T, P = 0.1) + 0.00281P, \ \lambda_S(T) = \frac{1}{6.6205 \times 10^{-4} T - 0.037037}. \tag{240}$$

Our result for the matrix thermal conductivity of sandstone at 298 K and at 0.1 MPa is 6.24 $W \cdot m^{-1} \cdot K^{-1}$. Sandstone is a sedimentary rock formed from a combination of many minerals settled down at high pressures. The major mineral component in almost all types of sandstones is quartz. Thermal conductivity of sandstone strongly depends on the properties of quartz. It is important to study of the mineral composition in sandstone. Using the thermal conductivity data of each mineral component we can estimate the matrix thermal conductivity as [124]

$$\lambda_S = \prod_{i=1}^{n} \lambda_i^{\phi_i}, \tag{241}$$

where λ_i and ϕ_i are the thermal conductivity and volume fraction of the *i*-th mineral constituent, respectively.

7.2. Application f the Porosity Dependence Models for Fluid-Saturated Sandstones

In this section we studied the applicability and predicting capabilities of the various theoretical and semiempirical models for the ETC of gas-, water-, and oil-saturated sandstones as a function of temperature, pressure, and porosity. For this purpose the models discussed in the sect. 5.3 for fluid-saturated rocks were applied to our experimental ETC data (see Appendix) for gas-, water- , and oil-saturated sandstones. Some models are failed to accurately represent ETC for large solid-fluid conductivity ratio, for example models developed in works [44, 236, and 477]. Several ETC models calculations are plotted in Figures 35 to 39 together with our single ETC data point for gas (air) saturated sandstone sample with porosity of 0.13. As one can see from this figures, the predicted values of ETC with Sugawara and Yoshizawa [224] and Volarovich [433] (Figure 34) models were excellent agree with the present measurements. Anand [171] prediction model $\lambda = \lambda_0(1-\phi)^4$ ($\lambda \propto \rho^4$) is low for gas-saturated sandstone by 20 %, while the values of ETC predicted by the $\lambda = \lambda_0(1-\phi)^3$ ($\lambda \propto \rho^3$ [433]) shows excellent agreement with the experimental values. Most theoretical models show higher predicted values (by 20-30 %) than the experiment. Figure 26 demonstrate the porosity behavior of ETC of gas-saturated sandstone sample at selected isotherm of 473 K and at 0.1 MPa predicted with Fricke model (Eqs. 89-95) [336] for dispersions of spheroids with various β. The structural parameter β is variable for various pore shapes (β=1 is the ideal mixing model, $\beta = 3/2$ spherical pores, $\beta = 5/3$ for needle–like pores). The optimal value of β for our ETC data for gas-saturated sandstone is 3.2. Therefore, for real porous rocks the values of β sufficiently higher than model values (3/2 or 5/3). The porosity dependence of the ETC predicted from Keller et al. [322] model is shown in Figure 36 for various values of the ratio *b*/*a* (where 2*b* is the pores size and 2*a* is the grain size). As one can see, Keller et al. [322] model shows good agreement with the measured values of ETC if pore and grain shapes ratio (or texture factor) (α_p/α_g) is 0.35 (or the ratio of pore and grain size is *b*/*a*=0.5). Figure 37 shows the temperature and pressure dependences of the texture factor (α_p/α_g) derived from our experimental ETC data. At low temperatures (α_p/α_g) are very slightly changes with pressure, while at high temperatures rapid changes are observed, especially at low pressures (see Figure 37 left).

At low temperatures and at pressures up to 100 MPa the effect of temperature on the texture factor is small, while at high temperatures (above 400 K) at high temperatures the rapid changes of (α_p/α_g) with T is observed, especially at high pressures (see Figure 37 right). Torquato [54] ETC model predictions for various values of the structural constants together with the values calculated from Eucken [368] model is shown in Figure 38.

Differences between various structural constants (face-centered cubic, body-centered cubic, and simple cubic) at porosities up to 40 % are negligible small, while at porosities between 50 and 90 % are within 20 %. Our ETC results are lower than the prediction values by 20-25 %. However, as Figure 39 (right) demonstrate, the prediction capability of the model considerable depends on the values of thermal conductivity of solid matrix, λ_S. If the values of λ_S reduce up to 3 W·m^{-1}·K^{-1} (by 20 %), the agreement between the model prediction and experiment is excellent (see also Figure 39).

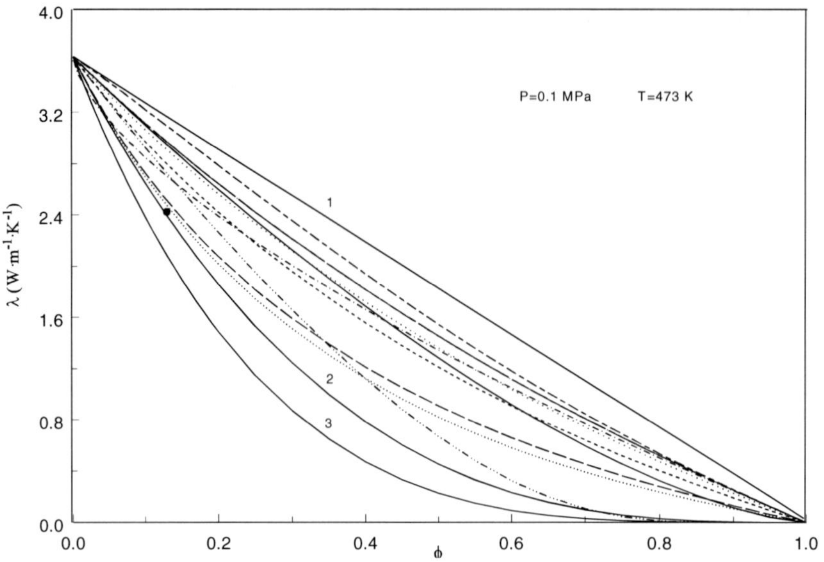

Figure 34. ETC of air-saturated sandstone as a function of porosity predicted from various theoretical models together with our experimental value for the porosity of 13 % at selected temperature of 473 K and pressure of 0.1 MPa. (– ··· – ··· –), Huang [172] (Eq. 162); (———), Waff [339]; (······), Fricke [336]; (– · – · –), Ribaud [409]; (——–—–), Russell [410]; (– – –), Odelevskii [327]; (——), Litovskii [480]; (—— ——), Buntebarth and Schopper [302]; (·········), Sugawara and Yoshizawa [224]; 2, Volarovich [433]; 3, Anand [171]; 1, arithmetic mean Eq. (50).

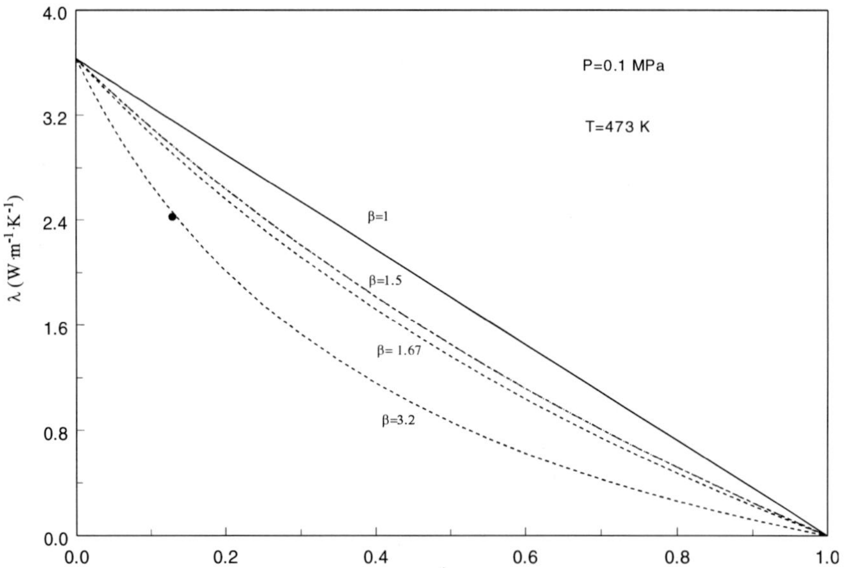

Figure 35. ETC of gas-saturated sandstone as a function of porosity predicted from Fricke [336] model (Eq. 95) for various values of structural parameter β together with our experimental value for the porosity of 13 % at selected temperature of 473 K and pressure of 0.1 MPa.

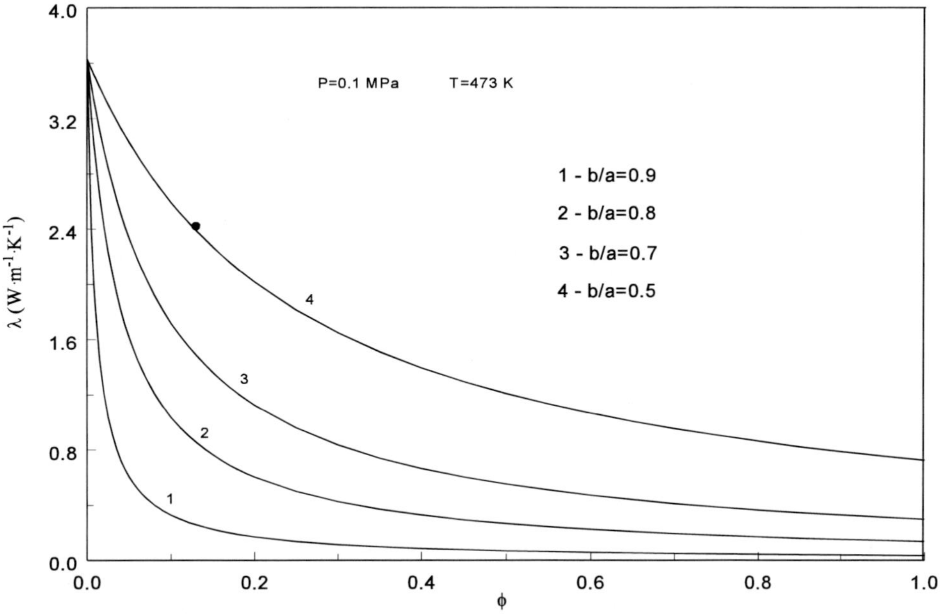

Figure 36. ETC of gas-saturated sandstone as a function of porosity **predicted from Keller's** [322] model for various values of ratio b/a together with our experimental value for the porosity of 13 % at selected temperature of 473 K and pressure of 0.1 MPa.

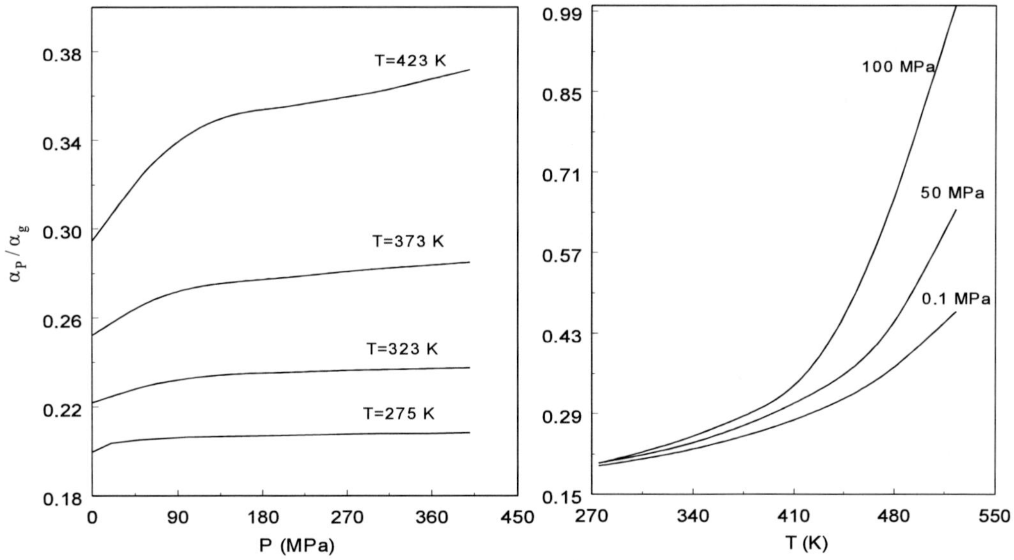

Figure 37. Calculated (from the measured ETC of gas-saturated sandstone [159]) values of the texture factor (α_P/α_g) in Keller's [322] model as a function of pressure (left) and temperature (right) at selected isotherms and isobars.

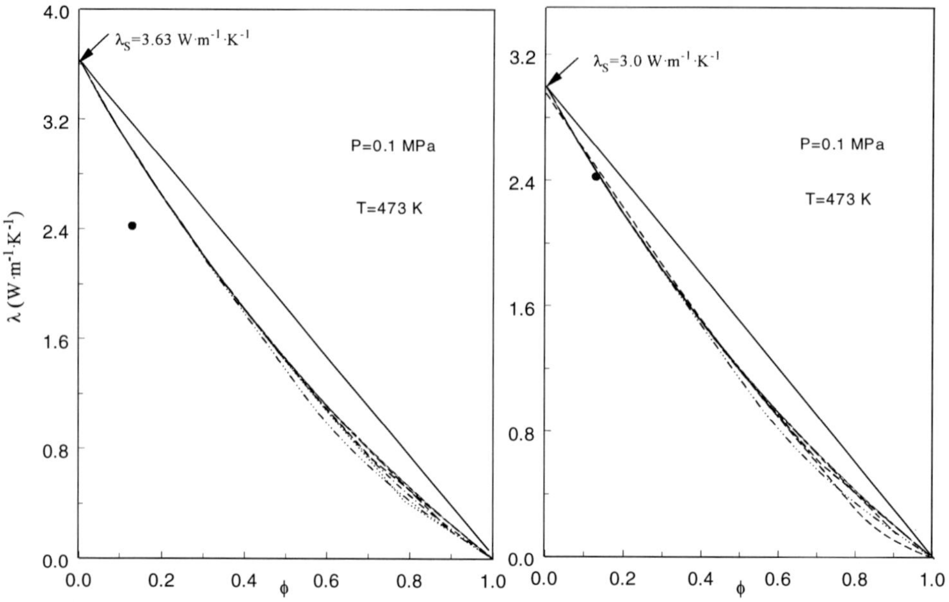

Figure 38. ETC of gas-saturated sandstone as a function of porosity predicted from Torquato [54] model for various values of the structural constants together with the experimental [159] value for the porosity of 13 % at selected temperature of 473 K and pressure of 0.1 MPa. (– ··· – ··· –), cubic lattice spheres; (———), fcc lattice of spheres; (······), square array of cylinders; (– · – · –), bcc lattice of spheres; (——–—–), hexagonal array of cylinders; (– – –), Eucken model (Eq. 201).

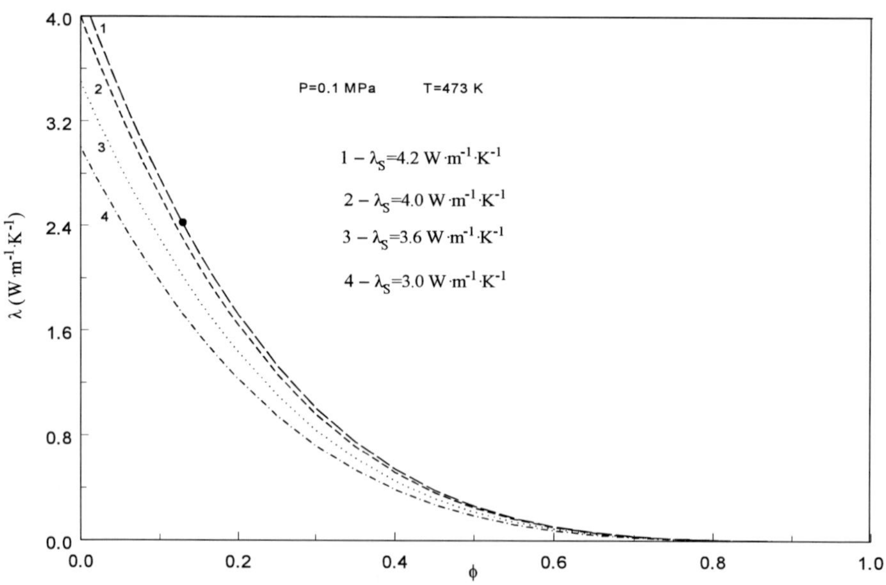

Figure 39. ETC of gas-saturated sandstone as a function of porosity predicted from the Anand's [171] model for various values of the matrix thermal conductivity together with the experimental value [159] for the porosity of 13 % at selected temperature of 473 K and pressure of 0.1 MPa.

The comparison results for oil-saturated sandstone are given in Figures 40 to 45. In these figures some selected frequently used ETC models predictions are plotted and compared with the measured ETC data for oil-saturated sandstone. Some models are failed to accurately represent ETC for oil-saturated sandstone, for example models by different authors [36, 74, 338, 339, 377, 369, 410, 477, and 478]. As one can see from Figure 40, the predicted values of ETC by geometric mean model, the models [166, 172, 224, 225, and 411] are excellent agree with our measured values of ETC for oil-saturated sandstone with porosity of 13 %. The predicted values of the ETC by other models [369, 377, 477, 479, and 480] are considerable higher than the present value (by 45-50 % and more), while the models by Hsu [36 and 338], Hashin and Shtrikman [363] and by Brown [369] lower than measured value (by about 45 %). The porosity dependence of the ETC predicted by Sugawara and Yoshizawa [224 and 225] model for various values of empirical parameter n shown in Figure 68 at two pressures (0.1 and 250 MPa) and at 294 K. As one can see, good agreement was achieved at n=6 for the oil-saturated sandstone.

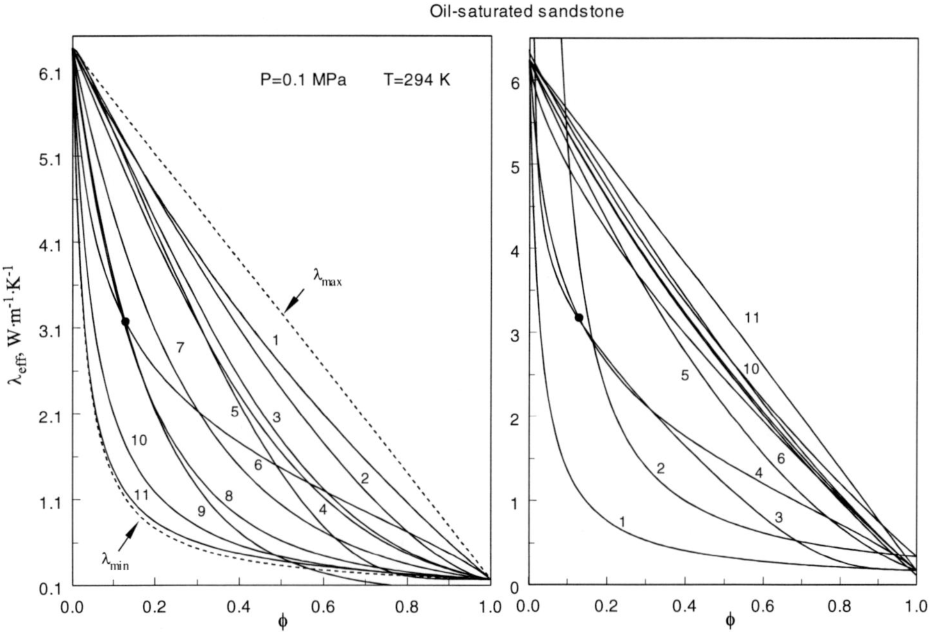

Figure 40. ETC of oil-saturated sandstone as a function of porosity predicted from various theoretical models together with the present experimental value for the porosity of 13 % at selected temperature of 294 K and at pressure of 0.1 MPa. Left: 1- Lichtenecker and Rother [479]; 2- Litovskii [480]; 3- Brown [369], second order; 4- Bruggeman [377]; 5-Sidorov [477]; 6-Geometric mean model; 8-Sugawara and Yoshizawa [224,225]; 9-Asaad [166]; 10- Hsu [36,338]; 11- Brown [369], first order; Right: 1-Ziman [338]; 2-Krupiczka [238]; 3- Huang [172] (Eq. 162); 4 – Chaudhary and Bhandari [411]; 5 – Roy et al. [102]; 6-Ribaud [409]; 7- Waff [339]; 8-Eucken [368]; 9-Russell [410]; 10-Maxwell [332]; 11-Krischer and Kröll [478].

Figure 42 shows the comparison between measured and predicted by Zimmerman [335] model values of the ETC for various types of pores (thin cracks, spherical pores, and needle–like). As one can see, for the thin-crack approximation the results much better (differences about 35 %) than for the spherical and needle–like pores (differences about 65-70 %). The porosity dependence of ETC predicted from the Keller et al. [322] model is shown in Figure 43 for various values of the ratio A_C / A_S (where A_C and A_S are the contact and cross-section areas, respectively). The ratio A_C / A_S is related with the pore 2b and grain 2a sizes as A_C / A_S=

$(1-b/a)^2$. Best result was found at A_C/A_S=0.1. The porosity dependence of the grain shapes ratio (or texture factor, α_p/α_g) at T=297 K and 0.1 MPa is shown in Figure 43 (right). Figure 44 demonstrate porosity dependence of the ETC of oil-saturated sandstone at selected constant temperature of 297 K and at 0.1 MPa predicted with the models by Cheng and Torquato [331] and Torquato and Rintoul [390]. In Figure 44 (right) the Cheng and Torquato [331] model plotted for various cubic lattice periodic arrays at fixed value of strength of the thermal resistance at interface between phase (solid-oil), R=20000.

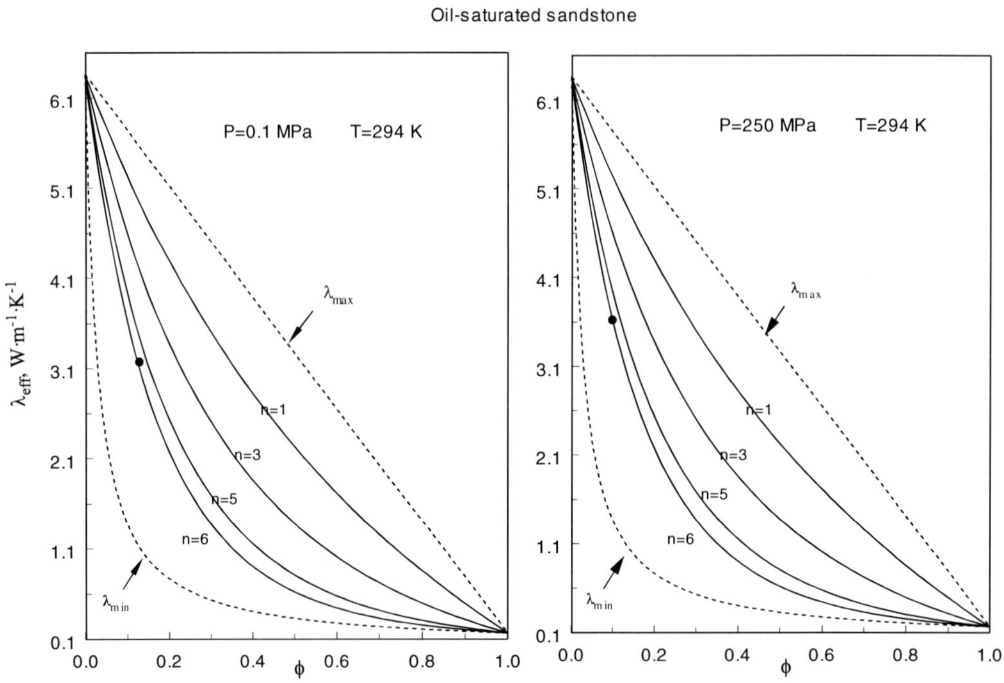

Figure 41. ETC of oil-saturated sandstone as a function of porosity predicted from Sugawara and Yoshizawa [224 and 225] model (Eq. 192) for various values of empirical parameter n together with the experimental value [292] for the porosity of 13 % at selected temperature of 294 K and at pressures of 0.1 MPa (left) and 250 MPa (right).

As this figure shows, the prediction values of ETC considerable higher than measured value. The prediction of ETC with model by Torquato and Rintoul [390] for various values of aforementioned three-point microstructural parameter ς_2, shown in Figure 44 (left). In this model we empirically assumed that ς_2 is proportional to porosity. As one can see, at low porosities, the predictive values of ETC are very small changes with ς_2 changing. The prediction capability of the Hsu [36 and 338] model is shown in Figure 45 for various values of the touching parameter $\gamma_c = c/a$. As we can note, best result for this model is achieved at $\gamma_c = 0.0$.

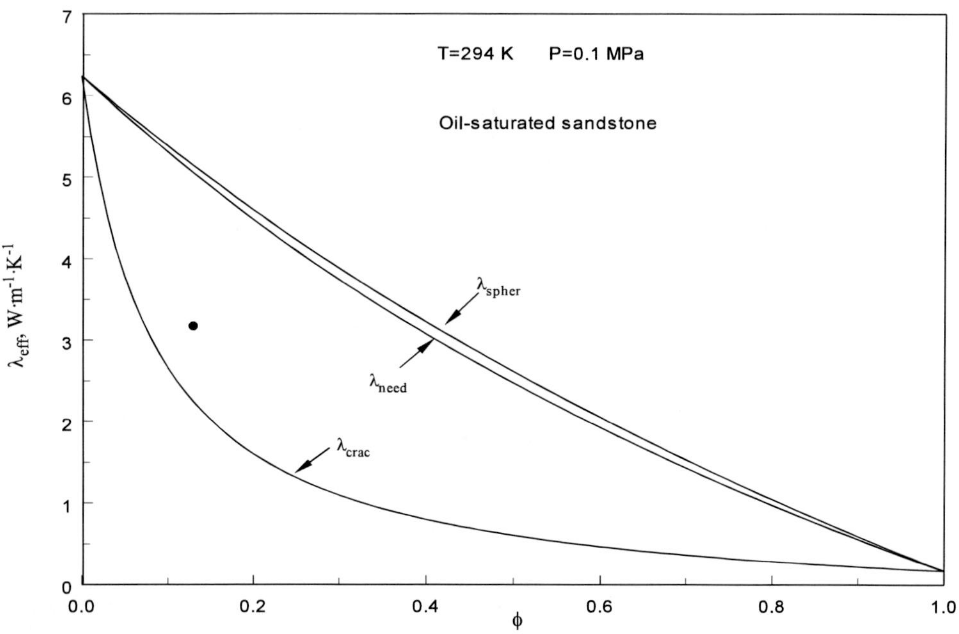

Figure 42. ETC of oil-saturated sandstone as a function of porosity **predicted from the Zimmerman's** [335] model (Eq. 237) for the various types of pores (thin cracks, spherical and needle–like pores) together with the present experimental value [292] for the porosity of 13 % at selected temperature of 294 K and at pressures of 0.1 MPa.

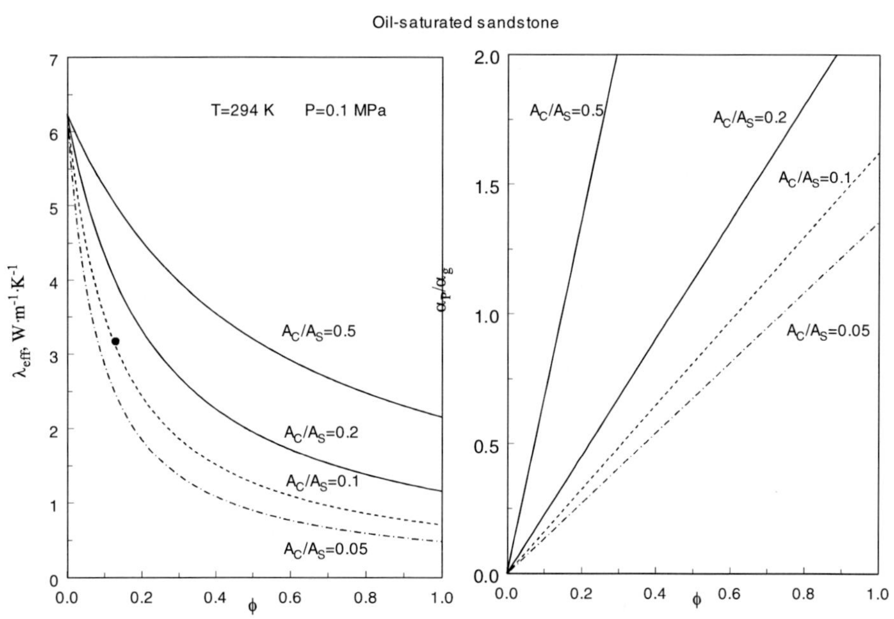

Figure 43. ETC of oil-saturated sandstone as a function of porosity predicted from the Keller's [322] model (left) for the various values of the ratio A_C / A_S (where A_C and A_S are the contact and cross-section areas, respectively) together with the experimental value [292] for the porosity of 13 % at selected temperature of 294 K and pressure of 0.1 MPa. Right: Porosity dependence of the grain shapes ratio (texture factor, α_p / α_g).

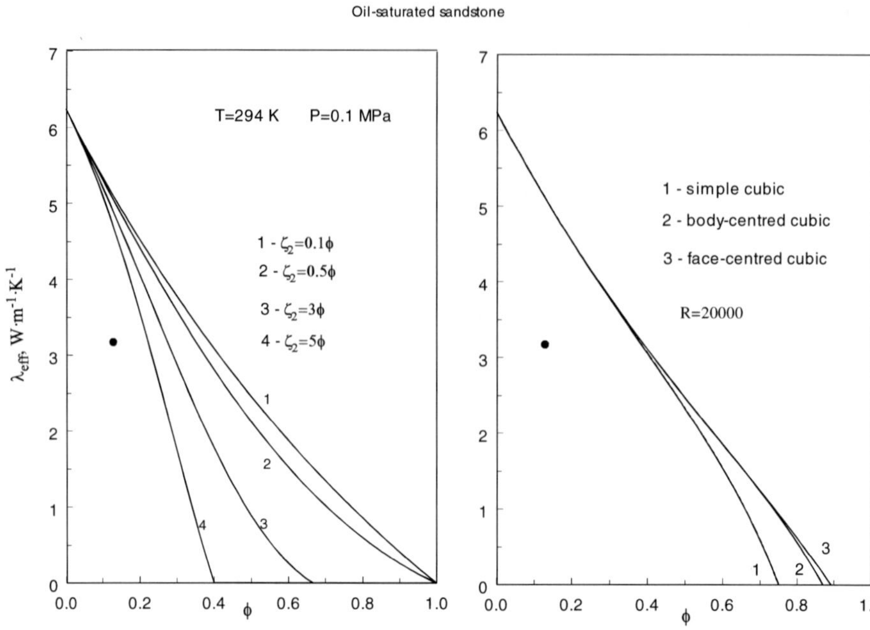

Figure 44. ETC of oil-saturated sandstone as a function of porosity predicted from Cheng and Torquato [331] (right, for the various aforementioned three-point microstructural parameter, ς_2) and Torquato and Rintoul [390] (left, for various cubic lattice periodic arrays at fixed value of the strength of thermal resistance at the interface between the phase, R=20000) models.

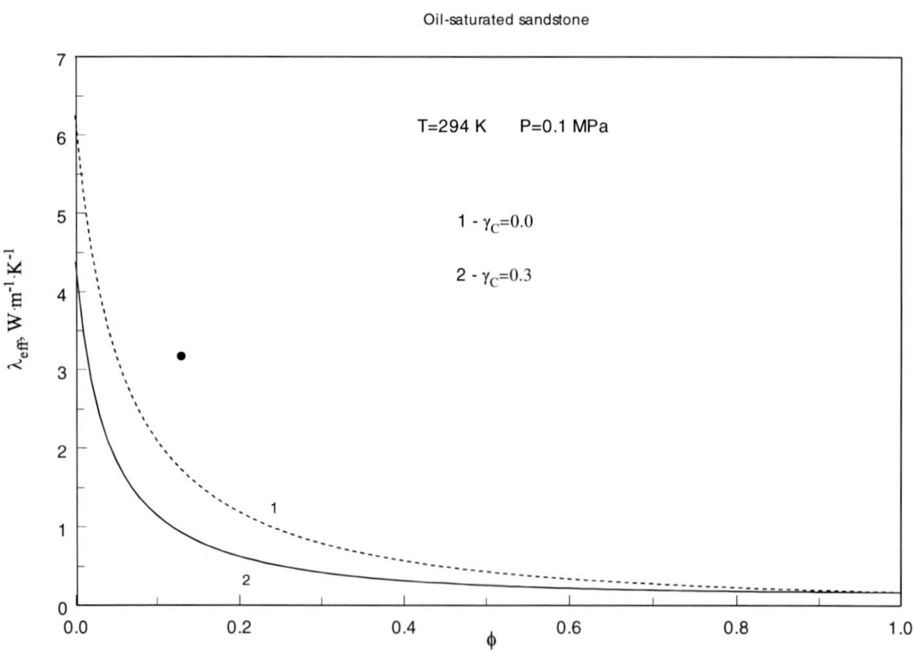

Figure 45. ETC of oil-saturated sandstone as a function of porosity predicted from Hsu et al. [36, 338] model (Eq. 108) for various values of the touching parameter $\gamma_c = c/a$ together with the experimental value [292] for the porosity of 13 % at selected temperature of 294 K and pressure of 0.1 MPa.

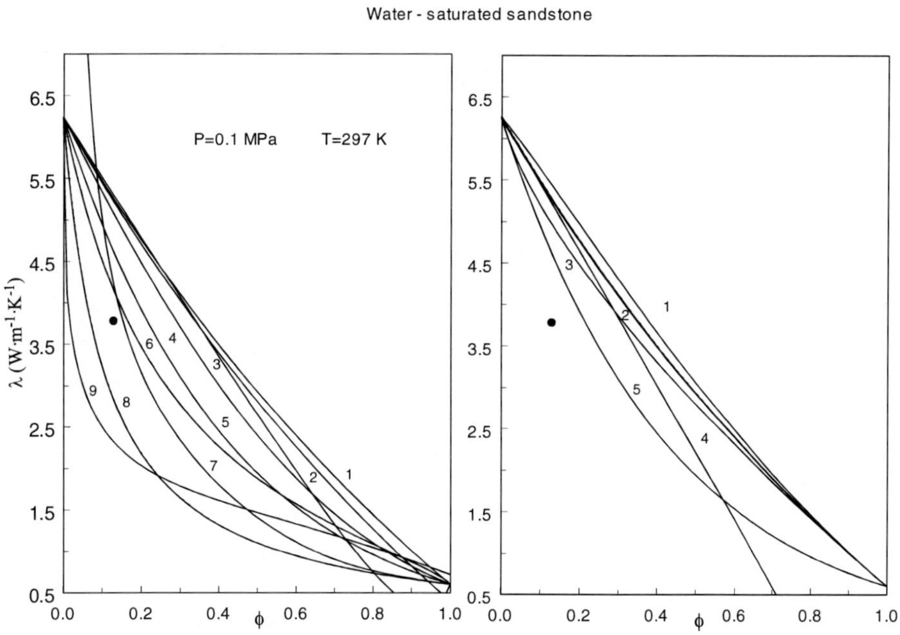

Figure 46. ETC of water-saturated sandstone as a function of porosity predicted from various theoretical models together with our experimental value. Left: 1- Odalevskii [327]; 2- Waff [339]; 3- Brown [369], first order; 4- Bruggeman [377]; 5- Asaad [166]; 6- Chan and Tien [340]; 7- Huang [172] (Eq. 162); 8 - Ziman [74]; 9- Chan and Jeffrey [221]; Right: 1-Russell [410]; 2- Rzhevsky and Novik [426]; 3 - Ribaud [409]; 4 – Mendel [328]; 5 - Intermediate model (Eq. 49).

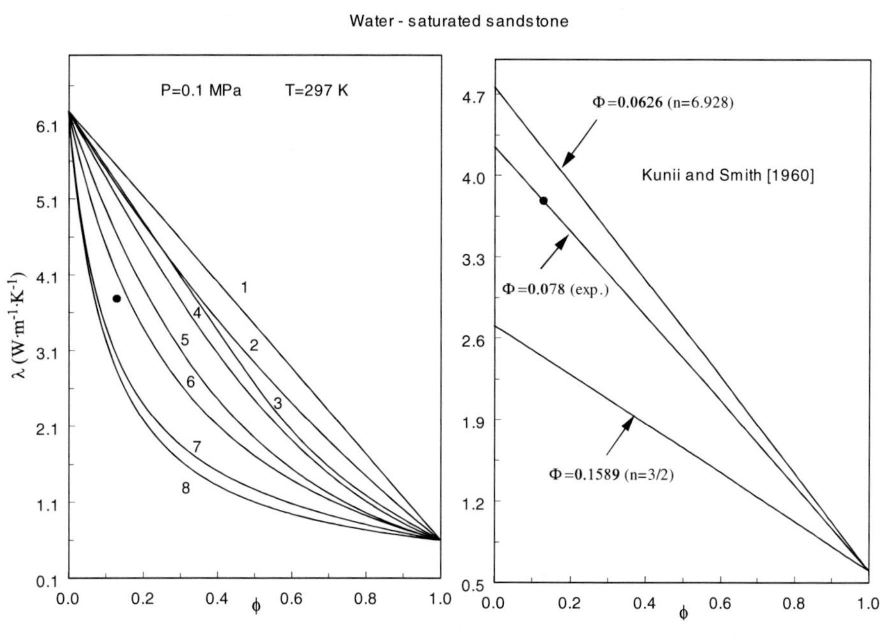

Figure 47. ETC of water-saturated sandstone as a function of porosity predicted from various mixing-law models (left) and Kunii and Smith [236] model (right) (Eq. 156) together with our experimental value [314]. Left: 1- arithmetic mean, first order; 2- Maxwell model (Eq. 52), second order; 3- arithmetic mean, second order; 4- geometric mean, second order; 5- geometric mean, first order; 6- Maxwell model, first order; 7- harmonic mean, second order; 8 - harmonic mean (Eq. 51), first order.

We also used our ETC data for water-saturated sandstone to examine most of the frequently used ETC models. The predictive capability of ETC models essentially depends on how they representing of the microstructure of the porous media. The results of application of some models to the ETC data of water-saturated sandstones are given in Figures 46 and 47. In these figures the wide used ETC models predictions are plotted and compared with our measured value for water-saturated sandstone with porosity of 13 % (single data point).

Our experimental result is close (deviation within 10 %) to the λ_{min} predicted by Walsh and Decker [169]. Therefore, the geometric mean model is the most realistic of these models, as is the only one that honors the fact that there are grain-to-grain contact paths through the material in the direction of lowest thermal conductivity. The main criticism of the geometric mean model is that it does not take into account the texture of the samples, and is therefore valid only for isotropic rocks. As one can see from Figure 46, the predicted values of ETC by Huang [172] are deviate from the measured value within 7.8 %. Majority of predicted values of ETC by various models are higher than measured value. For example, the value of ETC predicted by Asaad's [166] model higher than the present result by 21 %, while the models by Bruggeman [377]; Waff [339]; Odelevskii [327]; Brown [369]; and Eucken [368] are higher by 34 %.

The values of ETC predicted by two models Ziman [74] and Chan and Jeffrey [221] are lower than our measured value by 34 % and 64 %, respectively. Figure 46 (right) shows the comparison between the measured and predicted by Rzhevsky and Novik [426]; Lichtenecker and Rother [443]; Russell [410]; Mendel [328]; and Ribaud [409], and geometric mean model. The values of ETC predicted with these models for water-saturated sandstone higher than our experimental value by 21 to 34 %. The comparison various mixing-law models (parallel, series, first and second order geometric, first and second Maxwell, second order arithmetic and harmonic means models) together with measured value of ETC are presented in Figure 47 (left). The accuracy of the predicted values of ETC water –saturated sandstone by these models is varied within 7.9 % to 45 %. Best agreement is observed for first Maxwell model (7.9 %), the geometric and 2nd order harmonic mean are predict the ETC within 21 % to 26 %, while series and 2nd order geometric models are predict within 34 %. The agreement within 39 % is observed with the second Maxwell and second order arithmetic mean models, while parallel model prediction within 45 %. Kunii and Smith [236] developed a lumped parameter model for unconsolidated and consolidated materials such as sandstone. We compared the model developed by Kunii and Smith [236] with our experimental ETC data for water-saturated sandstone for various values of dimensionless geometric parameter Φ which is depend on the number of contact points, n (see Figure 47 right). This model with Φ =0.0626 and $n=4\sqrt{3}$ agree with the measured values of ETC for water-saturated sandstone within 10 %, while at Φ=0.1589 and n=3/2 the predicted value considerable (by 52 %) lower than experimental value. The optimal value of Φ (at $n=4\sqrt{3}$) derived from our ETC data is 0.078 (close to theoretical value 0.0626).

Figure 48 shows the comparison between the measured values of ETC and the values of λ_{max} and λ_{min} predicted with various models. As one can see, λ_{min} predicted with geometric model lower than λ_{min} by Walsh and Decker [169], while λ_{max} predicted by arithmetic model is higher than λ_{max} by Walsh and Decker [169] model. Walsh's model for the ETC considerable narrowing the boundary of the λ_{max} and λ_{min}. Our experimental result is close (deviation within 10 %) to the λ_{min} predicted by Walsh and Decker [169]. This model has been widely used by different authors [23, 27, 82, 124, 161, 364-366].

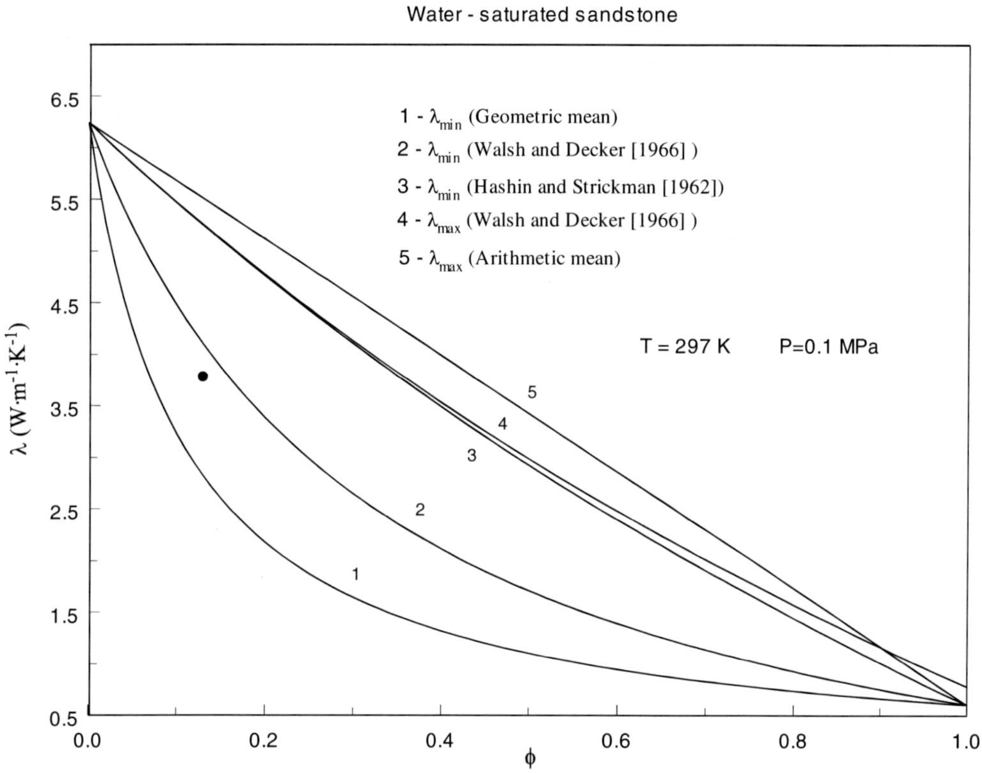

Figure 48. Maximum and minimum of the ETC of water-saturated sandstone as a function of porosity predicted from various models (Eqs. 49, 50, 164, 165, and 206).

The predicted ETC behavior for two models developed by Sugawara and Yoshizawa [224] and Buntebarth and Rueff [224] presented in Figure 49 for various model's parameters. In work [224] model the parameter A is the factor accounting for the pore geometry; $n>0$ ($\approx$ 0.5), determined by mode of packing, pore size, pore shape, and emissivity inside of the pore.

The parameter A becomes equal to porosity if the pores are considered cylindrical and parallel with the direction of heat flow. For more complicated structure A depends on porosity and cementation. The value of n for the sandstones is 2.5. But the value n=2.5 predict ETC water-saturated sandstone within 18 %. The experimental value of n is 4.5 for water – saturated sandstone, while for oil-saturated sandstone n=6.0 [292]. The porosity dependence of ETC predicted by Buntebarth and Rueff [185] is demonstrated in Figure 49 (right) for the various values of model's parameter, α , which is describe the geometry of the fluid phase in pores and varies with wetting properties of the liquid. At α =0, the model is reduced to parallel (arithmetic model). The optimal value of the parameter for water-saturated sandstone is 8.

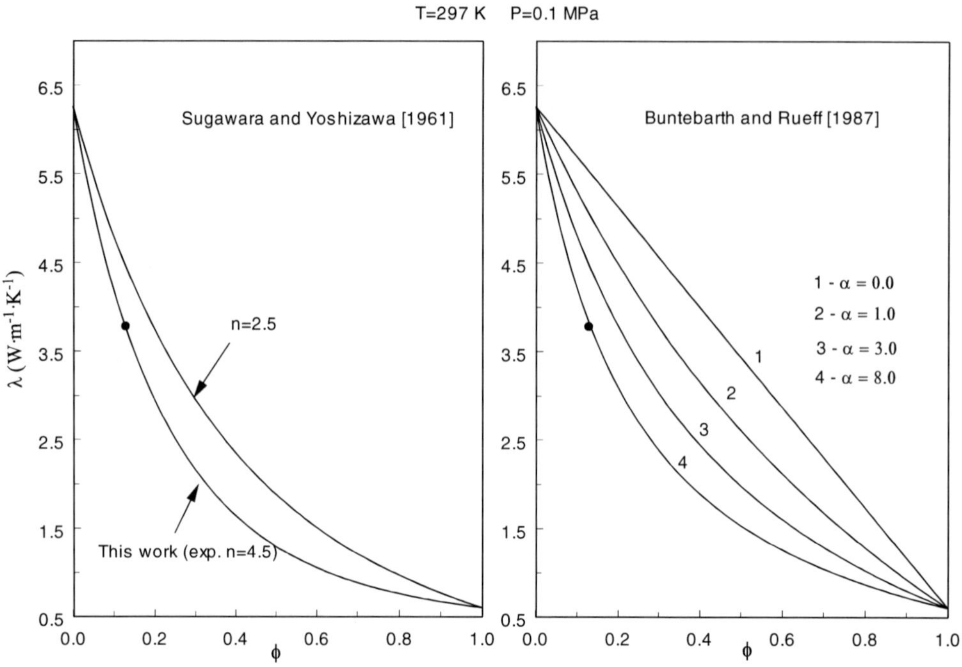

Figure 49. ETC of water-saturated sandstone as a function of porosity predicted from Sugawara and Yoshizawa [224 and 225] (left) and Buntebarth and Rueff [185] (right) models (Eq. 192) for various values of empirical parameter n and α together with our experimental values [314].

The porosity dependence of the ETC predicted from Keller et al. [322] model is shown in Figure 50 (left) for various values of the ratio A_C / A_S (where A_C and A_S are the contact and cross-section areas, respectively). The ratio A_C / A_S is related with the pore $2b$ and grain $2a$ sizes as $A_C / A_S = (1 - b/a)^2$. Best result was found at A_C / A_S=0.1. As was shown in our previous work [292], the optimal value of the ratio A_C / A_S for oil-saturated sandstone is the same, 0.1. The porosity dependence of the grain shapes ratio (or texture factor, α_P / α_g) at T=297 K and 0.1 MPa is shown in Figure 50 (right).

Figure 51 (left) shows the comparison between the measured and predicted by Zimmerman [335] model values of ETC for various types of pores (thin cracks, spherical pores, and needle–like). As one can see, for the thin-crack approximation the results much better (differences about 7 %) than for the spherical and needle–like pores (differences about 37 %). Sidorov [477] developed two different models for linear and nonlinear thermal flux. The prediction result for the ETC is presented in Figure 51 (right). The agreement between measured and predicted values of ETC is within 26 % for linear thermal flux and 42 % for the nonlinear thermal flux. Krupiczka [238] correlation model with B=-0.057 (author recommended value) agree with the measured values of ETC within 29 % (Figure 52, left). The porosity range where the model applicable is ($0.215 < \phi < 0.476$) slightly higher than for our measured sample (ϕ=0.13). Perhaps, this is one of the reason large discrepancies between predicted and measured values of ETC for water-saturated sandstone. Good result can be achieved if parameter B consider as empirical constant. The experimental values of the parameter B is -0.15. Large difference up to 53 % was found between measured and predicted by Schumann and Voss [398] model values of ETC for water-saturated sandstone (see Figure 52, right). Good agreement within 5 % is observed for the Hsu et al. [36] predictive model (see Figure 53, left) when touching parameter γ_c is 0 (the same as continuous fluid-phase model of Nozad et al. [310]). The experimental value of the γ_c is 0.01 for the 3-dimensional unit cell. At α_0=0 the Hsu's model is reduced to Zehner and Schunder [44] model. Hsu et al. [36] recommended value of α_0 is 0.002. Two different Maxwell's models are depicted in Figure 53 (right). As one can see, the model for spherical particles suspended in continues fluid good (within 7.8 %) predict the measured value of the ETC for water-saturated sandstone, while the model for random distributed spherical fluid-filled voids differ from the experiment by 42 %. Zehner and Schunder [44] model predicting results are shown in Figure 54 (left) for the shape parameter B correlation coefficients equal $B = C(1/\phi - 1)^{m_c}$ where C=1.25 and m_C=10/9.

The prediction accuracy of the present result is 7.9 %. The experimentally determined values of C and m_C are 1.15 and 1.0, respectively. At C=1.364 and m_C=1.055 the model reduced to Hsu et al. [36] model. Hadley [34] model with the empirical parameters ($f_0 = 0.8$ and α_p=0.02) predict the measured values of ETC within 18 % (Figure 54, right).

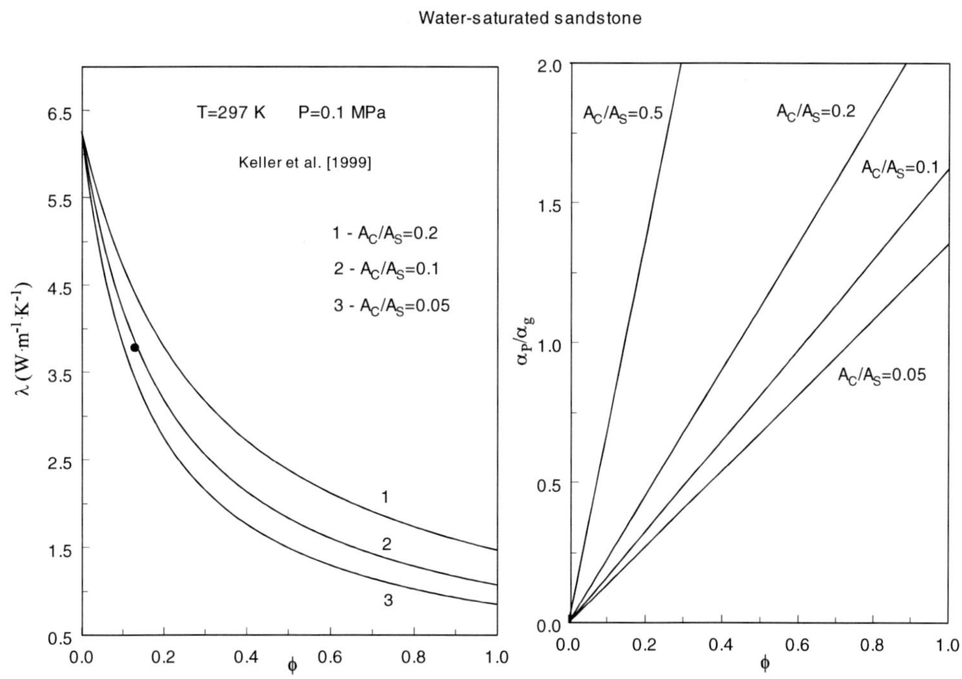

Figure 50. ETC of water-saturated sandstone as a function of porosity predicted from Keller et al. [322] model (left) for various values of the ratio of contact and cross-section areas, A_C / A_S. Right: Porosity dependence of the grain shapes ratio (texture factor, α_p / α_g).

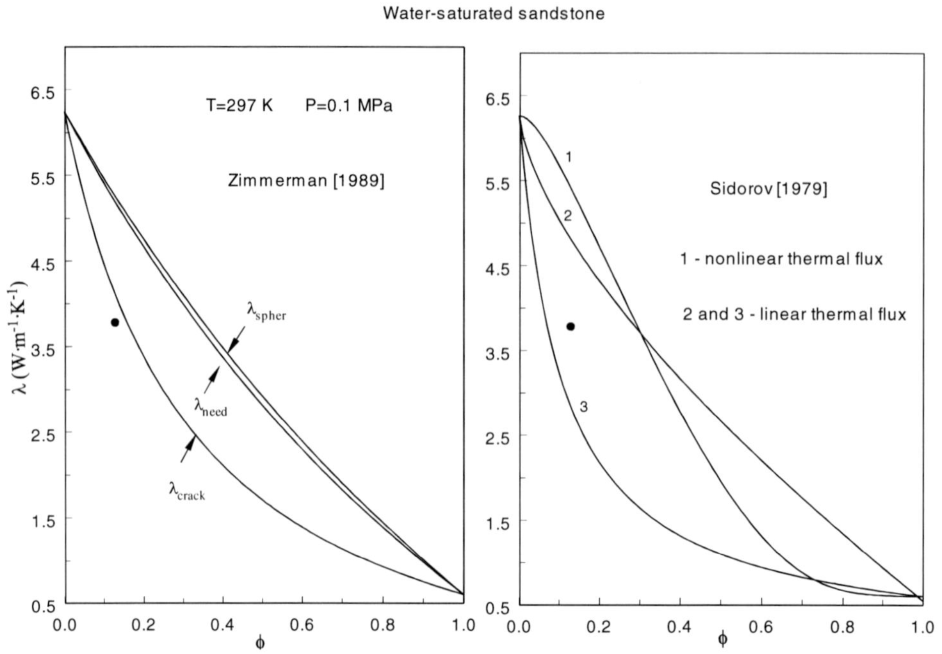

Figure 51. ETC of water-saturated sandstone as a function of porosity predicted from the Zimmerman's [335] model (left) for the various types of pores (thin cracks, spherical and needle–like pores) and Sidorov's [477] model (right) for linear and nonlinear thermal flux.

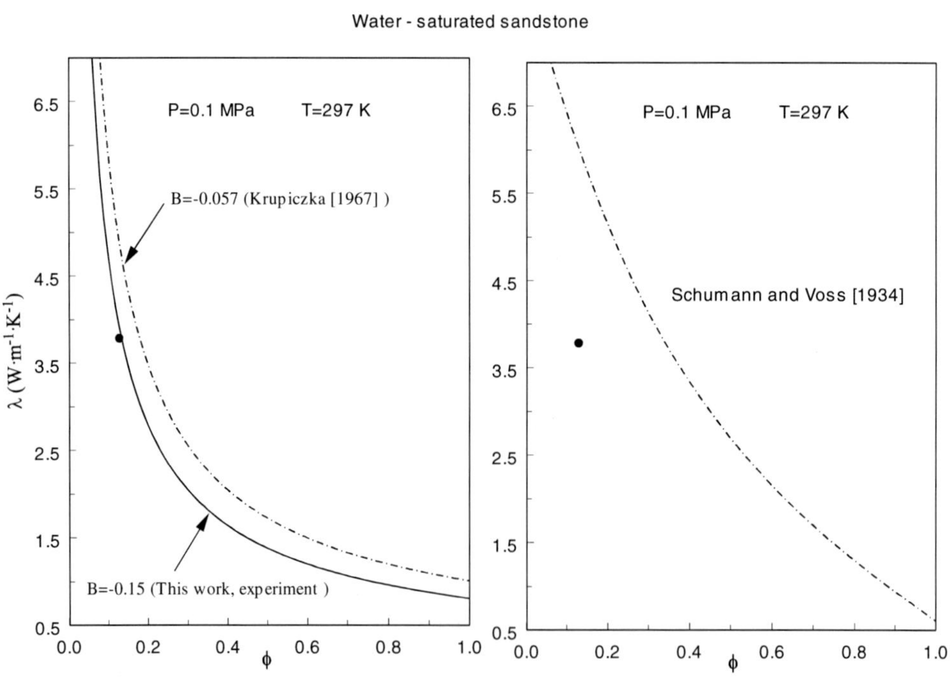

Figure 52. ETC of water-saturated sandstone as a function of porosity predicted from the Krupiczka's [238] (left) and Schumann and Voss [398] models (right) (Eqs. 153-155, and 161).

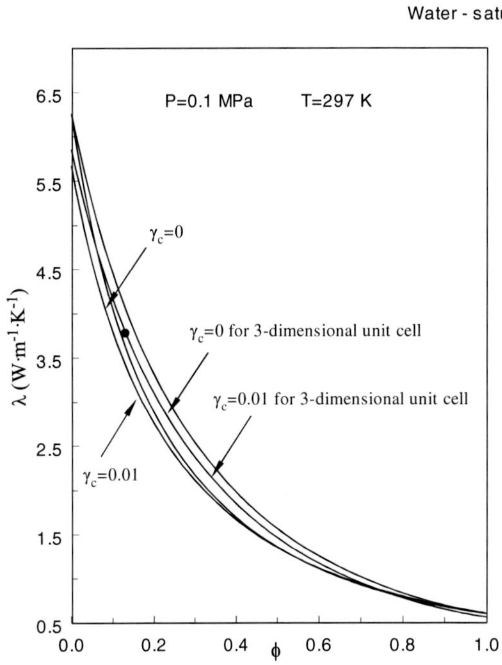

Figure 53. ETC of water-saturated sandstone as a function of porosity predicted from Hsu et al. [36 and 338] model (left) for various values of the touching parameter, $\gamma_c = c/a$, and Maxwell's models (1-for spherical fluid-filled voids; 2-for spherical particles suspended in a continued fluid).

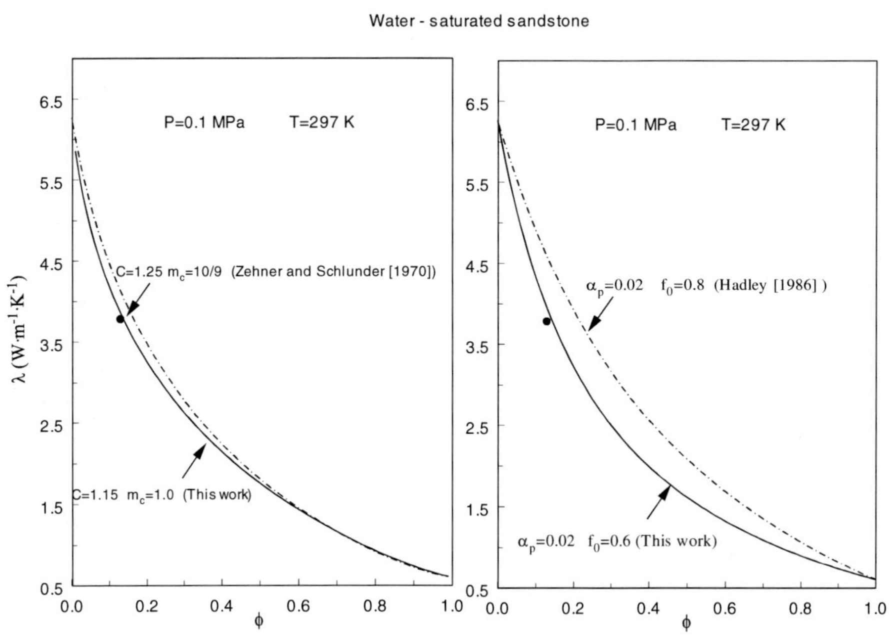

Figure 54. ETC of water-saturated sandstone as a function of porosity predicted from the Zehner and Schlünder [44] (left) and Hadley's [34] models (Eqs. 113, 114, 170-172) for various values of the model's parameters.

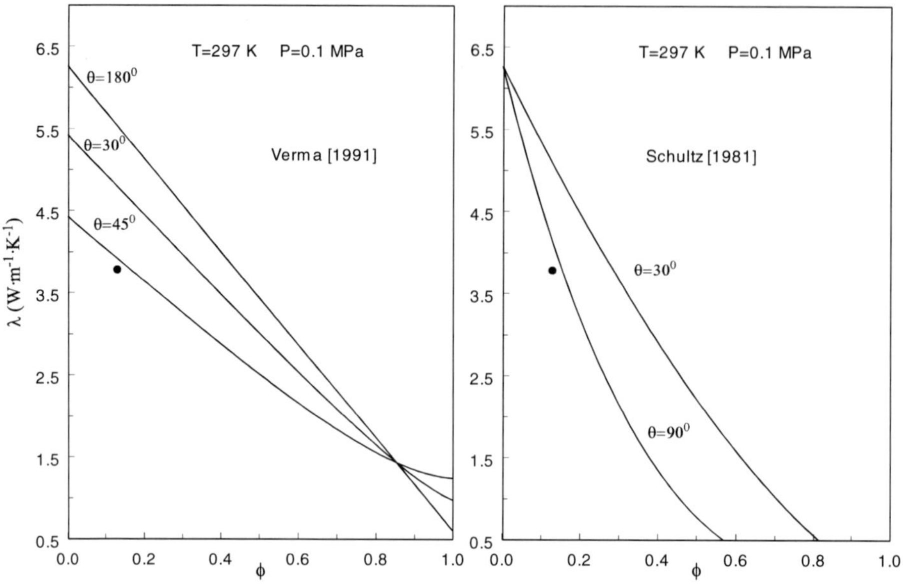

Figure 55. ETC of water-saturated sandstone as a function of porosity predicted from Verma et al. [255] (left) and Schultz [489] models (Eqs. 173-174) for various values of the model's parameters.

For water-saturated porous media (for metal matrix) the optimal values of the parameter f_0 are between 0.8 and 0.9. For water-saturated sandstone the optimal value of the parameter f_0 is 0.6. The model by Verma [255] (Figure 55, left) is excellent (2.6 %) agree with the present ETC data if θ=45° (sin45°=cos45°=$1/\sqrt{2}$). The best result for the Schultz [489] model was achieved at θ=90° (deviation between predicted and measured values is 7.9 %), while at θ=30° the agreement is within 34 % (Figure 55, right).

Conclusion

Available sources of data for the effective thermal conductivity of dry and fluid (gas, water, and oil)–saturated sandstones at high temperatures and high pressures have been collected and evaluated. A comprehensive review of the different contact (divided-bar, needle-probe, hot-wire, guarded hot plate, etc.) and contact-free (optical scanning, laser flash) measurement techniques were provided. The uncertainties of each method are presented. Various theoretical and semiempirical models for temperature, pressure, and porosity dependences of the ETC sandstones were applied for dry, gas-, water-, and oil –saturated sandstones. By means of the steady-state guarded parallel plate apparatus the λ, P, T relationships for the dry, gas-, water-, and oil-saturated sandstones with a various porosities were measured in the temperature range 273 K to 573 K and at pressures up to 400 MPa. The estimated uncertainty in ETC measurements is 4 % (expanded uncertainties at the 95 % confidence level with a coverage factor of k=2). The effect of pressure and temperature on the ETC behavior of dry, gas-, water-, and oil-saturated sandstones was studied. A sharp increase of the ETC was found for fluid-saturated sandstone sample at low pressures (between 0.1 and 100 MPa).

At high pressures (P>100 MPa) a weak linear dependence of the ETC with pressure was observed. The pressure effect is negligible small after the first 100 MPa where bridging of microcracks or improvement of grain contacts take place. The pressure and thermal coefficients were derived using the measured values of ETC as a function of temperature and pressure. The thermal coefficient, β_T, of ETC fluid –saturated sandstones is changed within $(0.42\text{-}0.60)\times10^{-3}$ K^{-1}, in the temperature range from 288 K to 520 K and at pressures up to 250 MPa, while the pressure coefficient, β_P ,of ETC for fluid-saturated sandstones is changes within $(0.045 \text{ to } 3.684)\times10^{-3}$ MPa^{-1} at the same temperature and pressure ranges. We examined the applicability various theoretical, empirical, and mixing-law models and their accuracy to predict of the ETC of fluid-saturated sandstones. The geometric mean model is effective for ETC prediction of sandstone. The effect of the structure (size, shape, and distribution of the pores) and mineralogical compositions (quartz content) on the ETC of various rocks types was discussed. The values of ETC provided in this work for fluid-saturated sandstones can be used to test and improve the existing theoretical models which can be applied to estimate thermal conductivities of all types of rocks at high temperatures and high pressures. The ETC of oil-saturated sandstone decreases with temperature at a rate approximately 20 % less than the rate for dry sandstone. The pressure rate of oil-saturated sandstone is about 12 % lower than the pressure rate for gas-saturated sandstone. The pores fluid pressure is primary cause of the less rapid decrease in gas –saturated sandstone ETC with pressure. The fabric model represents a good approach for the ETC of sandstone, based on the mineral composition, porosity, ant temperature. This model yield good estimates of ETC in complex fluid-rock systems. The results of the present study indicate that ETC of oil-saturated sandstone is determined for the most part by the conductivity of the saturating fluid. Porosity is second most important factor. Solid phase conductivity is the important correlation parameter.

THE APPENDIX

Table 1. Experimental thermal conductivity data of sandstones, λ ($W \cdot m^{-1} \cdot K^{-1}$), at high temperatures and high pressures

Sandstone (Buinaksk, density 2170 kg·m-3, ϕ =16.2%) [277]									
	Pressure, MPa								
T, K	0.1	50	100	150	200	250	300	400	
Gas-saturated (argon) [320]									
275	0.62	0.76	0.84	0.86	0.87	0.88	0.89	0.90	
323	0.64	0.77	0.85	0.89	0.90	0.90	0.92	0.92	
373	0.67	0.81	0.87	0.91	0.92	0.93	0.94	0.95	
423	0.70	0.84	0.91	0.93	0.90	0.95	0.95	0.98	

Table 1. (Continued)

Sandstone (Buinaksk, density 2170 kg·m-3, ϕ =16.2%) [277]									
T, K	Pressure, MPa								
	0.1	50	100	150	200	250	300	400	
Oil-saturated (castor) [277]									
275	1.64	1.81	1.84	1.86	1.87	1.89			
323	1.69	1.83	1.86	1.87	1.89	1.91			
373	1.70	1.87	1.90	1.91	1.92	1.96			
423	1.73	1.91	1.93	1.94	1.95	1.98			
Water-saturated [277]									
275	2.25	2.43	2.46	2.47	2.48	2.50			
323	2.26	2.44	2.46	2.48	2.49	2.51			
373	2.27	2.45	2.46	2.49	2.50	2.53			
423	2.30	2.48	2.48	2.52	2.53	2.54			

Sandstone (Solonchak, 3941 m, density 2180 kg·m^{-3}, porosity ϕ=13%) [291]

(air-saturated sample)										
	Pressure, MPa									
T, K	0.1	20	50	100	150	200	250	300	350	400
275	2.71	2.82	2.86	2.89	2.90	2.91	2.92	2.93	2.93	2.94
323	2.63	2.73	2.77	2.84	2.85	2.86	2.87	2.88	2.88	2.89
373	2.56	2.65	2.72	2.77	2.78	2.79	2.81	2.82	2.83	2.84
423	2.49	2.57	2.64	2.72	2.73	2.74	2.75	2.76	2.77	2.79
473	2.42	2.49	2.56	2.67	2.68	2.69	2.70	2.71	2.72	2.73
523	2.34	2.41	2.48	2.61	2.62	2.64	2.65	2.66	2.67	2.68

water-saturated sandstone [291]							
0.1 MPa		50 MPa		100 MPa		250 MPa	
T, K	λ, W·m^{-1}·K^{-1}	*T*, K	λ, W·m^{-1}·K^{-1}	*T*, K	λ, W·m^{-1}·K^{-1}	*T*, K	λ, W·m^{-1}·K^{-1}
281	3.81	279	3.91	284	4.02	281	4.11
289	3.81	307	3.90	296	3.98	290	4.11
297	3.78	318	3.88	328	3.92	304	4.05
305	3.77	345	3.81	350	3.90	335	4.02
325	3.75	364	3.80	398	3.79	362	3.94
347	3.72	388	3.77	417	3.78	392	3.90
368	3.66	405	3.72	445	3.71	408	3.88
391	3.65	437	3.69	464	3.76	441	3.79
398	3.62	460	3.62	495	3.62	463	3.76
417	3.57	478	3.61	509	3.62	478	3.72
454	3.53	505	3.55	-	-	490	3.70
472	3.49	521	3.54	-	-	516	3.66
498	3.44	-	-	-	-	-	-
518	3.41	-	-	-	-	-	-

oil-saturated sandstone [291]							
0.1 MPa		50 MPa		100 MPa		250 MPa	
T, K	λ, W·m^{-1}·K^{-1}	*T*, K	λ, W·m^{-1}·K^{-1}	*T*, K	λ, W·m^{-1}·K^{-1}	*T*, K	λ, W·m^{-1}·K^{-1}
288	3.18	291	3.44	294	3.54	294	3.62
318	3.15	308	3.42	316	3.50	309	3.61
344	3.11	322	3.40	334	3.45	356	3.53
356	3.09	329	3.37	352	3.43	370	3.48
388	3.02	362	3.34	364	3.40	394	3.45
428	2.96	378	3.31	392	3.38	418	3.39
470	2.91	399	3.25	406	3.33	427	3.37
500	2.87	428	3.23	435	3.31	467	3.31
515	2.83	463	3.15	457	3.24	492	3.25
-	-	495	3.12	481	3.23	505	3.22
-	-	520	3.06	507	3.16	519	3.21

Sandstone (Sukhokumsk, 4900 m, density 2280 kg·m^{-3}, porosity ϕ=7%) [290]

T, K	Pressure, MPa								
	0.1	50	100	150	200	250	300	350	400
Gas-saturated (argon) [290]									
273	2.06	2.18	2.20	2.24	2.25	2.26	2.27	2.28	2.29
323	1.91	2.10	2.13	2.16	2.17	2.19	2.20	2.21	2.22
373	1.83	2.03	2.07	2.10	2.11	2.13	2.14	2.15	2.16
423	1.76	1.97	2.07	2.04	2.06	2.08	2.09	2.10	2.10
473	1.71	1.92	1.97	2.00	2.01	2.04	2.05	2.06	2.05
523	1.66	1.88	1.94	1.96	1.97	2.00	2.01	2.02	2.02
Oil-saturated (castor) [290]									
273	2.16	2.24	2.29	2.34	2.37	2.40	2.42	2.44	2.46
323	2.07	2.14	2.21	2.25	2.28	2.31	2.33	2.36	2.34
373	1.99	2.06	2.13	2.19	2.21	2.24	2.26	2.29	2.31
423	1.93	2.00	2.09	2.13	2.15	2.19	2.20	2.23	2.25
473	1.87	1.94	2.05	2.08	2.10	2.13	2.15	2.18	2.20
523	1.83	1.90	2.01	2.04	2.06	2.09	2.11	2.14	2.16
Water-saturated [290]									
273	2.38	2.40	2.43	2.45	2.47	2.49	2.51	2.53	2.55
323	2.27	2.30	2.34	2.36	2.38	2.40	2.42	2.44	2.46
373	2.19	2.22	2.26	2.28	2.30	2.32	2.34	2.36	2.38
423	2.12	2.15	2.19	2.21	2.23	2.25	2.27	2.29	2.31
473	2.06	2.10	2.14	2.16	2.18	2.20	2.22	2.24	2.26
523	2.01	2.06	2.09	2.11	2.13	2.15	2.17	2.19	2.21

Sandstone (Gasha, 3420 m, density 2460 kg·m^{-3}, water content 0.49 %, quartz-20 %, Porosity, cement-12 %, feldspar-15 %, ϕ=10.5%) [285]

	Pressure, MPa											
T, K	0.1	5	10	15	20	30	50	70	90	110	130	150
gas-saturated (argon) [285]												
293	1.98	2.05	2.09	2.14	2.18	2.26	2.30	2.34	2.36	2.38	2.34	2.36
323	1.83	1.89	1.96	1.99	2.04	2.10	2.14	2.17	2.49	2.14	2.17	2.49
348	1.70	1.76	1.82	1.86	1.90	1.96	2.02	2.04	2.05	2.02	2.04	2.05
373	1.58	1.64	1.71	1.76	1.77	1.81	1.85	1.88	1.89	1.85	1.88	1.89
398	1.47	1.53	1.58	1.63	1.65	1.70	1.73	1.76	1.77	1.73	1.76	1.77
423	1.36	1.42	1.47	1.51	1.53	1.56	1.60	1.64	1.66	1.60	1.64	1.66
448	1.25	1.32	1.36	1.39	1.40	1.44	1.47	1.51	1.52	1.47	1.51	1.52
473	1.16	1.21	1.25	1.28	1.29	1.33	1.36	1.40	1.43	1.36	1.40	1.43
498	1.07	1.13	1.15	1.18	1.20	1.23	1.27	1.29	1.31	1.27	1.29	1.31
523	0.99	1.03	1.06	1.07	1.10	1.14	1.17	1.19	1.21	1.17	1.19	1.21
548	0.89	0.93	0.96	0.98	1.00	1.03	1.07	1.09	1.11	1.07	1.09	1.11
573	0.82	0.84	0.87	0.88	0.90	0.92	0.95	0.97	0.99	0.95	0.97	0.99
water- saturated [285]												
293	3.82	3.97	4.30	4.22	4.28	4.36	4.43	4.47	4.49	4.50	4.51	4.52
323	3.64	3.78	3.90	4.02	4.07	4.14	4.19	4.23	4.24	4.25	4.26	4.27
348	3.54	3.67	3.76	3.85	3.92	3.98	4.03	4.07	4.09	4.10	4.12	4.13
373	-	3.52	3.65	3.76	3.80	3.84	3.89	4.02	3.94	3.95	3.96	3.97
398	-	3.40	3.53	3.63	3.68	3.74	3.78	3.81	3.83	3.84	3.85	3.86
423	-	3.24	3.39	3.47	3.53	3.61	3.64	3.67	3.69	3.70	3.71	3.72
448	-	3.09	3.21	3.26	3.32	3.39	3.42	3.45	3.46	3.49	3.51	3.53
473	-	2.87	2.97	3.04	3.09	3.14	3.18	3.22	3.24	3.26	3.27	3.29
498	-	2.68	2.77	2.80	2.85	2.88	2.92	2.96	2.97	2.99	3.00	3.02
523	-	2.49	2.54	2.59	2.65	2.67	2.70	2.74	2.77	2.78	2.80	2.81
548	-	-	2.36	2.42	2.46	2.51	2.55	2.57	2.59	2.61	2.62	2.63
573	-	-	2.23	2.27	2.30	2.34	2.38	2.41	2.44	2.45	2.46	2.47
oil-saturated [285]												
293	2.59	2.68	2.77	2.84	2.88	2.94	2.98	3.02	3.04	3.06	3.08	3.09
323	2.42	2.52	2.61	2.68	2.71	2.77	2.82	2.86	2.88	2.89	2.90	2.91
348	2.27	2.37	2.46	2.51	2.55	2.61	2.66	2.70	2.72	2.73	2.75	2.76
373	-	2.23	2.31	2.37	2.42	2.46	2.51	2.53	2.56	2.58	2.59	2.60
398	-	2.08	2.16	2.21	2.26	2.30	2.36	2.38	2.41	2.43	2.44	2.46
423	-	1.93	2.02	2.07	2.10	2.14	2.20	2.22	2.24	2.26	2.28	2.29
448	-	1.80	1.86	1.91	1.94	1.98	2.04	2.08	2.10	2.11	2.13	2.14
473	-	1.55	1.71	1.76	1.79	1.83	1.85	1.91	1.94	1.96	1.97	1.98
498	-	1.53	1.60	1.63	1.65	1.69	1.73	1.76	1.79	1.81	1.83	1.84
523	-	1.40	1.45	1.48	1.50	1.55	1.56	1.58	1.60	1.63	1.65	1.66
548	-	-	1.32	1.36	1.39	1.42	1.44	1.47	1.49	1.50	1.52	1.53
573	-	-	1.20	1.23	1.25	1.28	1.31	1.34	1.36	1.37	1.38	1.39

Sandstone (North Kochubei, 4355 m, density 2340 kg·m^{-3}, water content 1.3 %, quartz-35 %, cement-6 to 7 %, porosity ϕ=12%) [285]

	Pressure, MPa											
T, K	0.1	5	10	15	20	30	50	70	90	110	130	150
gas-saturated (argon) [285]												
293	1.30	1.36	1.41	1.45	1.49	1.54	1.58	1.62	1.66	1.69	1.72	1.74
323	1.17	1.24	1.29	1.33	1.37	1.41	1.46	1.50	1.54	1.57	1.60	1.63
348	1.09	1.15	1.20	1.24	1.28	1.32	1.37	1.41	1.45	1.49	1.52	1.54
373	1.03	1.09	1.13	1.17	1.21	1.26	1.30	1.34	1.38	1.42	1.45	1.47
398	0.99	1.05	1.08	1.13	1.17	1.21	1.25	1.29	1.33	1.36	1.39	1.41
423	0.95	1.01	1.05	1.09	1.13	1.17	1.20	1.24	1.28	1.31	1.34	1.37
448	0.91	0.96	1.01	1.04	1.09	1.13	1.16	1.20	1.24	1.27	1.31	1.34
473	0.88	0.93	0.97	1.00	1.05	1.09	1.12	1.16	1.20	1.24	1.27	1.29
498	0.84	0.90	0.94	0.98	1.02	1.06	1.09	1.12	1.16	1.20	1.23	1.25
523	0.81	0.86	0.91	0.95	0.99	1.03	1.06	1.09	1.12	1.15	1.18	1.20
548	0.77	0.84	0.88	0.92	0.96	0.99	1.02	1.06	1.09	1.11	1.14	1.17
573	0.75	0.80	0.85	0.89	0.92	0.96	0.99	1.02	1.05	1.08	1.11	1.14
water- saturated [285]												
293	3.52	3.68	3.83	3.93	4.05	4.12	4.18	4.23	4.27	4.31	4.35	4.38
323	3.21	3.35	3.52	3.64	3.75	3.84	3.91	3.96	4.01	4.06	4.09	4.13
348	2.95	3.10	3.24	3.38	3.50	3.59	3.67	3.73	3.78	3.83	3.86	3.90
373	-	2.88	3.02	3.15	3.26	3.35	3.42	3.49	3.55	3.59	3.64	3.68
398	-	2.72	2.84	2.96	3.05	3.14	3.21	3.28	3.33	3.37	3.43	3.47
423	-	2.58	2.72	2.82	2.90	3.00	3.07	3.12	3.16	3.20	3.25	3.29
448	-	2.42	2.54	2.67	2.76	2.84	2.90	2.95	3.00	3.05	3.09	3.13
473	-	2.27	2.40	2.51	2.59	2.67	2.74	2.80	2.85	2.90	2.94	2.97
498	-	2.14	2.25	2.37	2.44	2.51	2.54	2.65	2.71	2.75	2.80	2.84
523	-	1.98	2.12	2.21	2.30	2.38	2.44	2.51	2.56	2.61	2.65	2.69
548	-	-	1.99	2.09	2.17	2.24	2.31	2.36	2.42	2.47	2.51	2.56
573	-	-	1.87	1.98	2.05	2.12	2.18	2.23	2.28	2.33	2.38	2.42
oil-saturated [285]												
293	2.46	2.64	2.84	2.95	3.06	3.14	3.21	3.26	3.31	3.34	3.38	3.42
323	2.20	2.37	2.54	2.67	2.78	2.87	2.94	2.99	3.05	3.10	3.10	3.17
348	2.04	2.19	2.36	2.48	2.59	2.67	2.74	2.80	2.84	2.89	2.93	2.96
373		2.06	2.23	2.33	2.44	2.51	2.57	2.63	2.68	2.73	2.77	2.81
398		1.97	2.12	2.23	2.32	2.39	2.45	2.51	2.56	2.60	2.64	2.67
423		1.91	2.04	2.14	2.23	2.30	2.36	2.41	2.46	2.50	2.53	2.57
448		1.85	1.99	2.09	2.17	2.23	2.30	2.35	2.40	2.44	2.48	2.51
473		1.81	1.95	2.03	2.11	2.19	2.25	2.30	2.35	2.39	2.43	2.45
498		1.76	1.90	1.99	2.07	2.14	2.20	2.25	2.31	2.35	2.38	2.41
523		1.71	1.86	1.95	2.03	2.10	2.16	2.22	2.27	2.31	2.35	2.38
548			1.82	1.92	2.00	2.06	2.12	2.18	2.23	2.28	2.32	2.34
573			1.80	1.89	1.97	2.03	2.09	2.15	2.19	2.25	2.28	2.31

Sandstone- quartz, (Duzlak, 3360 m, density 2700 kg·m^{-3}, water content 0.60 %, quartz-20 %, cement-12 %, porosity ϕ=6.9 %) [285]

	Pressure, MPa											
T, K	0.1	5	10	15	20	30	50	70	90	110	130	150
Gas-saturated (argon) [285]												
293	2.22	2.28	2.35	2.39	2.44	2.51	2.57	2.63	2.68	2.72	2.77	2.82
323	2.08	2.13	2.18	2.24	2.30	2.37	2.43	2.50	2.54	2.58	2.63	2.68
348	1.97	2.01	2.07	2.12	2.18	2.25	2.32	2.38	2.43	2.47	2.53	2.58
373	1.84	1.89	1.95	2.01	2.07	2.13	2.21	2.27	2.31	2.36	2.41	2.47
398	1.71	1.77	1.83	1.89	1.95	2.01	2.09	2.15	2.21	2.27	2.31	2.35
423	1.62	1.67	1.71	1.78	1.85	1.92	1.99	2.06	2.11	2.16	2.21	2.24
448	1.48	1.55	1.61	1.67	1.74	1.80	1.87	1.95	2.00	2.05	2.10	2.14
473	1.41	1.46	1.52	1.57	1.63	1.71	1.78	1.85	1.91	1.95	2.00	2.04
498	1.30	1.36	1.42	1.48	1.54	1.61	1.69	1.76	1.81	1.86	1.90	1.94
523	1.20	1.28	1.34	1.39	1.45	1.52	1.59	1.66	1.71	1.77	1.81	1.84
548	1.12	1.18	1.25	1.30	1.36	1.43	1.50	1.57	1.62	1.68	1.72	1.75
573	1.01	1.09	1.16	1.21	1.27	1.34	1.42	1.48	1.53	1.58	1.62	1.66
Water- saturated [285]												
293	4.44	4.54	4.66	4.76	4.83	4.94	5.01	5.07	5.11	5.14	5.20	5.24
323	3.92	4.07	4.18	4.24	4.36	4.48	4.57	4.62	4.67	4.71	4.76	4.80
348	3.56	3.69	3.82	3.91	3.99	4.12	4.22	4.28	4.34	4.39	4.42	4.48
373	-	3.40	3.51	3.60	3.69	3.81	3.90	3.96	4.01	4.07	4.12	4.16
398	-	3.23	3.32	3.40	3.47	3.57	3.67	3.73	3.77	3.81	3.86	3.90
423	-	3.06	3.16	3.24	3.32	3.41	3.49	3.56	3.61	3.64	3.68	3.71
448	-	2.88	2.97	3.04	3.13	3.22	3.29	3.35	3.40	3.45	3.51	3.55
473	-	2.62	2.73	2.81	2.88	2.98	3.07	3.12	3.16	3.22	3.27	331
498	-	2.32	2.43	2.51	2.57	2.68	2.77	2.84	2.90	2.95	3.01	3.05
523	-	2.02	2.13	2.22	2.29	2.38	2.47	2.53	2.58	2.64	2.70	2.75
548	-	-	1.88	1.96	2.05	2.15	2.26	2.31	2.36	2.43	2.47	2.52
573	-	-	1.63	1.72	1.80	1.93	2.01	2.06	2.11	2.16	2.22	2.26
oil-saturated [285]												
293	3.26	3.37	3.48	3.52	3.61	3.68	3.75	3.81	3.86	3.90	3.95	4.01
323	2.95	3.06	3.17	3.24	3.32	3.39	3.45	3.51	3,56	3.62	3.66	3.72
348	2.73	2.83	2.92	3.02	3.08	3.15	3.22	3.29	3.33	3.40	3.46	3.51
373	-	2.63	2.71	2.78	2.84	2.92	2.99	3.06	3.12	3.18	3.23	3.27
398	-	2.42	2.51	2.57	2.65	2.71	2.77	2.82	2.88	2.95	3.01	3.06
423	-	2.18	2.27	2.36	2.41	2.50	2.56	2.62	2.67	2.73	2.80	2.86
448	-	2.03	2.10	2.17	2.24	2.33	2.38	2.44	2.49	2.55	2.62	2.66
473	-	1.82	1.92	1.98	2.04	2.13	2.20	2.26	2.32	2.36	2.42	2.47
498	-	1.67	1.74	1.82	1.87	1.95	2.03	2.08	2.12	2.19	2.24	2.30
523	-	1.53	1.60	1.66	1.72	1.78	1.85	1.91	1.96	2.02	2.06	2.13
548	-	-	1.46	1.63	1.65	1.69	1.70	1.76	1.81	1.86	1.90	1.97
573	-	-	1.32	1.38	1.44	1.59	1.57	1.63	1.68	1.72	1.77	1.81

Sandstone (Tarki, 1990 m, density 2310 kg·m^{-3}, water content 0.89 %, quartz and feldspar -35 %, cement-6%, porosity ϕ=13.9 %) [285]

	Pressure, MPa											
T, K	0.1	5	10	15	20	30	50	70	90	110	130	150
Gas-saturated (argon) [285]												
293	1.45	1.59	1.73	1.83	1.89	2.01	2.11	2.14	2.17	2.20	2.23	2.24
323	1.40	1.55	1.69	1.78	1.84	1.95	2.07	2.10	2.14	2.17	2.14	2.20
348	1.37	1.51	1.64	1.74	1.81	1.91	2.00	2.04	2.10	2.12	2.14	2.16
373	1.29	1.43	1.58	1.70	1.75	1.85	1.95	2.00	2.06	2.07	2.10	2.12
398	1.24	1.39	1.55	1.67	1.72	1.88	1.92	1.97	2.05	2.06	2.08	2.09
423	1.21	1.37	1.53	1.63	1.70	1.79	1.90	1.95	2.00	2.04	2.06	2.06
448	1.19	1.34	1.49	1.59	1.67	1.75	1.87	1.93	1.97	2.00	2.03	2.05
473	1.14	1.31	1.45	1.56	1.63	1.72	1.85	1.90	1.95	1.98	2.01	2.03
498	1.10	1.29	1.44	1.54	1.62	1.70	1.83	1.89	1.94	1.97	1.99	2.00
523	1.08	1.26	1.41	1.52	1.61	1.69	1.82	1.88	1.93	1.95	1.97	1.99
548	1.06	1.24	1.39	1.51	1.59	1.68	1.81	1.87	1.91	1.94	1.96	1.98
573	1.04	1.22	1.37	1.50	1.58	1.67	1.80	1.87	1.90	1.94	1.95	1.97
Water- saturated [285]												
293	3.93	4.11	4.28	4.39	4.48	4.67	4.81	4.94	4.99	5.05	5.08	5.13
323	3.82	4.04	4.16	4.29	4.39	4.54	4.73	4.82	4.90	4.95	5.01	5.03
348	3.74	3.87	4.04	4.16	4.28	4.44	4.59	4.73	4.83	4.88	4.94	4.96
373	-	3.80	3.94	4.08	4.20	4.38	4.52	4.67	4.75	4.79	4.86	4.89
398	-	3.72	3.88	4.03	4.15	4.32	4.45	4.61	4.69	4.74	4.79	4.84
423	-	3.69	3.87	3.99	4.12	4.25	4.41	4.58	4.65	4.70	4.75	4.78
448	-	3.65	3.79	3.94	4.04	4.23	4.39	4.55	4.60	4.64	4.70	4.72
473	-	3.61	3.75	3.89	4.02	4.19	4.35	4.49	4.57	4.63	4.65	4.68
498	-	3.58	3.72	3.86	3.97	4.16	4.33	4.46	4.54	4.60	4.63	4.65
523	-	3.56	3.71	3.85	3.96	4.15	4.32	4.43	4.52	4.57	4.61	4.64
548	-	-	3.69	3.82	3.93	4.12	4.31	4.41	4.50	4.55	4.60	4.63
573	-	-	3.68	3.81	3.91	4.09	4.29	4.38	4.47	4.53	4.58	4.61
oil-saturated [285]												
293	2.72	2.92	3.09	3.18	3.27	3.39	3.48	3.57	3.63	3.69	3.73	3.76
323	2.62	2.79	2.96	3.09	3.19	3.32	3.41	3.49	3.56	3.61	3.65	3.67
348	2.52	2.70	2.82	2.99	3.07	3.23	3.32	3.43	3.49	3.53	3.57	3.59
373	-	2.62	2.78	2.88	2.99	3.14	3.24	3.32	3.42	3.46	3.49	3.52
398	-	2.57	2.72	2.84	2.95	3.06	3.20	3.29	3.39	3.41	3.45	3.48
423	-	2.54	2.67	2.79	2.91	3.02	3.15	3.26	3.34	3.37	3.41	3.43
448	-	2.51	2.64	2.76	2.87	2.98	3.12	3.20	3.29	3.34	3.38	3.39
473	-	2.48	2.60	2.73	2.84	2.95	3.09	3.18	3.25	3.30	3.33	3.35
498	-	2.46	2.58	2.70	2.81	2.91	3.07	3.15	3.22	3.27	3.31	3.34
523	-	2.44	2.57	2.68	2.78	2.90	3.05	3.14	3.21	3.25	3.28	3.31
548	-	-	2.55	2.67	2.76	2.90	3.05	3.12	3.18	3.23	3.26	3.28
573	-	-	2.54	2.66	2.75	2.88	3.03	3.10	3.16	3.21	3.24	3.26

Sandstone (Aktash, 2977 m, porosity ϕ=5%) density is 2.18 g·cm^{-3}, weakly cemented, weakly carbonated, quartzed, color is grey, moderate grained, and porosity is 5 %. The sample come from the Aktash, Dagestan, Russia, borehole #1, depth 2977 m, 80-90 % forming material has crystalline structure [297]

	Pressure (MPa)					
T, K	0.1	50	100	150	200	250
273	2.01	2.05	2.08	2.10	2.12	2.14
323	1.93	1.96	1.99	2.01	2.03	2.05
373	1.88	1.91	1.93	1.94	1.96	1.98
423	1.82	1.85	1.87	1.88	1.90	1.92

Sandstone (Sukhakumsk ϕ=7%) [522]									
	Pressure, MPa								
T, K	0.1	50	100	150	200	250	300	350	400
275	2.06	2.18	2.20	2.24	2.25	2.26	2.27	2.28	2.29
323	1.91	2.10	2.13	2.16	2.17	2.19	2.20	2.21	2.22
373	1.83	2.03	2.07	2.10	2.11	2.13	2.14	2.15	2.16
423	1.76	1.97	2.07	2.04	2.06	2.08	2.09	2.10	2.10
473	1.71	1.92	1.97	2.00	2.01	2.04	2.05	2.06	2.05
523	1.66	1.88	1.94	1.96	1.97	2.00	2.01	2.02	2.02

	Gas-saturated sandstone (Kochubei ϕ=12%)						
	Pressure, MPa						
T, K	0.1	20	50	100	150	200	250
275	2.24	2.91	3.12	3.34	3.39	3.4	3.41
323	2.40	2.99	3.18	3.37	3.42	3.43	3.44
373	2.56	3.06	3.23	3.39	3.45	3.46	3.47
423	2.70	3.12	3.27	3.42	3.47	3.48	3.49
473	2.83	3.18	3.32	3.44	3.49	3.50	3.51
523	2.96	3.24	3.36	3.46	3.52	3.53	3.54

	Water-saturated sandstone (Kochubei ϕ=12%)						
	Pressure, MPa						
T, K	0.1	20	50	100	150	200	250
275	2.42	2.61	2.81	2.92	2.94	2.98	3.05
323	2.54	2.73	2.90	3.01	3.04	3.08	3.13
373	2.65	2.84	2.99	3.1	3.13	3.17	3.21
423	2.75	2.94	3.06	3.17	3.21	3.25	3.28
473	2.84	3.03	3.13	3.24	3.28	3.32	3.35
523	2.93	3.12	3.20	3.31	3.35	3.39	3.41

ACKNOWLEDGMENTS

(I.M.A.) thank the Applied Chemicals and Materials Division at the National Institute of Standards and technology (NIST) for the opportunity to work as a Guest Researcher at NIST during the course of this research.

REFERENCES

[1] Revil, A. Thermal conductivity of unconsolidated sediments with geophysical applications. *J. Geophys. Res.* 2000, *105*, 16749-16768.

[2] Somerton, W.H.; Keese, J.A.; Chu, S.C. Thermal behavior of unconsolidated oil sands. *Soc. Pet. Eng. J.* 1974, *14*, 513-521.

[3] Orr, F.M.; Sliva, M.K.; Lien, C. L. Equilibrium phase compositions of CO/sub 2//hydrocarbon mixtures-part 1: measurement by a continuous multiple-contact experiment. *Soc. Pet. Eng. J.* 1982, *22*, 272-280.

[4] Holm, L.W.; Josendal, V.A. Effect of oil composition on miscible-type displacement by carbon dioxide. *Soc. Pet. Eng. J.* 1982, *22*, 87-98.

[5] Doscher, T.M.; El-Arabi, M. Scaled model experiments show how CO_2 might economically recover residual oil. *Oil Gas J.* 1982, *80*, 144-158.

[6] Stalkup, F.I. *Miscible Displacement*, (Soc. Petrol. Eng. Monograph, Dallas, TX, 8), 1983.

[7] Stalkup, F.I. Status of miscible displacement. *J. Petrol. Tech.* 1983, *1*, 815.

[8] Boberg, T.C. *Thermal method of oil recovery*, (Somerset: John Wiley), 1988.

[9] Burger, J.; Sourieau, P.; Combarnous, M. *Thermal Methods of Oil Recovery*, (Houston: Gulf Publishing Co.), 1986.

[10] Butler, R.M. *Thermal recovery of oil and bitumen*, (Univ. of Calgary, Calgary, Alberta, Old Tappan, NJ, United States: Prentice Hall Inc.), 1991.

[11] Čermak, V.; Rybach, L. Thermal conductivity and specific heat of minerals and rocks. In: *Landolt and Börnstein V/1: Physical Properties of Rocks,* ed., G. Angenheiser, (Berlin: Springer), 1982, vol.1, pp. 305-343.

[12] Ramey, H.J.; Brigham, W.E.; Chen, H.K.; Atkinson, P.G.; Arihara, N. *Thermodynamic and hydrodynamic properties of hydrothermal systems*, (Stanford, Stanford University, California, USA), 1974.

[13] Yamano, M.S.; Uyeda, S.; Aoki, Y.; Shipley, T.H. Estimates of heat flow derived from gas hydrates. *Geology*. 1982, *10*, 339-343.

[14] Gallagher, K. Thermal conductivity of sedimentary and basement rocks from the Eromanga and Cooper basins, South Australia. *Exploration Geophys.* 1987, *18*, 381-392.

[15] Gallagher, K.; Ramsdale, M.; Lonergan, L.; Morrow, D. The role of thermal conductivity measurements in modeling thermal histories in sedimentary basins. *Marine and Petroleum Geology*. 1997, *14*, 201-214.

[16] Ungerer, P.; Burrus, J.; Doligez, B.; Chenet, P.Y.; Bessis, F. Basin evaluation by integrate two-dimensional modeling of heat transfer, fluid flow, hydrocarbon generation, and migration. *AAPG Bulletin* 1990, *74*, 309-335.

[17] Waples, D.W. Maturity modeling: thermal indicators, hydrocarbon generation, and oil cracking. In: *The Petroleum System-From Source to Trap*, eds., L.B. Magoon and W.G. Dow, AAPGM, 1994, *60*, 285-306.

[18] Schön, J.H. *Physical properties of rocks: Fundamentals and principles of petrophysics.* Handbook of geophysics exploration, eds., K. Helbig and S. Treitel, (Oxford: Pergamon), 1996, vol. 18.

[19] Eckert, E.R.; Faghri, M. A general analysis of moisture migration caused by temperature differences in an unsaturated porous medium. *Int. J. Heat Mass Transfer* 1980, *23*, 1613-1623.

[20] Dayan, A.; Merbaum, A. H.; Segal, I. Temperature distributions around buried pipe networks in soil with a temperature dependent thermal conductivity. *Int. J. Heat Mass Transfer* 1984, *27*, 409-417.

[21] Beniwal, R.S.; Singh, R.; Chaudhary, D.R. Heat losses from a salt-gradient solar pond. *Appl. Energy* 1985, *19*, 273-285.

[22] Sasaki, A.; Aiba, S.; Fukuda, H. A study on the thermophysical properties of a soil. *Transaction ASME* 1987, *109*, 232-237.

[23] Sass, J.H.; Lachenbruch, A.H.; Munroe, R. Thermal conductivity of rocks from measurements on fragments and its application to heat-flow determinations. *J. Geophys. Res.* 1971, *76*, 3391-3401.

[24] Boldizsár, T.; Gózon, J. The geothermic flow at Komló-Zobák. *Acta Techn. Hung.* 1963, *43*, 467-476.

[25] Kappelmeyer, O.; Haenel, R. *Geothermic with special reference to application*, (Gebrueder Borntraeger, Berlin and Stuttgart), 1974, 238 p.

[26] Luchenbruch, A.H.; Sass J.H. *Heat flows in the United State, in the earth's crust*, ed., J.G. Heacock, (Washington DC: AGU), 1977, pp. 626-675.

[27] Sass, J.H.; Lachenbruch, A.H.; Munroe, R.; Greene, G.; Moses, T. Heat flow in the western United States. *J. Geophys. Res.* 1971, *76*, 6376-6413.

[28] Williams, C.F.; Sass, J.H. The role of temperature-dependent thermal conductivity in heat transfer at the Geysers. In: *Proc. Int. Association of Seismology and Physics of the Earth's Interior*, 27th General Assembly, (Wellington, New Zealand), 1994.

[29] Crane, R. A.; Vachon, R. I. A prediction of the bounds on the effective thermal conductivity of granular materials. *Int. J. Heat Mass Transfer* 1977, *20*, 711-723.

[30] Côté, J.; Konrad, J.M. A generalized thermal conductivity model for soils and construction materials. *Can. Geochem. J.* 2005, *42*, 443-458.

[31] Demirci, A.; Görgülü, K.; Durutük, Y.S. Thermal conductivity of rocks and its variation with uniaxial and triaxial stress. *Int. J. Rock Mech. and Mining Sciences* 2004, *41*, 1133-1138.

[32] Cheremisinoff, N. P. *Handbook of Ceramics and Composites*, (New York: Marcel Dekker), 1990, vols. 1-3.

[33] Giechrist, J.D. *Extraction Metallurgy*, (London: Pergamon Press), 1989.

[34] Hadley, G.R. Thermal conductivity of packed metal powders. *Int. J. Heat Mass Transfer* 1986, *29*, 909-920.

[35] Hasan, A. Optimising insulation thickness for buildings using life cycle cost. *Applied Energy* (Amsterdam: Elsevier), 1999, *3*, 115-124.

[36] Hsu, C.T.; Cheng, P.; Wong, K.W. A Lumped – parameter model for stagnant thermal conductivity of spatially periodic porous media. *Transaction ASME* 1995, *117*, 264-269.

[37] Ma, Y.T.; Yu, B.M.; Zhang, D.M.; Zou, M.Q. A self-similarity model for effective thermal conductivity of porous media. *J. Phys. D: Applied Phys*. 2003, *36*, 2157-2164.

[38] Mirsa, K.; Shrotriya, A.K.; Singh, R.; Chaudhary D.R. *J. Phys. D: Applied Phys*. 1994, *27*, 732-735.

[39] Naidu, A.D.; Singh, D.N. Field probe for measuring thermal resistivity of soils. *J. Geothech. Geoenvironmental Eng*. 2004, 213-216.

[40] Pande, R.N.; Saxena, N.S.; Chaudhary, D.R. Measurement of effective thermal conductivity of bilding – construction materials at different interstitial air pressures. *Ind. J. Thechnology* 1984, *22*, 66-69.

[41] Rao, M.V.B.B.G.; Singh, D.N. A generalized relationship to estimate thermal resistivity of soils. *Can. Geotech. J*. 1999, *36*, 767-773.

[42] Waite, W.F.; de Martin, B.J.; Kirby, S.H.; Pinkston, J.; Ruppel, C.D. Thermal conductivity measurements in porous mixtures of methane hydrate and quartz sand. *Geophys. Res. Lett.* 2002, *29*, 82-1-82-4.

[43] Whitaker, S.; Chou, W.T. –H. Drying granular porous media-Theory and Experiment. *Drying Technology* 1983, *1*, 3-33.

[44] Zehner, P.; Schlünder, E.U. Thermal conductivity of granular materials at moderate temperatures. *Chem. Ingr.-Tech*. 1970, *42*, 933-941.

[45] Özkahraman, H.T.; Selver, R.; Işik, E.C. Determination of the thermal conductivity of rock from P-wave velocity. *Int. J. Rock Mech. and Mining Sciences* 2004, *41*, 703-708.

[46] Biancheria, A. The effect or porosity on thermal conductivity of ceramic bodies. *Trans. Amer. Nucl. Soc*. 1966, *9*, 15-19.

[47] Caren, R.P.; Cunnington, G.R. Heat transfer in multilayer insulation systems. *Chem. Eng. Prog. Symp. Series* 1968, *64*, 67-81.

[48] Batchelor, G.K. Transport properties of two-phase materials with random Structure. *Annu. Rev. Fluid Mech.* 1974, *6*, 227-255.

[49] Gruescu, C.; Giraud, A.; Homand, F.; Kondo, D.; Do, D.P. Effective thermal conductivity of partially saturated porous rocks. *Int. J. Solid Struc*. 2007, *44*, 811-833.

[50] Sommerton, S.W.; Robers, J. Computer simulation of a natural convection driven packed bed of an oxide coating. *ASME Proc. National Heat Transfer Conference*, 1988, HTD-Vol. 96, pp. 379-384.

[51] Staicu, D.M.; Jeulin, D.; Beauvy, M.; Laurent, M.; Berlanga, C. Effective thermal conductivity of heterogeneous materials: calculation methods and application to different microstructures. *High Temperatures – High Pressures* 2001, *33*, 293-301.

[52] Sievers, A.J. In: *Solar Energy Conversion-Solid-State Physics Aspects*, ed. by B.O. Serahin, Topics Appl. Phys., (Berlin: Springer), 1979, *30*, 57.

[53] Tien, C.L.; Cunnington, G.R. Cryogenic Insulation *Heat Transfer*. In: *Advances in Heat Transfer* 1973, 9, 349-417.

[54] Torquato, S. Thermal conductivity of disordered heterogeneous media from microstructure. *Rev. Chem. Eng*. 1987, *4*, 151-204.

[55] *Hale, D.K. The physical properties of composite materials. J. Mater. Sci., 1976,* 11, *2105-2145.*

[56] Watt, J.P.; Daviers, G.F.; O'Connel, R.J. The elastic properties of composite materials. *Rev. Geophys.* 1976, *14*, 541-563.
[57] Giraud, A.; Gruescu, C.; Homand, F.; Kondo, D. Effective thermal conductivity of transversely isotopic media with arbitrary oriented ellipsoidal inhomogeneities. *Int. J. Solid Struc.* 2007, *44*, 2627-2647.
[58] Calmidi, V.V. *Transport Phenomena in High Porosity Metal Foam*. Ph.D. Thesis, (Boulder: University of Colorado), 1998.
[59] Kooi, H. Spatial variability in subsurface warming over the last three decades; insight from repeated borehole temperature measurements in the Netherlands. *Earth Planet Sc. Let.* 2008, *270*, 86-94.
[60] Ingham, D.B; Bejan, A.; Mamut, E.; Pop, I. *Emerging Technologies and Techniques in Porous Media*, (London: Kluwer Academic Publishers), 2003.
[61] Sayari, A.; Jaroniec, M. *Nonoporous Materials IV*, (New York: Elsevier Science Publisher Co.), 2005.
[62] Pan, N.; Zhong, W. *Fluid Transport Phenomena in Fibrous Materials*, (Cambridge, UK: Woodhead Publishing Ltd.), 2006.
[63] Saravacos, G.D.; Maroulis, Z.B. *Transport Properties of Foods,* (New York: Marcel Dekker Inc.), 2004.
[64] Brucker, K. A.; Majdalani, J. Effective thermal conductivity of common geometric shapes. *Int. J. Heat Mass Transfer* 2005, *48*, 4779-4796.
[65] Meyer, M.J.; McGee, H.W. Oil and gas fields accompanied by geothermal anomalies in Rocky Mountain region. *AAPG Bulletin* 1985, *69*, 933-945.
[66] Issler, D.R.; Snowdon, L.R. Hydrocarbon generation kinetics and thermal modeling, Beaufort-Mackenzie Basin. *Bulletin of Canadian Petroleum Geology 1990,* 38, *1-16.*
[67] Bullard, E.C. Heat flow in South Africa. *Proc. R. Soc.London, Series A* 1939, *173*, 474–502.
[68] O'Brien, J.J.; Lerche I. The influence of salt domes on paleotemperature distributions. *Geophys*. 1984, *49*, 2032–2043.
[69] Ljubimova, E.A.; Maslennikov, A.I.; Ganiev, Yu.A. The method of prediction of the deep Earth temperature on the bases of thermal conductivity of rocks at high temperatures and pressures at water- and oil-containing. In: *Earth Heat Field" Proc. All Union Conf. "Methodologicaal Problems of Geothermy", Geothermal Methods,* (Makhachkala: Geothermal Research Institute of the Russian Academy of Sciences), 1979, Vol. II, pp.4-11.
[70] Aichlmayr, H.T.; Kulacki, F.A. The effective thermal conductivity of saturated porous media. *Advances in Heat Transfer*, eds., G. Greene et al., (New York: AP), 2006, vol. 39, pp. 377-460.
[71] Kaviany, M. *Principles of heat transfer in porous media*. Second edition, (New York: Springer), 1991.
[72] Somerton, W.H. *Thermal properties and temperature related behavior of rock/fluid system*, (New York: Elsevier), 1992.
[73] Peierls, R. *Quantum theory of solids*, (Moscow: IL), 1956.
[74] Ziman, J.M. *Electrons and phonons. The theory of transport phenomena in solids*, (Oxford: Clarendon Press), 1962.
[75] Spitzer, D.P. Lattice thermal conductivity of semiconductors: A Chemical bond approach. *J. Phys. Chem. Solids.* 1970, *31*, 19-40.

[76] Roufosse, M.; Klemens, P. G. Lattice thermal conductivity of minerals at high temperatures. *J. Geophys. Res.* 1974, *79*, 703-705.
[77] Hofmeister, A.M. Mantle values of thermal conductivity and the geotherm from phonon lifetimes. *Science* 1999, *283*, 1699-1706.
[78] Honda S.; Yuen D.A. Interplay of variable thermal conductivity and expansivity on the thermal structure of oceanic lithosphere II. *Earth Planets Space* 2004, *56*, e1-e4.
[79] Birch, F.; Clark, H. The thermal conductivity of rocks and its dependence upon temperature and composition. Part II. *Am. J. Science* 1940, *238*, 613-635.
[80] Sekiguchi, K. A method for determining terrestrial heat flow in oil basins areas. *Tectonophysics* 1984, *103*, 67-69.
[81] Carslaw H.S.; Jaeger J.C. *Conduction of heat in solids*, 2nd edition, (Oxford: Oxford University Press), 1959, pp. 345-375.
[82] Woodside, W.; Messmer, J. Thermal conductivity of porous media. II. Consolidated rocks. *J. Appl. Phys.* 1961, *32*, 1688-1699.
[83] Beck, A.E. A steady state method for the rapid measurements of the thermal conductivity of rocks. *J. Sci. Instr.* 1957, *3*, 186-189.
[84] Parker, J.W.; Jenkins, J.R.; Butler, P.C.; Abbott, G.I. Flash method of determining thermal diffusivity, heat capacity, and thermal conductivity. *J. Appl. Phys.* 1961, *32*, 1679-1684.
[85] Mongelli, F. Thermal conductivity measurements of some Apulian limestones by the "cut core" method, (Atti 18th Convegno Annuale di Associazione Geofisica Italiana), 1969, pp.137-154.
[86] Tye, R.P. *Thermal conductivity*, (London: AP), 1969.
[87] Singh, R. Thermal conduction through porous systems. In: *Cellular and Porous Materials. Thermal Properties Simulation and Prediction*, eds., A. Ochsner, G.E. Murch, M.J.S. de Lemos, (Weinheim: Wiley-VCH), 2008, pp.199-265.
[88] Jessop, A.M. *Thermal Geophysics*, (Amsterdam: Elsevier), 1990.
[89] Hofmeister, A.M.; Branlund, J.M.; Pertermann, M. Properties of rocks and minerals-thermal conductivity of the Earth. Treatise in Geophysics, G. Schubert, ed. in chief, Mineral Physics (G.D. Price, ed.), (The Nitherlands: Elsevier), 2007, vol. 2, pp. 543-577.
[90] Seki, J.P.; Cheng, K.C.; Fukusako, S. Mesurement of thermal conductivity of Alberta oil sands. In: *Thermal Conductivity 17,* (New York: Plenum Press), 1982, p.635.
[91] Ljubimova, E.A. *Termika Zemli I Luny* (*in Russian*), (Moscow: Nauka), 1968, p. 256.
[92] Hammerschmidt, U.; Sabuga, W. Transient hot wire (THW) method: Uncertainty assessment. *Int. J. Thermophys.* 2000, *21*, 1255-1278.
[93] Middleton, M.F. A transient method of measuring the thermal properties of rocks. *Geophysics*. 1993, *58*, 357-365.
[94] Beck, A.E. *Geophysics* 1977, *41*, 9-14.
[95] MidttØmme, K.; Roaldset, E. Thermal conductivity of sedimentary rocks: uncertainties in measurement and modeling. In: *Muds and Mudstones: Physical and Fluid Flow Properties*, eds., A.C. Aplin, A.J. Fleet, and J.H.S. Macquaker, (London: Geological Society), Special Publications, 1999, Vol. 158, pp.45-60.
[96] Ross, R.G.; Andersson, P.; Sundqvist, B.; Bäckström, G. Thermal conductivity of solidsand liquids under pressure. *Rep. Prog. Phys.* 1984, *47*, 1347-1402.

[97] Li, Y.Z.; Madhusudana, C.V.; Leonardi, E. Experimental investigation of thermal contact conductance: Various of surface microhardness and rougness. *Int. J. Thermophys.* 1998, *19*, 1691-1704.
[98] Dumont, I.; Moyne, C.; Degiovanii, A. Thermal contact resistance: Experiment versus theory. *Int. J. Thermophys.* 1998, *19*, 1681-1690.
[99] Strack, K.M.; Ibrahim, A.W.; Keller, G.V.; Stoyer, C.H. A method for the determination of the thermal conductivity of sandstone using a variable state approach. *Geophys. Prospect* 1982, *30*, 454-469.
[100] Pribnow, D.; Davis, E. E.; Fisher, A. T. Borehole heat flow along the eastern flank of the Juan de Fuca Ridge, including effects of anisotropy and temperature dependence of sediment thermal conductivity. *J. Geophys. Res.* 2000, *105*, 13449-13456.
[101] Beck, A.E. Methods for determining thermal conductivity and thermal diffusivity. In: *Handbook of Terrastrial Heat Flow Density Determination*, eds., R. Haenel, L. Rybach, and L. Stegena, (Dordrecht: Kluwer), 1988, pp.87-124.
[102] Roy, R.F.; Beck, A.E.; Touloukian, Y.S. Thermophysical Properties of Rocks. In: *Physical Properties of Rock and Minerals*, eds., Y.S. Touloukian, W.R. Judd and C.Y. Ho, (New York: McGraw-Hill), 1981, Vol. II-2, pp. 409-502.
[103] Davis, E.E. Ocean heat flow density, In: *Handbook of Terrestrial Heat –Flow Density Determination*, eds., R. Haenel, L. Rybach, L. Stegena, (Dordrecht: Kluwer Academic Publishers), 1988, pp.223-260.
[104] Mongelli, F., Geothermal prospecting for geothermal fields. In: *Geophysical aspect of the energy problem*, eds., A. Rapolla, G.V. Keller, and D.J. Moore, (Amsterdam: Elsivier), 1989, pp.83-99.
[105] Von Herzen, R.P.; Maxwell, A.E. The measurement of thermal conductivity of deep sea sediments by a needle-probe method. *J. Geophys. Res.* 1959, *64*, 1557-1563.
[106] Sass, J.H.; Stone, C.; Munroe, R.J. Thermal conductivity determinations on solid rock-a comparison between a steady-state divided –bar apparatus and a commercial transient line – source device. *J. Volcanology and Geother. Res.* 1984, *20*, 145-153.
[107] Horai, K. Thermal conductivity of rock-forming minerals. *J. Geophys. Res.* 1971, *76*, 1278-1308.
[108] Horai, K. Thermal conductivity of Hawaiian Basalt: A new interpretation of Robertson and Peck's Data. *J. Geophys. Res.* 1991, *96*. 4125-4132.
[109] Horai, K.; Susaki, J. The effect of pressure on the thermal conductivity of silicate rocks up to 12 kBar. *Phys. Earth Planet. Inter.* 1989, *55*, 292-305.
[110] Kiyohashi, H.; Deguchi, M. Derivation of a correlation formula for the effective thermal conductivity of geological porous materials by three-phase geometric-mean model. *High Temperatures - High Pressures* 1998, *30*, 25-38.
[111] Vosteen, H.D.; Schellschmidt, R. Influence of temperature on thermal conductivity, thermal heat capacity and thermal diffusivity for different types of rock. *Phys. Chem. Earth.* 2003, *28*, 499-509.
[112] Xu, Y.; Shankland, T.J.; Linhardt, S.; Rubie, D.C.; Langenhorst, F.; Kiasinski, K. Thermal diffusivity and conductivity of olivine, wadsleyite, and ringwoodite to 20 GPa and 1373 K. *Physics of the Earth and Planetary Interiors* 2004, *143-144*, 321-326.
[113] Pribnow, D. F.C. *Ein Vergleich von Bestimmungsmethoden der Wärmeleitfähigkeit unter Berücksichtigung von Gesteinsgefügen und Anisotropien*, (Düsseldorf: VDI-Verlag GmbH), 1994.

[114] Popov, Y.A.; Pribnow, D.; Sass, J.H.; Williams, C.F.; Burkhardt, H. Characterization of rock thermal conductivity by high – resolution optical scanning. *Geothermics* 1999, *28*, 253-276.

[115] Pribnow, D.; Sass, J.H. Determination of thermal conductivity from deep boreholes. *J. Geophys. Res.* 1995, *100*, 9981-9994.

[116] Honarmand, H. *3ohrlochsonden zur Bestimmung der Wärmeleitfähigkeit in größeren Tiefen*, (Berlin: Verlag fur Wissenschaft und Forschung), 1993.

[117] Suleiman, B.M. Thermal conductivity of saturated samples using the hot-disk technique. In *Proc. 4th Int. Conf. Heat Transfer, Thermal Engineering and Environment*, (Elounda, Greece), 2006, pp. 140-145.

[118] Suleiman, B.M.; Borjesson, L.; Berastegui, P. Effect of cobalt doping on the thermal conductivity of YBa2Cu3O7-d superconductor. *Phys. Rev. B.* 1996, *53*, 5901-5906.

[119] Gustafsson, S.E. The transient plane-source technique: for measurements of thermal conductivity and diffusivity of solid materials. *Rev. Sci. Instr.* 1991, *62*, 797-804.

[120] Gustavsson, M.; Gustafsson, S.E. On the use of transient plane source sensors for studying materials with directions dependent properties. In: 26th ITCC in Boston, 2001, pp. 195-199.

[121] Krishnaiah, S.; Singh, D.N.; Jadhav, G.N. A methodology for determining thermal properties of rocks. *Int. J. Rock Mechanics and Mining Sci.* 2004, *41*, 877-882.

[122] Sass, J.H.; Lachenbruch, T.H.; Moses, T.H.; Morgan, P. Heat flow from a scientific research well at Cajon Pass, California. *J. Geophys. Res.* 1992, *97*, 5017-5030.

[123] Brigaud, F.; Vasseur, G. Mineralogy, porosity and fluid control on thermal conductivity of sedimentary rocks. *Geophys. J.* 1989, *98*, 525-542.

[124] Brigaud, F.; Chapman, D. S.; Le Douaran, S. Estimating thermal conductivity in sedimentary basins using lithologic data and geophysical well logs. *Bull. Amer. Assoc. Petrol. Geol.* 1990, *74*, 1459-1477.

[125] Johansen, Ø. *Thermal conductivity of soils*. Ph.D. Thesis, (Hanover: NTH), 1977.

[126] Farouki, O.T. *Thermal conductivity of soils. Cold Region Research and Engineering Lab*. Monograph, 1981, Vol. 81-1.

[127] Beck, A.E.; Beck, J.M. On the measurement of the thermal conductivities of rocks by observations on a divided bar apparatus. *Transactions - American Geophysical Union.* 1958, *39*, 1111–1123.

[128] Gourley, P. Determination of thermal conductivity of rock chip samples by the divided bar method. *Proc. North Dakota Academy of Science*, 1974, *28*, 8.

[129] Jessop, A.M. The effect of environment on divided bar measurements. In: *Tectonophysics, Proc. Sympos. Geothermal Problems, Madrid Spain,* 1969, (Amsterdam, Netherlands: Elsevier), 1970, vol. 10, pp.39–49.

[130] Schilling, F.R. A transient technique to measure thermal diffusivity at elevated temperatures. *Eur. J. Mineral.* 1999, *11*, 1115-1124.

[131] Höfer, M., Schilling, F.R., Heat transfer in quartz, orthoclase, and sanidine at elevated temperatures. *Phys. Chem. Minerals* 2002, *29*, 571-584.

[132] Gibert, B.; Schilling, F.R.; Tommasi, A.; Mainprice, D. Thermal diffusivity of olivine single-crystal and polycrystalline aggregates at ambient conditions-A comparison. *Geophys. Res. Lett.* 2005, *30*, 1029.

[133] Cahill, D.; Watson, S.K.; Pohl, R.O. Lower limit of thermal conductivity of disordered solids. *Phys. Rev. B.* 1992, *46*, 6131-6140.

[134] Chen, F.; Shulman, J.; Xue, Y.; Chu, C.W.; Nolas, G.S. Thermal conductivity measurements under hydrostatic pressure using the 3 omega method. *Rev. Sci. Inst.* 2004, *75*, 4578-4584.

[135] Parker, J.W.; Jenkins, J.R. A flash method for determining thermal diffusivity over a wide temperature range. WADD Technical Report 65-91, *Directorate of Materials and Processes*. 1961.

[136] Blumm, J.; Lemerchand, S. Influence of test conditions on the accuracy of laser flash measurements. *High Temp. – High Press*. 2002, *34*, 523-528.

[137] Azumi, T.; Takahashi, Y. Novel finite pulse-width correction in flash thermal diffusivity measurement. *Rev. Sci. Instrum*. 1981, *52*, 1411-1413.

[138] Whittington, A.G.; Hofmeister, A.M.; Nabelek, P.I. Temperature dependent thermal diffusivity of the earth's crust and implications for magmatism. *Nature* 2009, *458*, 319-321.

[139] Holt, J.B. Thermal diffusivity of olivine. *Earth and Planetary Sci*. 1975, *27*, 404-408.

[140] Buttner, R.; Zimanowski B.; Blumm, J.; Hagemann, L. Thermal conductivity of a volcanic rock material (olivine-melilite) in the temperature range between 288 and 1470 K. *J. Volcanology and Geotherm. Res*. 1998, *80*, 293-302.

[141] Hofmeister, A.M. Thermal diffusivity of garnets at high temperatures. *Phys. Chem. Minerals* 2006, *33*, 45-62.

[142] Braüer, H.; Dusza, L.; Schulz, B. New laser flash equipment LFA 427. *Interceramic* 1992, *41*, 489-492.

[143] Mehling, H.; Hautzinger, G.; Nilsson, O.; Fricke, J.; Hofmann, R.; Hahn, O. Thermal diffusivity of semitransparent materials determined by the laser-flash method applying a new mathematical model. *Int. J. Thermophys*. 1998, *19*, 941-949.

[144] Hofmann, R.; Hahn, O.; Raether, F.; Mehling, H.; Fricke, J. Determination of thermal diffusivity in diathermic materials by the laser-flash technique. *High Temp.–High Press*. 1997, *29*, 703-710.

[145] Hanley, E.J.; DeWitt D.P.; Taylor R.E. The thermal transport properties at normal and elevated temperature of eight representative rocks. In: *Proc. 7th Symp. Thermophys. Prop.,* ed., A. Cezairlyan, (New York: ASME), 1977, pp.386-391.

[146] Taylor, R.E. Critical evaluation of the flash method for measuring thermal diffusivity, (PRF 6764, NTIS PB 225 591/7AS), 1973.

[147] Taylor, R.E. Improvements in data reduction for the flash diffusivity method. In: *Proc. 13th Int. Conf. Thermal Conductivity*, 1973.

[148] Strack, K.M.; Ibrahim, A.W.; Keller, G.V.; Stoyer, C.H. A method for the determination of the thermal conductivity of sandstone using a variable state approach. *Geophys. Prospect* 1982, *30*, 454-469.

[149] Kubičár L.; Vretenár V.; Boháč V.; Tiano P. Thermophysical properties of sandstone by pulse transient method. *Int. J. Thermophys*. 2006, *27*, 220-234.

[150] Kubičár, L. In: *Comprehensive Analytical Chemistry*, ed. G. Svehla, (Amsterdam: Elsevier), 1990, p.360.

[151] Schatz, J.F.; Simmons, G. Thermal conductivity of Earth materials at high temperatures. *J. Geophys. Res*. 1972, *77*, 6966-6983.

[152] Schatz, J.F.; Simmons, G. Method of simultaneous measurements of radial and lattice thermal conductivity. *J. Appl. Phys*. 1972. *43*, 2588-2594.

[153] Zaug, J.; Abransom, E.; Brown, J.M.; Slutsky, L.J. Elastic constant, equations of state and thermal diffusivity at high pressure. In: *High-Pressure*, eds., Y. Syono and M.H. Manghnani, (Washington DC: Terra/AGU), 1992. pp. 157-166.

[154] Chai, M.; Brown, J.M.; Slutsy, L.J. Thermal diffusivity of mantle minerals. *Phys. Chem. Miner.* 1996, *23*, 470-475.

[155] Pangilinan, G.I.; Ladouceur H.D.; Russell, T.P. All-optical technique for measuring thermal properties of materials at static high pressure. *Rev. Sci. Inst.* 2000, *71*, 3846-3852.

[156] Hofmeister, A.M. Thermal conductivity of spinels and olivines from vibrational spectroscopy: Ambient conditions. *Amer. Mineralogist* 2001, *86*, 1188-1208.

[157] Shankland, T.J.; Nitsan, U.; Duba, A.G. Optical absorption and radiative heat transfer in olive at high temperatures. *J. Geophys. Res*. 1979, *75*, 1603-1610.

[158] Klemens, P.G. *Thermal conductivity*, (New York: AP), 1969.

[159] Abdulagatova, Z.Z.; Abdulagatov, I.M.; Emirov, S.N. Effect of temperature and pressure on the thermal conductivity of sandstone. *Int. J. Rock Mechanics and Mining Sci*. 2009, *46*, 1055-1071.

[160] Sukharev, G.M.; Vlasova, S.P.; Taranukha, Yu. K. *Dok. Akad. Nauk SSSR*, 1967, *177*, 3-11.

[161] Woodside, W.; Messmer, J. Thermal conductivity of porous media. II. Consolidated rocks. *J. Appl. Phys.* 1961, *32*, 1699-1706.

[162] Seipold, U. Temperature dependence of thermal transport properties of crystalline rocks-a general law. *Tectonophysics* 1998, *291*, 161-171.

[163] Seipold, U. Der Warmetransport in kristallinen Gesteinen unter den Bedingungen der kontinentalen Kruste. *Scientific Technical Report STR01/13* (Potsdam: GFZ), 2001.

[164] Birch, F.; Clark, H. The thermal conductivity of rocks and its dependence upon temperature and composition. Part I. Am. *J. Science*, 1940, *238*, 529-558.

[165] Birch, F. Thermal conductivity and diffusivity. In: *Handbook of Physical Constants*, eds., F. Birch, F.J. Schairer and H. C. Spicer, (New York: GSA), Special Paper 36, 1942, pp. 243-266.

[166] Asaad, Y. *A study of the thermal conductivity of fluid-bearing porous rocks*. Ph.D. Thesis, (Berkley: University of California), 1955.

[167] Susaki, J.; Horai, K. Thermal conductivity of rocks at upper mantle condition. In: *Abstracts of the 29th Int. Geological Congress, Kyoto, Japan, August 24–September 3, 1992,* vol. 29, p.58.

[168] Clark, S. P., *Thermal conductivity. In: Handbook of Physical Constants,* ed., S.P. Clark, Memoir, (New York: GSA), 1966, *97*, 461-482.

[169] Walsh, J.B.; Decker, E.R. Effect of pressure and saturating fluid on the thermal conductivity of compact rock. *J. Geophys. Res*. 1966, *71*, 3053-3061.

[170] Tikhomirov, V. Conductivity of rocks and their relationship with density, saturation and temperature. *Russian Neftyanoe Khozyaistvo* 1968, *46*, 151-161.

[171] Anand, J. *Thermal conductivity of fluid saturated rocks at elevated pressures and temperatures.* M.S. Thesis, (Berkeley: University of California), 1971.

[172] Huang, J.H. Effective thermal conductivity of porous rocks. *J. Geophys. Res*. 1971, *76*, 6420-6427.

[173] Schatz, J.F. *Thermal conductivity of earth materials at high temperatures.* Ph.D. Thesis (Cambridge: MIT), 1971.

[174] Gomaa, E.E. *Thermal behavior of partially liquid saturated porous media*. Ph.D. Thesis, (Berkeley: University of California), 1973.

[175] Desai, P.D.; Navarro, R.A.; Hasan, S.E.; Ho, C.Y.; DeWitt, D.P.; West, T.R. *Thermophysical properties of selected rocks*. CINDAS Report 23. (West Lafayette: Purdue University, Indiana), 1974.

[176] Beck, A. E. An improved method of computing the thermal conductivity of fluid-filled sedimentary rocks. *Geophysics* 1976, *41*, 133-144.

[177] Özbek, H. *Thermal conductivity of multi-fluid saturated porous media*. Ph.D. Thesis, (Berkeley: University of California), 1976.

[178] Ito, Y.; Saito, T.; Nanun, M. *Chinetsu*. 1977, *14*, 21-34.

[179] Ljubimova, E.A. Thermal history of the Earth with consideration of the variable conductivity of its mantle. *Geophys. J. R. Astron Soc*. 1958, *1*, 115-134.

[180] Ljubimova, E.A.; Maslennikov, A.I.; Ganijev, Yu. A. Thermal conductivity of dry, wet and oil-saturated rocks at slightly high temperatures and pressures. *Geophys. Res. Bull.* 1977, *15*, 59-66.

[181] Ljubimova, E.A.; Maslennikov, A.I.; Ganijev, Yu.A. Thermal conductivity of rocks at high temperatures and pressures in water- and oil-saturated conditions. *Russian Izv. Akad. Nauk SSSR, ser. Fizika Zemli* 1979, *9*, 87-93.

[182] Sorokin, J.A.; Dzavadov, L.N.; Genshaft, J.S. Thermal conductivity of calcite, hornblende and crystalline slate for pressures up to 8 GPa. *Russian Izv. Akad. Nauk SSSR, ser. Fizika Zemli* 1982, *8*, 88-92.

[183] Kristiansen, J.I. The transient cylindrical probe method for determination of thermal parameters of earth materials. *Geoskrifter, Aarhus Univ.* 1982, *18*, 1-154.

[184] Schärli, U.; Rybach, L. On the thermal conductivity of low –porosity rocks. *Tectonophysics* 1984, *103*, 307-313.

[185] Buntebarth, G.; Rueff, P. Laboratory thermal conductivities applied to crustal conditions. *Rev. Brasil Geofisica* 1987, *5*, 103-109.

[186] Buntebarth, G. The thermal properties of KTB Oberpfalz VB core sample at elevated temperature and pressure. *Sci. Drilling* 1991, *2*, 73-80.

[187] Wang, H.F.; Bonner, B.P.; Carlson, S.R.; Kowallis, B.J.; Heard, H.C. Thermal stress cracking in granite. *J. Geophys. Res.* 1989, *94*, 1742-1758.

[188] Diment, W.H.; Pratt, H.R. Thermal conductivity of some rock-forming minerals: Tabulation. (USGS Open File Report), 1988, p.690.

[189] Davis, B.W. Thermal conductivities of reservoir rocks, (Chevron Oil Field Res. Co. Techn. Mem., TM84000293), 1984, p. 32.

[190] Kukkonen, I.T.; Jokinen, J.; Seipold, U. Temperature and pressure dependencies of thermal transport properties of rocks: Implications for uncertainties in thermal lithosphere models and new laboratory measurements of high-grade rocs in the central Fennoscandian shield. *Surv. Geophys*. 1999, *20*, 33-59.

[191] Huenges, E.; Burkhard, H.; Erbas, K. Thermal conductivity profile of the KTB pilot corehole. *Scientific Drilling* 1990, *1*, 224-230.

[192] Pribnow, D.; Umsonst, T. Estimation of thermal conductivity from mineral composition: Influence of fabric and anisotropy. *Geophys. Res. Lett.* 1993, *20*, 2199-2202.

[193] Pribnow, D.; Williams, C.F.; Burkhardt, H. Log-derived estimate for thermal conductivity of crystalline rocks from the 4 km KTB Vorbohrung. Geophys. *Res. Lett.* 1993, *20*, 1155-1158.

[194] Mohanty, S. *Experimental determination and theoretical prediction of effective thermal conductivity of porous media.* Ph.D. Thesis (Austin: University of Texas at Austin), 1993.

[195] Seipold, U.; Gutzeit, W.; Stromeyer, D. Application of pulse measuring techniques with cylindrical sample symmetry for determining the thermal conductivity of rocks under high pressures. In: *Theoretical and Experimental Investigations of Physical Properties of Rocks and Minerals under Extreme P,T-Conditions,* eds., H. Stiller, M.P. Volarovich, T.S. Lebedev, H. Vollstaedt and D. Freund, (Berlin: Federal Republic of Germany, Akad.-Verlag), 1979, pp.155–162.

[196] Seipold, U. Simultaneous measurements of thermal diffusivity and thermal conductivity under high pressure using thermal pulses of finite length. *High Temperatures – High Pressures* 1988, *20*, 609-613.

[197] Seipold, U. Pressure and temperature dependence of thermal transport properties for granites. *High Temperatures – High Pressures* 1990, *22*, 541-548.

[198] Seipold, U. The variation of thermal transport properties in the earth's crust. *J. Geodynam.* 1995, *20*, 145-154.

[199] Seipold, U. Measurements of thermal conductivity and thermal diffusivity of serpentinities at high pressures up to 500 MPa. *High Temperatures – High Pressures* 1995/1996, *27/28*, 147-155.

[200] Seipold, U.; Mueller, H.–J.; Tuisku, P. Principle differences in the pressure dependence of thermal and elastic properties of crystalline rocks. *Phys. Chem. Earth.* 1998, *23*, 357-360.

[201] Seipold, U.; Huenges, E. Thermal properties of gneisses and amphibolites-high pressure and high temperature investigations of KTB-rock samples. *Tectonophysics* 1998, *291*, 173-178.

[202] Seipold, U. Investigation of the thermal transport properties of amphibolites. I. Pressure dependence. *High Temperatures – High Pressures* 2002, *34*, 299-306.

[203] Popov, Yu.; Tertychnyi, V.; Bayuk, I.; Korobkov, D. To estimation of rock thermal conductivity on logging data. In: *Proc. Int. Conference "The Earth's Thermal Field and Related Research Methods"* (Moscos: MSGPU), 2002, pp. 228-231.

[204] Popov, Yu.; Romushkevic, R. Thermal conductivity of sedimentary rocks of oil-gas fields. In: *Proc. Int. Conference "The Earth's Thermal Field and Related Research Methods"* (Moscow: MSGPU), 2002, pp. 219-223.

[205] Pribnow, D.; Williams, C. F.; Sass, J. H.; Keating, R. Thermal conductivity of water – saturated rocks from KTB pilot hole at temperatures of 25 to 300 °C. *Geophys. Res. Lett.* 1996, *23*, 391-394.

[206] Somerton, W.H.; Mossahebi, M. Ring heat source probe for rapid determination of thermal conductivity of rocks. *Rev. Scientific Instr.* 1967, *38*, 1368–1371.

[207] Morin, R.; Silva, A.J. High pressure-high temperature laboratory apparatus for the measurement of deep sea sediment physical properties. *Ocean Engineering* 1983, *10*, 481–487.

[208] Silva, A.J.; Morin, R.H. The sensitivity of sediment physical properties to changes in temperature, pressure, and porosity. In: *Thermal Modeling in Sedimentary Basins, Proc.*

Meeting: 1st IFP Exploration Research Conference, Carcans, France, June 3–7, 1985, ed., J. Burrus, (Paris: Ed. Technip.), 1985, vol. 44, pp.289-309.

[209] Sakvarelidze, E.; Mamulia, N. Determination of rock heat conduction based on the regular heat regime method in wide temperature range. *Bulletin of the Georgian Academy of Sciences* 1998, *158*, 237–239.

[210] Sclater, J.G.; Christie, P.A.F. Continental stretching: An explanation of the post Mid-Cretaceous subsidence of Central North Sea Basin. *J. Geophys. Res.* 1980, *85(B7)*, 3711–3739.

[211] Horai, K.; Simmons, G. Thermal conductivity of rock-forming minerals. *Earth and Planetary Science Lett.* 1969, *6*, 359–368.

[212] Cull, J.P. Thermal conductivity probes for rapid measurements in rock. *J. Phys. E: Scientific Instr.* 1974, *7*, 771–774.

[213] Deming, D. Estimation of the thermal conductivity anisotropy of rock with application to the determination of terrestrial heat flow. *J. Geophys. Res.* 1994, *99(B11)*, 22087–22091.

[214] Beck, A.E.; Anglin, F.M.; Sass, J.H. Analysis of heat flow data: *In situ* thermal conductivity measurements. *Can. J. Earth Sciences* 1971, *8*, 1–19.

[215] Falvey, D.; Middleton, M. Passive continental margins: Evidence for a pre-breakup, deep crustal metamorphic subsidence mechanism. Oceanologica Acta 1981, *4*, 103–114.

[216] Touloukian, Y.S.; Liley, P.E.; Saxena, S.C. *Thermal Properties of Matter, Thermal Conductivity, Non-Metallic Liquids and Gases.* (New York-Washington: IFI/Plenum), 1970, vol. 3.

[217] Touloukian, Y.S.; Powell, R.W.; Ho, C.Y.; Klemens, P.G. *Thermal Properties of Matter, Thermal Conductivity, Non-Metallic Solids*, (New York-Washington: IFI/Plenum), 1970, vol. 2.

[218] Baldwin, B.; Butler, C.O. Compaction curves. *AAPG Bulletin* 1985, *69*, 622–626.

[219] de Vries, D.A.; Peck, A.J. On the cylindrical probe method of measuring thermal conductivity with special reference to soils. *Australian J. Phys.* 1958, *11*, 255–271.

[220] Clark, S.P. Heat conductivity in the mantile. In: *The Earth's Crust and Upper Mantle*, Geophys. Monog. Ser., ed. by P.J. Hart, (Washington DC: AGU), 1969. vol.13, pp.622-626.

[221] Chan, T.; Jeffrey, J.A. Scale and water-saturated effects for thermal properties of low-porosity rock. *AEG, Proc. 24th US Symp. Rock Mechanics*, 1983, pp.287-301.

[222] Anand, J.; Somerton, W.H.; Gomaa, E.E. Predicting thermal conductivities of formations from other known properties. *Soc. Petr. Eng. J.* 1973, *13*, 267-273.

[223] Keese, J. *Thermal conductivity of unconsolidated oil sands.* M.S. Thesis, (Berkeley: Univ. California), 1973.

[224] Sugawara, A.; Yoshizawa, Y. An investigation on the thermal conductivity of porous materials and its applications to porous rocks. *Austral. J. Phys.* 1961, *14*, 469-480.

[225] Sugawara, A.; Yoshizawa, Y. An experimental investigation on the thermal conductivity of consolidated porous materials. *J. Appl. Phys.* 1962, *33*, 3135-3138.

[226] Niven, C.D. Thermal Conductivity of Some Sedimentary Rocks. *Can. J. Res. A.* 1940, *18*, 132-137.

[227] Zierfuss, H.; Viliet, G.V. Laboratory measurements of heat conductivity of sedimentary rocks. *AAPG Bull.* 1956, *40*, 2475-2488. [227a]. Khan, A.M. *A thermoelectric method*

for measurement of steady–state thermal conductivity of rocks. MS Thesis, (Berkeley: University California), 1964.

[228] Edmondson, T. *Thermal diffusivity of sedimentary rocks subjected to simulated overburden pressure.* M.S. Thesis, (Berkeley: University California), 1961.

[229] Bridgman, P.W. The thermal conductivity and compressibility of several rocks under pressure. *Am. J. Sci.* 1924, *7*, 81-102.

[230] Somerton, W.H. Some thermal characteristics of porous rocks. *J. Pet. Tech.* 1958, *213*, 61-64.

[231] Troschke, B.; Burkhardt, H. Ermittlung der gesteinswarmeleitfahig-keit von sedimentgesteinen aus messungen am Bohrklein, Tagungsband d.4. *Geothermischen Fachtagung der Geothermischen Vereinigung*, 1997, *4*, 217-223.

[232] Troschke, B.; Burkhardt, H. Thermal conductivity models for two-phase systems. In: *Phys. Chem. Earth. Proceedings of Meeting: Relationship Between Various Properties in Sedimentary and Crystalline Rocks, Vienna, Austria, 1997,* eds., E. Huenges, I. Kukkonen and J. Urai, (Oxford-New York-Toronto: Pergamon), 1998, *23*, 351-355.

[233] Deguchi, M.; Kiyohashi, H.; Enomoto, H. Experimental study on the thermal conductivity of porous rocks containing water and air. *J. Mining and Material Proc. Inst. Jap.* 1995, *111*, 537-542.

[234] Deguchi, M.; Kiyohashi, H.; Suzuki, S. Effect of moisture on thermal conductivity of porous rocks. In: *12th Japan Symp. Thermophys. Properties* 1991, pp. 275-278.

[235] Bayuk, I.; Tertychnyi, V.; Popov, Yu. Theoretical modeling as a key for interpretation of experimental data on rock thermal conductivity. In: *Proc. Int. Conference "The Earth's Thermal Field and Related Research Methods"*, (Moscow: MSGPU), 2002, pp.12-17.

[236] Kunii, D.; Smith, J.M. Heat transfer characteristics of porous rocks. *J. AIChE.* 1960, *6*, 71-78.

[237] de Vries, D.A. *Mededelingen Landbouwhogeschol Wageningen*, 1957, *52*, 373-389.

[238] Krupiczka, R. Analysis of thermal conductivity in granular materials. *Int. Chem. Engr.* 1967, *7*, 122-144.

[239] Sammartino, S.; Siitari-Kauppi, M.; Meunier, A.; Sardini, P.; Bouchet, A.; Tevissen, E. An imaging method for the porosity of sedimentary rocks: adjustment of the PMMA method-example of a characterization of calcareous shale. *J. Sediment Res.* 2002, *72*, 937-943.

[240] André, S.; Degiovanni, A. A theoretical study of the transient coupled conduction and radiation heat transfer in glass: phonic diffusivity measurements by the flash technique. *Int. J. Heat Mass Transf.* 1995, *38*, 3401-3412.

[241] Prats, M.; O'Brien, S.M. *J. Petroleum Tech.* 1975, *97*, 114-119.

[242] Tihen, S.S.; Carpenter, H.C.; Sohns, H.W. *NBS Special Publ.* 1968, *302*, 529-536.

[243] Nottenburg, R.; Rajeshwar, K.; Rosenvold, R.; DuBow, J. Measurement of thermal conductivity of Green River oil shales by the thermal comparator technique. *Fuel.* 1978, *57*, 789-795.

[244] Woodside, W. *ASTM-STP.* 1957, *217*, 49-61.

[245] Mirkovich, V.V. Comparative method and choice of standards for thermal conductivity determinations.*J. Am. Ceram. Soc.* 1965, *48*, 387-391.

[246] Powell, R.W. In: *Thermal conductivity*, ed., R.P. Tye, AP. 1969, Vol. 2, Chap. 6, p. 276.

[247] Messmer, J.H. The thermal conductivity of porous media. IV. Sandstones. The effect of temperature and saturation. In: *Proc. 5th Conference on Thermal Conductivity*. (Colorado Denver: University of Denver), 1965, Vol. 1, pp. 1-29.

[248] Blackwell, J.H. The axial-flow error in the thermal conductivity probe. *Can. J. Phys.* 1956, *34*, 412-417.

[249] Somerton, W.H.; Boozer, G.D. Thermal characteristics of porous rocks at elevated temperatures. *AIME Trans*. 1960, *219*, 418-422.

[250] Morinl, R.; Silva, A.J. The effects of high pressure and high temperature on some physical properties of ocean sediments. *J. Geophys. Res*. 1984, *89*, 511-526.

[251] Ratcliffe, E.H. Thermal conductivity of fused and crystalline quartz. *Brit. J. Appl. Phys*. 1959, *10*, 22-25.

[252] MacDonald, K.; Simmons, G. Temperature coefficient of the thermal conductivities of ocean sediments. *Deep Sea Res*. 1972, *19*, 669-671.

[253] Singh, R.; Beniwal, R.S.; Chaudhary, D.R. Thermal conduction of multi-phase systems at normal and different interstitial air pressures. *J. Phys. D: Appl. Phys*. 1987, *20*, 917-922.

[254] Singh, A.K.; Singh, R.; Chaudhary, D.R. Prediction of effective thermal conductivity of moist porous materials. *J. Phys. D: Appl. Phys.* 1990, *23*, 698-702.

[255] Verma, L.S.; Shrotriya, A.K.; Singh, R.; Chaudhary, D.R. Thermal conduction in two-phase materials with spherical and non-spherical inclusions. *J. Phys. D: Appl. Phys*. 1991, *24*, 1515-1526.

[256] Hartmann, A.; Rath, V.; Clauser, C. Thermal conductivity from core and well log data. *Int. J. Rock Mech. Mininig Sci*. 2005, *42*, 1042-1055.

[257] Surma, F.; Geraud, Y. Porosity and thermal conductivity of the Soultz-sous-Forets granite. *Pure Appl. Geophys*. 2003, *160*, 1125-1136.

[258] Popov, Yu. A. Optical scanning technology for nondestructive contactless measurements of thermal conductivity and diffusivity of solid matter. In: *Experimental Heat transfer, Fluid Mechanics and Thermodynamics, Proc. 4th world congress on experimental heat transfer, Fluid Mechanics and Thermodynamics*, eds., Giot M., Mayinger F., Celeta G.P., (Belgium, Brussels), 1997, vol. 1, pp.109-116.

[259] Sepaskhah, A.R.; Boersma, L. Thermal conductivity of soils as a function of temperature and water content. *Soil. Sci. Soc. Am. J*. 1979, *43*, 439-444.

[260] Moyne, C.; Batsale, J.C.; Degiovanni, A.; Maillet, D. Thermal conductivity of wet porous media: Theoretical analysis and experimental measurements. In: *Thermal Conductivity-21*. ed., J. Gremers, (New York: Plenum Press), 1990, pp.109-120.

[261] Zeng, S.Q.; Hunt, A.; Greif, R. Geometric Structure and thermal conductivity of porous medium silica aerogel. *J. Heat Transfer* 1995, *117*, 1055-1058.

[262] Huenges, E.; Burkhard, H.; Erbas, K. Thermal conductivity profile of the KTB pilot corehole. *Scientific Drilling* 1990, *1*, 224-230.

[263] Buntebarth, G. The thermal properties of KTB Oberpfalz VB core sample at elevated temperature and pressure. *Sci. Drilling* 1991, *2*, 73-80.

[264] Grubbe, K.; Haenel, R.; Zoth, G. Determination of the vertical component of thermal conductivity by line source methods. *Zeitschrift für Geologie und Paläontologie*. Teil I H (1/2), 1983, pp. 49-56.

[265] Dobrynin, V.M. Effect of overburden pressure on some properties of sandstones. *Soc. Petrol. Eng. J.* 1962, *12*, 360-366.

[266] Aichlmayr, H.T.; Kulacki, F.A. A transient technique for measuring the effective thermal conductivity of saturated porous media with a constant boundary heat flux. *J. Heat Transfer* 2006, *128*, 1217-1220.
[267] Ofuchi, K.; Kunii, D. Heat transfer characteristics of packed beds with stagnant fluids. *Int. J. Heat Mass Trnasfer* 1965, *8*, 749-757.
[268] King, M. Thermal conductivity measurements on saturated rocks at permafrost temperatures. *Can. J. Earth Sci.* 1979, *16*, 73-79.
[269] Davis, E.E.; Seemann, D.A. Anisotropic thermal conductivity of Pleistocene turbidite sediments of the northern Huann de Fuca Ridge, Proc. Ocean Drill. *Program, Sci. Results* 1994, *139*, 559-564.
[270] Dorofeeva, R.P.; Shapova, M.V. Thermal conductivity studies of underwater boreholes in lake Baikal and relationship with other geophysical properties. In: *Proc. Int. Conference "The Earth's Thermal Field and Related Research Methods"*, (Moscow: MSGPU), 2002, pp.43-45.
[271] Antipov, V.; Afonina, T.; Badalov, O.; Bezrukova, E.; Bychinsky, V.; Bukhariov, A.; Williams, D.; Gelety, V.; Gvozdkov, A.; Golubev, V.; Goreglyad, A.; Gorokhov, I.; Dmitriev, A.; Dorofeeva, R.; Duchkov, A. Late Cenozoic paleoclimatic record from Lake Baikal from the data of investigation of 600 m core of a deep drilling. *Geology and Geophysics* 2000, *41*, 3-32.
[272] Sukharev, G.M.; Sterlenko, Z.V. Thermal properties of water and oil saturated sandstones. *Dok. Akad. Nauk SSSR* 1970, *194*, 683-685.
[273] Mücke G., *Bergbau Arch.* 1964, *25*, 35-58.
[274] Emirov S.N., Tsomaeva T.A., Mursalov B.A., Study of the thermal conductivity of water- and oil-saturated sandstones at high temperatures and high pressures. In: *Geothermic. Geothermal Energetic.* (Makhachkala: Geothermal Research Institute of the Dagestan Scientific Center of the Russian Academy of Sciences), 1994, pp.172-176.
[275] Emirov, S.N.; Tsomaeva, T.A.; Askerov, S.Ya. Thermal conductivity of fluid-saturated sandstones at high temperatures and high pressures. *Int. Geophys. J.* 1997, *19*, 68-71.
[276] Emirov, S.N.; Gairbekov, Kh.A.; Ramazanova, E.N.; Kulikov, S.N.; Suleimanov, M.Zh. In: *Proc. Int. Conference "The Earth's Thermal Field and Related Research Methods"*, (Moscow: MSGPU), 2002, p.56.
[277] Emirov, S.N.; Ramazanova, E.N. Thermal conductivity of fluid-saturated rocks at high temperatures and high pressures. *Fizika*, 2007, *13*, 77-80.
[278] Emirov, S.N. *Experimental study of the thermal conductivity of semiconductors and rocks at high temperatures and high pressures.* Ph.D. Thesis, (Moscow: Moscow Power Engineering Institute), 1997.
[279] Emirov, S.N. Thermal conductivity of certain rocks under high pressures and temperatures. In: *High Pressure Investigations in Geosciences*, (Berlin: Akademie-Verlag, KAPG), Chapter III, 1989, *1*, 123-126.
[280] Kurbanov, A.A. Thermal conductivity of rocks at various P-T conditions. In: *Geothermic. Geothermal Energetic.* (Makhachkala: Dagestan Scientific Center of the Russian Academy of Sciences), 1994, pp.172-176.
[281] Kurbanov, A.A. Device for thermal conductivity measurements. A.C. 779670 (SSSR), 1980, N42.

[282] Kurbanov, A.A. Combined effect of temperature and pressure on the fluid-saturated rocks. In: *Proc. VII Int. Conference "New Approaches in Earth Science*", Moscow, 2005, p.176.

[283] Kurbanov, A.A. Study of the simultaneous effect of temperature, pressure, and fluid-saturation on the thermal conductivity of rocks. In: *Proc. Int. Geophysical Conference*, Moscow, 2003, pp.56-59.

[284] Kurbanov, A.A. Thermal conductivity of gas-, water-, and oil-saturated rocks at deep underground conditions. *Russian Izv. AN SSSR, ser. Fizika Zemli.* 1988, 107-112.

[285] Kurbanov, A.A. *Thermophysical properties of fluid-saturated rocks at PT-conditions.* (Makhachkala: Dagestan Scientific Center of the Russian Academy of Sciences), 2000.

[286] Kurbanov, A.A. *Study of the thermophysical properties of fluid-saturated rocks at various temperatures and pressures.* Ph.D. Thesis, (Moscow: Earth Physics Institute of the Russian Academy of Sciences), 2007.

[287] Amirkhanov, Kh.I.; Suetnov, V.V.; Gairbekov, Kh.A.; Kurbanov, A.A. In: *Thermal Field of the Earth.* (Makhachkala: Dagestan Scientific Center of the Russian Academy of Sciences), 1979, v.2, pp.22-25.

[288] Gairbekov, Kh.A.; Kurbanov, A.A. Measurements of the thermal conductivity of rocks at PT-conditions and estimating of the heat loses. In: *Thermomechanics of the geothermal systems.* (Makhachkala: Dagestan Scientific Center of the Russian Academy of Sciences), 1990, pp.62-70.

[289] Tsomaeva, T.A.; Emirov, S.N. Thermal conductivity of sandstones at high temperatures and high pressures. In: *First All Russian Petrography Meeting.* (Ufa: 1995), Vol.3, p. 136.

[290] Ramazanova, A.E.; Emirov, S.N. In: *Proc. 2th Int. Conference "Renewal Energy. Problems and Perspective*s" (Makhachkala: Geothermal Research Institute), 2010, pp.153-157.

[291] Abdulagatova, Z.Z. Ph.D. Thesis, (Makhachkala: Geothermal Research Institute of the Russian Academy of Sciences), 2010.

[292] Abdulagatova, Z.Z.; Emirov, S.N.; Abdulagatov, I.M. Effect of pressure, temperature, and oil-saturation on the thermal conductivity of sandstone up to 250 MPa and 520 K. *J. Pet. Sci. Eng.* 2010, *73*, 141-155.

[293] Abdulagatova, Z.Z.; Emirov, S.N.; Abdulagatov, I.M., Effect of temperature and pressure on the thermal conductivity and speed of sound in andezite rocks. In: *Ultrasonic and Thermodynamic Properties of Substances*, (Kursk: KU), 2006, *33*, 5-24.

[294] Abdulagatov, I.M.; Emirov, S.N.; Tsomaeva, T.A.; Gairbekov, Kh.A.; Askerov, S.Ya.; Magomaeva, M.A. Thermal conductivity of fused quartz and quartz ceramic at high temperatures and high pressures. *J. Phys. Chem. Solids* 2000, *6*, 779-787.

[295] Abdulagatov, I.M.; Emirov, S.N.; Gairbekov, Kh.A.; Magomaeva, M.A.; Askerov, S.Ya.; Ramazanova, E.N. Effective thermal conductivity of fluid-saturated porous mica ceramics at high temperatures and high pressures. *Ind. Eng. Chem. Res.* 2002, *41*, 3586-3593.

[296] Abdulagatov, I.M.; Emirov, S.N.; Tsomaeva, T.A.; Gairbekov, Kh.A.; Askerov, S.Ya. Thermal conductivity of the porous glasses at high temperatures and high pressures. *Russ. High Temperature.* 1998, *36*, 401-409.

[297] Abdulagatov, I.M.; Emirov, S.N.; Abdulagatova, Z.Z.; Askerov, S.Ya. Effect of pressure and temperature on the thermal conductivity of rocks. *J. Chem. Eng. Data* 2006, *51*, 22-33.
[298] Ljubimova, E.A.; Maslennikov, A.I.; Ganiev, Yu.A. Thermal properties of water- and oil-saturated sedimentary rocks. *Acta Geophys. Polonica* 1977, *25*, 273-286.
[299] Ljubimova, E.A.; Maslennikov, A.I.; Ganiev Yu.A. The thermal conductivity of dry, wet-, and oil-saturated rocks under of slightly high temperatures and pressures. *Geophys. Res. Bull.* Hyderabad 1977, *15*, 59-66.
[300] Maslennikov, A.I.; Ganiev, Yu.A. The effect of pressure and temperature on the thermal conductivity rocks. In: *"Physical Processes in Mining"*, (Leningrad: LGI), 1977, pp.137-140.
[301] Jamalova, A.S. Deep heat flow in Dagestan region. In: *Thermophysical Properties of Rocks*, (Moscow: Nauka), 1969, Chap. IV, N4, pp.1-126.
[302] Buntebarth, G.; Schopper, J.R. Experimental and theoretical investigations on the influence of fluids, solids and interactions between them on thermal properties of porous rocks. *J. Phys. Chem. Earth.* 1997, *23*, 1141-1146.
[303] Fuchs, S.; Förster, A. Rock thermal conductivity of mesozoic geothermal aquifers in the northeast German basin. *Chemie der Erde- Geochem.* 2010, *70*, 13-22.
[304] Devyatkov, E.D.; Petrova, A.R.; Smirnov, I.A. Fused quartz as a standard material at thermal conductivity measurements. *Solid State Physics* 1960, *11*, 738-747.
[305] Cheng, Sh.; Li, Zh.; Ge, X.; Li, J.; Zhang, Y.; Liang, X.; Song, Y.; Feng, H. In-situ measurements of the effective thermal conductivity of oil-sandstone samples under simulated conditions. *High Temperatures - High Pressures* 1990, *22*, 641-648.
[306] Xie, H.; Gu, H.; Fujii, M.; Zhang, X. An experimental study on thermal and electrical properties of packed carbon nanofibers. *Int. J. Thermophys.* 2006, *27*, 244-257.
[307] Yagi, S.; Kunii, D. Studies on effective thermal conductivities in packed beds. *AIChE J.* 1957, *3*, 373-381.
[308] Deissler, R.G.; Eian, C.S. Investigation of effective thermal conductivities of powders, (Tech. Report NACA RM E52C05, National Advisory Committee for Aeronautics), 1952.
[309] Deissler, R.G.; Boegli, J.S. An investigation of ETC of powders in various gases. *ASME Trans.* 1958, *80*, 1417-1425.
[310] Nozad, S.; Carbonell, R.G.; Whitaker, S. Heat conduction in multiphase systems. I. Theory and experiments for two-phase systems. *Chem. Eng. Sci.* 1985, *40*, 843-855.
[311] Nozad, S.; Carbonell, R.G.; Whitaker, S. Heat conduction in multiphase systems. II. Experimental method and results for three-phase systems. *Chem. Eng. Sci.* 1985, *40*, 857-863.
[312] Aichlmayr, H.T. *The effective thermal conductivity of saturated porous media.* Master's Thesis, (The University of Minnesota), 1999.
[313] Maqsood, A.; Kamran, K. Thermophysical properties of porous sandstones: Measurements and comparative study of some representative thermal conductivity models. *Int. J. Thermophys.* 2005, *26*, 1617-1632.
[314] Abdulagatova, Z.Z.; Abdulagatov, I.M.; Emirov, S.N. Effective thermal conductivity of water-saturated sandstone at high temperatures and high pressures. *J. Appl. Geophys.* 2011 (in press).

[315] Zierfuss, H. Heat conductivity of some carbonate rocks and clayey sandstones. *Am. Assoc. Pet. Geol. Bull.* 1969, *53*, 251-260.

[316] Clark H., The effect of simple compression and wetting on the thermal conductivity of rocks. *Transaction of the AGU*, 1941, pp.543-544.

[317] Boldizsar, T.; Gozon, J. The geothermic flow at Komlo-Zobak. *Acta Techn. Hung.* 1963, *43*, 467-476.

[318] Babaev, A.Yu.; Gairbekov, Kh.A.; Mursalov, B.A.; Tsomaeva, T.A. Effect of different fluid-saturation on thermal conductivity and speed of sound of sedimentary rocks. In: *Geothermic. Geothermal Energetic.* (Makhachkala: Geothermal Research Institute of the Dagestan Scientific Center of the Russian Academy of Sciences) 1994, pp.177-181.

[319] Batyrmurzaev, A.S.; Guseinov, A.A.; Alibekov, G.I. Physical properties of quartz *Russian Doklady Akademii Nauk* 1996, *348*, 641-643.

[320] Osadchii, V.G.; Ishchenko, P.N.; Popov, Yu.A. Thermal conductivity *Russian Izvestiya Akademii Nauk SSSR, Seriya Geologicheskaya* 1989, *5*, 130-145.

[321] Schoen, J.H. Physical Properties of Rocks. Fundamentals and principles of petrophysics. In: *Handbook of Geophysical Exploration, Section I, Seismic Exploration*, ed., K. Helbig, S. Treitel, (Pergamon Press, Inc.), 1996, vol. 18.

[322] Keller, Th.; Motschmann, U.; Engelhard, L. Modelling the poroelasticity of rocks and ice. *Geophys. Prospect* 1999, *47*, 509-526.

[323] Parrott, J.E.; Stuckes, A.D. *Thermal Conductivity of Solids*, (London: Pion Limited), 1975.

[324] Kiyohashi, H. Review of studies on effective thermal conductivity of three –phased materials. In: *Proc. 13th Japan Symp. Thermophys. Prop*. 1992, pp. 101-104.

[325] Zarichnyak, Yu.P.; Novikov, V.V. Effective conductivity of heterogeneous systems with disordered structure. *Inzh. Fiz. Zhurnal* 1978, *34*, 648-655.

[326] Midttømme, K.; AAgaard, P. Termiske egenskaper I sedimentoere bergarter. Rapport fra institutt for geologi og bergteknikk. Rapport fra institutt for geologi og bergteknikk (Norwegian University of Technology, NTH), 1994, p.29.

[327] Odelevskii, V.I. Calculation of the generalized conductivity of heterogeneous systems. *Russ. J. Tech. Phys.* 1951, *21*, 667-678.

[328] Mendel, A.M. Relation between thermal conductivity of rocks and structure of the pores. *Russ. J. Geol. Prospect* 1997, *1*, 112-119.

[329] Dul'nev, G.N. Thermal conductivity of mixtures and composite materials, (Leningrad: Energiya), 1974.

[330] Vacquier, V.; Mathieu, Y.; Legendere, E.; Blondin, E. Experiment on estimating thermal conductivity of sedimentary rocks from oil well logging. *Bull. Amer. Assoc. Petrol. Geol.* 1988, *72*, 758-764.

[331] Cheng, H.; Torquato, S. Effective conductivity of periodic arrays of spheres with interfacial resistance. *Proc. R. Soc. London A*. 1997, *453*, 145-161.

[332] Maxwell, C.J. A Treatise on Electrical and Magnetism, 3rd ed., (Oxford: Clarenton Press), 1904. pp. 440-445.

[333] Yu, F.; Wei, G.; Zhang, X.; Chen, K. Two effective thermal conductivity models for porous media with hollow spherical agglomerates. *Int. J. Thermophys.* 2006, *27*, 293-303.

[334] Ghaffari, *A. Model for predicting thermal conductivity of rock/fluid systems*, Ph.D. Thesis (Berkeley: University of California), 1980.

[335] Zimmerman, R.W. Thermal conductivity of fluid saturated rocks. *J. Pet. Sci. and Eng.* 1989, *3*, 219-227.

[336] Fricke, H.A mathematical treatment of the electric conductivity and capacity of disperse systems. *Phys. Rev.* 1924, *24*, 575-587.

[337] Boomsma, K.; Poulikakos, D. On the effective thermal conductivity of a three-dimensionally structured fluid-saturated metal foam. *Int. J. Heat Mass Transfer* 2001, *44*, 827-836.

[338] Hsu, C.T.; Cheng, P.; Wong, K.W. Modified Zehner-Schlunder models for stagnant thermal conductivity of porous media. *Int. J. Heat Mass Transfer* 1994, *37*, 2751-2759.

[339] Waff, H.S. Theoretical considerations of electrical conductivity in partially molten mantle and implications for geothermometry. *J. Geophys. Res.* 1974, *79*, 4003-4010.

[340] Chan, C.K.; Tien, C.L. Conductance of packed spheres in vacuum. *Trans. ASME.* 1973, *95*, 302-308.

[341] Dul'nev, G.N.; Volkov, D.P.; Malarev, V.I. Thermal conductivity of moisture porous materials. *Inzh. Fiz. Zhurnal* (*Russian*), 1989, *56*, 281-291.

[342] Touloukian, Y.S.; Judd, W.R.; Roy, R.F., eds., *Physical Properties of Rock and Materials*, (New York: Hemisphere), 1989.

[343] Woodside, W. Calculation of the thermal conductivity of porous media. *Can. J. Phys.* 1958, *36*, 815-823.

[344] Cosenza, P.; Guérin, R.; Tabbagh, A. Relationship between thermal conductivity and water content of soils using numerical modeling. *Eur. J. Soil. Sci.* 2003, *54*, 581-588.

[345] Kohout, M.; Collier, A.P.; Štepánek, F. Effective thermal conductivity of wet particle assemblies. *Int. J. Heat Mass Trans.* 2004, *47*, 5565-5574.

[346] Cathles, L.M.; Ciprianni, F.; Manhardt, P. *Basin-Theory, Templates and Example Manual*, (New York: Cornell Univ. Ithaca), 1993, p.55.

[347] Luo, M.; Wood, J.R.; Cathles, L.M. Prediction of thermal conductivity in reservoir rocks using fabric theory. *J. Appl. Geophys.* 1994, *32*, 321-334.

[348] Clauser, C.; Huenges, E. Thermal conductivity of rocks and minerals. In: *Rock Physics and Phase Relations. A Handbook of physical constants, rock physics and phase relations*, ed., T.J. Ahrens, (Washington D.C.: AGU), 1995, vol. 3, pp. 105-126.

[349] Beardsmore, G.R.; Cull, J.P. *Crustal Heat Flow: A Guide to Measurement and Modeling,* (Cambridge: Cambridge University Press), 2001.

[350] Deguchi, M.; Kiyohashi, H.; Enomoto, H. A correlation equation derived from three phase geometric mean model and analysis of the effective thermal conductivity of porous rocks: Studies on the thermal conductivity of porous rocks: Second report. *J. Mining and Materials Proc. Inst. Japan*, 1995, *111*, 1007–1013.

[351] Eckstein, Y.; Dahl, P.S.; Vitaliano, C.J. Petrographic and physical factors controlling thermal conductivity of granitic rocks in Illinois deep holes UPH 1, 2, and 3. *J. Geophys. Res.* 1983, *88(B9)*, 7381–7385.

[352] Gibiansky, L.; Torquato, S. Rigorous connection between physical properties of porous rocks. *J. Geophys. Res.* 1998, *103(B10)*, 23911–23923.

[353] Jones, F.W.; Pascal, F. Numerical model calculations of the effects of grain sizes and orientations on the thermal conductivities of composites. *Geothermics*, 1994, *23*, 365–371.

[354] Kukkonen, I.T.; Peltoniemi, S. Relationships between thermal and other petrophysical properties of rocks in Finland. In: *Physics and Chemistry of the Earth. Relationship*

between Various Properties in Sedimentary and Crystalline Rocks, eds., E. Huenges, I. Kukkonen, J. Urai, (Oxford-New York-Toronto: Pergamon), 1998, vol. 23(3), pp.341–349.

[355] Luo, M.; Wood, J.R.; Cathles, L.M. Prediction of thermal conductivity in reservoir rocks using fabric theory. *J. Appl. Geophys.* 1994, *32*, 321-334.

[356] Midttomme, K.; Roaldset, E. The effect of grain size on thermal conductivity of quartz sands and silts. *Petroleum Geoscience*, 1998, *4*, 165–172.

[357] Murali, S. Geothermal methods. In: *Proceedings of Meeting: Geophysical Technology for Coal Prospecting, Hyderabad, India, January 20–February 10, 1988*, eds., B.V.S. Murty, G. Nandakumar, P.C. Reddy, and S.M.V. Rao, (Hyderabad, India: Osmania University Centre for Exploration Geophysics), 1988, pp.111–115.

[358] Nobes, D.C.; Villinger, H.; Davis, E.E.; Law, L.K. Estimation of marine sediment bulk physical properties at depth from seafloor geophysical measurements. *J. Geophys. Res.* 1986, *91*, 14033-14043.

[359] Simmons, G. Anisotropic thermal conductivity. *J. Geophys. Res.* 1961, *66*, 2269–2270.

[360] Vasseur, G.; Brigaud, F.; Demongodin, L. Thermal conductivity estimation in sedimentary basins. In: *Tectonophysics, Proc. of Meeting: European Geophysical Society, 15th General Assembly, Symposium on Heat Flow and Lithospheric Thermal Regimes, Copenhagen, Denmark, April 23–27, 1990*, eds. N. Balling and E.R. Decker, (Amsterdam, Netherlands: Elsevier), 1995, vol. 244(1/3), pp.167–174.

[361] Drury, M.J.; Jessop, A.M. The estimation of rock thermal conductivity from mineral content: an assessment of techniques. *Zbl. Geol. Paleont. Teil I*, 1983, H1/2, 35-48.

[362] Horai, K.; Baldridge, S. Thermal conductivity of nineteen igneous rocks-I-Application of the needle probe method to the measurement of the thermal conductivity of rocks. *Phys. Earth Planet Int.* 1972, *5*, 151-156.

[363] Hashin, Z.; Shtrikman, S.A. A variational approach to the theorey of the effective magnetic permeability of multiphase materials. *J. Appl. Phys.* 1962, *33*, 3125-3131.

[364] Demongodin, L.; Pinoteau, B.; Vasseur, G.; Gable, R. Thermal conductivity and well logs: a case study from the Paris basin. *Geophys. J. Int.* 1991, *105*, 675-691.

[365] McKenna, T.E.; Sharp, J.M.; Lynch, L.F. Thermal conductivity of Wilcox and Frio sandstones in South Texas (Gulf of Mexico Basin). *AAPG Bulletin* 1996, *80*, 1203-1215.

[366] Balling, N.; Kristiansen, J.I.; Breiner, N.; Poulsen, K.D.; Rasmussen, R.; Saxov, S. *Geothermal measurements and subsurface temperature modeling in Denmark,* (Geoskrifter: Department of Geology, Aarhus University), 1981, vol. 16.

[367] Midttømme, K.; Saettem, J.; Roaldset, E. Thermal conductivity of unconsolidated marine sediments from Voring Basin, (Nordic Petroleum Technology, Series II), 1998, pp. 145-197.

[368] Eucken, A. Forsch. *Gehiete Ingenieurue B3*, *Forschnung*, 1932, *353*, 16.

[369] Brown, W.F. Solid mixture permitivities. *J. Chem. Phys.* 1955, *23*, 1514-1517.

[370] Kaganer, M.G. Contact heat transfer in granular material in vacuum. *J. Eng. Phys.* 1966, *11*, 19-30.

[371] Yovanovich, M.M. Thermal contact resistance across elastically deformed spheres. *J. Spacecraft Rockets* 1967, *4*, 119-122.

[372] Ogniewicz, Y.; Yovanovich, M.M. Effective conductivity of regularly packed spheres: Basic cell model with constriction. In: *Progress in Astronautics and Aeronautics,* (Reston: VA, AIAA), 1977, vol. 60, pp. 209-228.

[373] Rayleigh, L. On the influence of obstacles arranged in rectangular order upon the properties of a medium. *Philos. Mag.* 1892, *34*, 481-502.

[374] Meredith, R.E.; Tobias, C.W. Resistance to potential flow through a cubical array of spheres. *J. Appl. Phys.* 1960, *31*, 1270-1273.

[375] Sanagani, A.S.; Acrivos A. On the thermal conductivity and permeability of regular arrayes of spheres. *Proc. R. Soc. London A* 1981, *385*, 216-225.

[376] Brailsford, A.D.; Major, K.G. The thermal conductivity of aggregates of several phases, including porous materials. *J. App. Phys.* 1964, *15*, 313-319.

[377] Bruggeman, D.A.G. Calculation of difference physical constants of heterogeneous substances. I. Dielectric constant and conductivity of media of isotopic substances. *Ann. Phys. Ser. 5*, 1935, *24*, 636-679.

[378] Frey, G.S. Uber die elektrische leitfahigkeit binarer aggregate. *Z. Electrochem. Angew. Phys. Chem.* 1932, *38*, 260-273.

[379] Dul'nev, G.N.; Zarichniak, Yu.P. *Thermophysical properties of composites,* (Leningrad: Energy), 1974.

[380] Alishaev, M.G. The effective thermal conductivity of porous medium filled with the fluid. *Inzh. Fiz. Zhur.* 1983, *14*, 143-144.

[381] Cheng, C.H.; Toksöz, M.N. Elastic wave propagation in fluid-filled borehole and synthetic acoustic logs. *Geophysics* 1981, *46*, 1042-1053.

[382] Cheng, C.H.; Toksöz, M.N. Inversion of seismic velocities for the pore aspect ratio distribution of a rock. *J. Geophys. Res.* 1979, *84*, 7534-7543.

[383] Prasad, V.; Kladias, B.A.; Tian, Q. Evaluation of correlations for stagnant thermal conductivity of liquid saturated porous beds of spheres. *Int. J. Heat Mass trans.* 1989, *32*, 1793-1796.

[384] Saez, A.E.; Perfetto, J.C.; Rusinek, I. Prediction of effective diffusivities in porous media using spatially periodic models. *Transp. Porous Media* 1991, *6*, 143-157.

[385] Beniwal, R.S.; Singh, R.; Pande, R.N.; Kumar, V.; Chaudhary, D.R. *Ind. J. Pure Appl. Phys.* 1985, *23*, 289-297.

[386] Adnani, P.; Catton, I.; Raffray, A.R.; Abdou, M.A. Effective thermal conductivity of binary mixtures at high solid to gas conductivity ratios. *Chem. Eng. Comm.* 1993, *120*, 45-58.

[387] Yu, B.; Cheng, P. Fractal models for the effective thermal conductivity of bidispersed porous media. *J. Thermophysics and Heat Transfer* 2002, *16*, 22-29.

[388] Lee, S.L.; Yang, J.H. Modeling of effective thermal conductivity for a nonhomogeneous anisotropic porous medium. *Int. J. Heat Mass Transfer* 1998, *41*, 931-937.

[389] Lipton, R.; Vernescu, B. Composites with imperfect interfaces. *Proc. R. Soc. London A* 1996, *452*, 329-358.

[390] Torquato, S.; Rintoul, M.D. Effect of the interface on the properties of composite media. *Phys. Rev. Lett.* 1995, *75*, 4067-4070.

[391] Miloh, T.; Benveniste, Y. On the effective conductivity of composites with ellipsoidal inhomogeneities and high highly conducting interfaces. *Proc. R. Soc., London A* 1996, *455*, 2687-2706.

[392] Singh, K.J.; Singh R., Chaudhary, D.R. Heat conduction and a porosity correction term for spherical and cubic particles in a simple cubic packing. *J. Phys. D: Appl. Phys.* 1998, *31*, 1681-1687.

[393] Verma, L.S.; Shrotriya, A.K.; Singh, R.; Chaudhary, D.R. Thermal conduction in two-phase materials with spherical and non-spherical inclusions. *J. Phys. D: Appl. Phys.* 1991, *24*, 1515-1526.

[394] Berryman, J.G. Thermal conductivity of porous media. *J. Phys. Lett.* 2005, *86*, 032905.

[395] Schulz, B. Thermal conductivity of porous and highly porous materials. *High Temperatures – High Pressures* 1981, *13*, 649-660.

[396] Burger, H.C. *Phys. Zr.* 1915, *20*, 73-76.

[397] Kimura, M. *Chem. Eng. (Japan)* 1957, *21*, 472.

[398] Schumann, J.E.; Voss, V. Heat flow through granulated materials. *Fuel Sci. Practice* 1934, *13*, 249-256.

[399] Wilhelm, R.H.; Johnson, W.C.; Wynkoop, R.; Collier, C.W. Reaction rate, heat transfer, and temperature distribution in fixed bed catalytic converters. *Chem. Eng. Progr.* 1948, *44*, 105-116.

[400] Gemant, A. The thermal conductivity of soils. *J. Appl. Phys.* 1950, *21*, 750-752.

[401] Swift, D.L. The thermal conductivity of spherical metal powders including the effect of oxide coating. *Int. J. Heat Mass Transfer* 1966, *9*, 1061-1074.

[402] Jaguaribe, E.F.; Beasley, D.E. Modeling of the effective thermal conductivity and diffusivity of a packed bed with stagnant fluid. *Int. J. Heat Mass Transfer* 1984, *27*, 399-407.

[403] Misra, K.; Shrotriya, A.K.; Singh, R.; Chaudhary, D.R. Porosity correction for thermal conduction in real two-phase systems. *J. Phys. D: Appl. Phys.* 1994, *27*, 732-735.

[404] Reiter, M.A.; Tovar, R.J.C. Estimates of terrestrial heat flow in Northern Chihuahua, Mexico. *Can. J. Earth Sci.* 1982, *22*, 1503–17.

[405] Cheng, S.C.; Vachon, R.I. The prediction of the thermal conductivity of two and three phase solid heterogeneous mixtures. *Int. J. Heat Mass Transfer* 1969, *12*, 249-264.

[406] Goss, R.D.; Combs, J.; Timur, A. Prediction of thermal conductivity in rocks from other physical parameters and from standard geophysical well logs. In: *Transactions of Meeting: SPWLA 10th Annual Logging Symp.*, (Houston, Texas, United States: Society of Professional Well Log Analysts), 1975, pp.1-21.

[407] Houbolt, J.J.; Wells. P.R.A. Estimation of heat flow in oil wells based on a relation between heat conductivity and sound velocity. *Geologie en Mijnbouw* 1980, *59*, 215-224.

[408] Wyble, D.O. Effect of applied pressure on the conductivity, porosity and permittivity of sandstones. *Pet. Trans. AIME.* 1958, *213*, 430-432.

[409] Ribaud, M. Theoretical study of thermal conductivity of porous and pulverienlent materials. *Chal. Et Ind.* 1937, *18*, 36-43.

[410] Russell, H.W. Principles of heat flow in porous insulators. *J. Amer. Ceram. Soc.* 1935, *18*, 1-5.

[411] Chaudhary, D.R.; Bhandari Thermal conductivity of two phase porous materials. *Brit. J. Appl. Res.* (*J. Physics – D*) 1969, *2*, 609-610.

[412] Bogomolov, V.Z. *Trans. Phys. Astron.*, (Inst. Agric. Press), 1941.

[413] Kaufman, B.I. *Thermal Conductivity of Building Materials*, (Moscow), 1955.

[414] Litovsky, E.Ya.; Shapiro, M. Gas pressure and temperature dependences of thermal conductivity of porous ceramic materials: Part 1, Refractors and Ceramics with porosity below 30 %. *J. Am. Ceram. Soc.* 1992, *75*, 3425-3439.

[415] Dachanov, V.N.; Djakonov, D.J. *Termicheskie issledovaniya skvazhin.* (Moscow: Nedra), 1952, pp.16-29.

[416] Budiansky, B. Thermal and thermoelastic properties of isotropic composites. *J. Composite Materials* 1970, *4*, 286-295.

[417] Robertson, D.E.; Peck, D.L. Thermal conductivity of vesicular basalt from Hawaii. *J. Geophys. Res.* 1974, *79*, 4875-4888.

[418] Blackwell, D.D.; Steele, J.L. Thermal conductivity of rocks-measurement and significance. In: *Thermal History of Sedimentary Basins*: Method and Case Histories, eds., N.D. Naeser, T.H. McCulloh, (New York: Springer), 1989, pp.13-36.

[419] Doveton, J.H.; Forster, A.; Merriam, D.F. Predicting thermal conductivity from petrophysical logs: a midcontinet Paleozoic case study. In: *Proc. IAMG'97*, ed., V. Pawlowsky-Glahn, (Barcelona: CIMNE), 1977, pp.212-217.

[420] Evans, T.R. Thermal properties of North Sea rocks. *Log Analyst.* 1977, *18*, 3-12.

[421] Decker, E.R.; Sahai, S.K.; Choquette, P.W. The thermal conductivity of porous sedimentary rock: Clean sandstones, limestones and dolomites. *EOS*, 1981, *62*, 392-401.

[422] Nosal, E.A. Predicting thermal conductivity of sandstones and carbonates. *EOS*, 1981, *62*, 392-398.

[423] Adler, D. Electrical conductivity in disordered systems. *Solid State Commun.* 1973, *12*, 9-12.

[424] Williams, C.F.; Anderson, R.A. Thermophysical properties of the earth's crust: in situ measurements from continental and oceanic drilling. *J. Geophys. Res.* 1990, *95*, 9209-9236.

[425] Dea, S. *Thermal conductivity estimation for oil sands from well logging data.* MS Research Report, (Berkeley: University of California), 1976.

[426] Rzhevsky, V.; Novik, G. *The Physics of Rocks,* (Moscow: Mir), 1971.

[427] Brace, W.F. Some new measurements of liner compressibility of rocks. *J. Geophys. Res.* 1965, *70*, 391-398.

[428] Hughes, D.S.; Maurette, C. Elastic wave velocities in granite. *Geophysics.* 1956, *21*, 277-284.

[429] Hughes, D.S.; Maurette, C. Détermination des vitesses d'ondes élastiques dans diverses roches en fonction de la pression et de la temperature. *Rev. Inst. Franc. Petrole Ann. Combust. Liquides* 1957, *12*, 730-738.

[430] Birch, F. The velocity of compressional waves in rocks to 10 kbars. *J. Geophys. Res.* 1960, *65*, 1083-1103.

[431] Birch, F. The velocity of compressional waves in rocks to 10 kbars. *J. Geophys. Res.* 1961, *66*, 2199-2224.

[432] Simmons, G. Velocity of shear waves in rocks to 10 kbars. *J. Geophys. Res.* 1964, *69*, 1123-1130.

[433] Volarovich, M.P., ed., *Physical properties of minerals and rocks at high thermodynamic parameters*, Handbook, 2nd edition, (Moscow: Nedra), 1988.

[434] Somerton, W.H.; Selim, M. Additional thermal data for porous rocks-thermal expansion and heat of reaction. *AIME Petr. Trans.* 1961, *222*, 249-253.

[435] Anderson, D.L. *Theory of the Earth*, (Boston: Blackwell Scientific Publ.), 1989, Chap.5.

[436] Gotze, H.J.; Huenges, E.; Kukkonen, I.; Ural, J. eds., *Physics and Chemistry of the Earth*. (New York: Pergamon Press), 1998.

[437] Anderson, D.L.; Kanamori, H. Shock-wave equations of state for rocks and minerals1contribution. *J. Geophys. Res*. 1968, *73*, 6477-6502.

[438] Davies, G. F.; Anderson, D.L. Revised shock-wave equations of state for high-pressure phases of rocks and minerals contribution. *J. Geophys. Res*. 1971, *76*, 2617-2627.

[439] Eucken, A. Allgemeine Geseteztmassig keitne für das Warmeleitvermogen Versciediner Stoffarten und Aggregatzustande. Forsch. Gebiete Ingenieur Ausgabe A, *V.D.I. Forschnung*, 1940, *11*, 6-10.

[440] Clauser, C. Opacity-the concept of radiative thermal conductivity, In: *Handbook of Terrestrial Heat–Flow Density Determination,* R. Haenel, L. Rybach, L. Stegena, eds., (Dordrecht: Kluwer Academic Publishers), 1988, pp.143-165.

[441] Beck, A.E.; Darbha, D.M.; Schloessin, H.H. Lattice conductivities of single-crystal and polycrystalline materials at mantle pressures and temperatures. *Phys. Earth Planet. Inter*. 1978, *17*, 35-53.

[442] McBirney, A.R. Conductivity variation and terrestrial heat heat-flow distribution. *J. Geophys. Res*. 1963, *68*, 6323-6329.

[443] Lee, Y.; Deming, D. Evaluation of thermal conductivity temperature correlations applied in terrestrial heat flow studies. *J. Gophys. Res*. 1998, *103*, 2447-2454.

[444] Vulis, L.; Putseluiko, A. On the determination of the temperature dependence of thermal conductivity. *Zh. Tech. Fiziki*. 1956, *26*, 76-87.

[445] Gore, D.; Flucker, E. Experimental determination of heat transfer coefficients of porous solid-liquid systems. *Status Report on API Grant-in Aid No. 43, Sept. 15*, 1958.

[446] Francl, J.; Kingery, D.W. Thermal conductivity: IV, Apparatus for determining thermal conductivity by a comparative method. *Am. Ceramic Society* 1954, *37*, 80-84.

[447] Buntebarth, G. Thermal models of cooling. In: *Equilibrium and kinetics in contact metamorphism: the Ballachulish igneous complex and its aureole*, eds., G. Voll, D. Pattison, *J. Topel and F. Seifert*, (New York: Springer), 1991, 377.

[448] Leibfried, G.; Schlömann, E. *Nachr. Ges. Wiss. Goett. Math.-Phys. Kl.* 1954, *2*, 71-93.

[449] Lee, D.W.; Kingery, W.D. Thermal Conductivity: XIV, Conductivity of Multicomponent Systems. *J. Amer. Ceram. Soc*. 1960, *43*, 594-605.

[450] Roufosse, M.; Klemens, P.G. Thermal conductivity. *Phys. Rev. B*. 1973, *7*, 5379-5386.

[451] Zoth, G.; Hänel, R. In: *Handbook of Terrestrial Heat –Flow Density Determination*, eds., R. Haenel, L. Rybach, L. Stegena, (Dordrecht: Kluwer Academic Publishers), 1988.

[452] Chapman, D.S.; Keho, T.H.; Bauer, M.S.; Picard, M.D. Heat flow in the Uinta Basin determined from bottom hole temperature (BHT) data. *Geophysics*. 1984, *49*, 453-466.

[453] Funnell, R.; Chapman, D.; Allis, R.; Amstrong, P. Thermal state of the Taranaki Basin, New Zealand. *J. Geophys. Res*. 1996, *101*, 25197-25215.

[454] Jõeleht, A.; Kirsimäe, K.; Shogenova, A.; Šliaupa, S.; Kukkonen, I.T.; Rasteniene, V.; Zabele, A. Thermal conductivity of Cambrian siliciclastic rocks from the Baltic basin. *Proc. Estonian Acad. Sci. Geol*. 2002, *51*, 5-15.

[455] Osako, M.; Ito, E.; Yoneda, A. Simultaneous measurements of thermal conductivity and thermal diffusivity for garnet and olivine under high pressure. *Phys. Earth Planet. Inter.* 2004, *143-144*, 311-320.

[456] Dzhavadov, L.N. Measurement of thermophysical properties of dielectrics under pressure. *High Temp.- High Press.* 1975, *7*, 49-54.

[457] Lebedev, T.S.; Korchin, B.Yu.; Savenko, V.I. *Petrophysical studies at high PT-parameters*, (Kiev: Naukova Dumka), 1988, p. 248.

[458] Lebedev, T.S.; Korchin, V.A.; Savenko, B.Ya.; Shapovalov, V.I.; Shepel, S.I. *Physical properties of minerals and matter under PT-conditions of the lithosphere*, (Kiev: Naukova Dumka), 1986.

[459] Lebedev, T.S., ed., *Thermophysical properties of rocks*, (Moscow: Nedra), 1987.

[460] Osako, M.; Ito, E. Simultaneous thermal diffusivity and thermal conductivity measurements of mantle materials up to 6 GPa. *Rev. High Pressure Sci. Technol.* 1998, *7*, 110-112.

[461] Seipold, U.; Schilling, F. R. Heat transport in serpentinites. *Tectonophysics* 2003, *370*, 147-162.

[462] Katsura, T. Thermal diffusivity of olivine under upper mantle conditions. *Geophys. J. Int.* 1995, *122*, 63-69.

[463] MacPherson, W.R.; Schloessin, H.H. Lattice and radiative thermal conductivity variations through high P,T polymorphic structure transitions and melting points. *Phys. Earth Planet. Inter.* 1982, *17*, 35-53.

[464] Edmondson, T. *Thermal diffusivity of sedimentary rocks subjected to simulated overburden pressure,* M.S. Thesis, (Berkeley: University of California), 1961.

[465] Fujisawa, H.; Fujii, N.; Mizutani, H.; Kanamori, H.; Akimoto, S. Thermal diffusivity of Mg_2SiO_4, Fe_2SiO_4, and NaCl at high pressures and temperatures. *J. Geophys. Res.* 1968, *73*, 4727-4733.

[466] Lebedev, T.S.; Shapavalov, V.I. *Temperature dependence of the thermal conductivity coefficient. Petrophysical studies at high PT-conditions*, (Kiev: Nukovo Dumka), 1988.

[467] Bäckström, G. Measurements of the thermophysical properties of of solids under high pressure. In: *Proc. 7th Symp. Thermophys. Prop.*, ed., A. Cezairlyan, (New York: ASME), 1977, pp.169-180.

[468] Brydsten, U.; Gerlich, D.; Bäckström, G. Thermal conductivity of single-crystal NaCl under uniaxial compression. *J. Phys. C: Solid State Physics*. 1983, *16*, 143-146.

[469] Alm, O.; Bäckström, G. Thermal conductivity of KCl up to 19 kbar. *J. Phys. Chem. Solids*. 1974, *35*, 421-424.

[470] Shimozura, D. Elasticity of rocks and some related geophysical problems. *Jap. J. Geophys*. 1960, *2*, 1-85.

[471] Lemmon, E.W.; Jacobsen, R.T. Viscosity and Thermal Conductivity Equations for Nitrogen, Oxygen, Argon, and Air. *Int. J.* Thermophys. 2004, *25*, 21-69,

[472] Aliev, N.A.; Magaramova, F.G. *Complex semiconductors*, (Baku: Azerbajan Academy of Sciences), 1962.

[473] Averkin, A.A.; Zhaparov, Zh.Zh.; Stil'bans, A.S. *Solid State Physics* (*Russian*), 1971, *5*, 2236-2245.

[474] Neizer, V.G. *Experimental study of the thermodynamic properties of poliorganic liquids*, Ph.D. Thesis, (Minsk: Belarus University), 1970.

[475] Kestin, J.; Sengers, J.V.; Kamgar-Parsi, B.; Levelt Sengers, J.M.H. Thermophysical properties of fluid H_2O. *J. Phys. Chem. Ref. Data*. 1984, *13*, 175-183.

[476] Nieto, de Castro, C.A.; Li S. F.; Nagashima, A.; Trengove, R.D.; Wakeham, W.A. *J. Phys. Chem. Ref. Data* 1986, *15*, 1073-1086.

[477] Sidorov, A.M. Effective thermal conductivity of porous rocks. *Geology and Geophysics* 1979, *10*, 87-94.

[478] Krischer, O.; Kröll, K. *Die wissenschaftlichen grundlagen der trocknungstechnik*, (Berlin-Göttingen-Heidelberg), 1956, Bd.1.

[479] Lichtencker, K.; Rother, K. Die Herleitung des logarithmischen Mischungs-gesetzes aus allegemeinen Prinzipien der stationaren Stromung. *Phys. Z.* 1931, *32*, 255-260.

[480] Litovskii, E.Ya. Interpolation formula for the expression of thermal conductivity of solid materials on porosity. *Russian Izv. Akad. Nauk, ser. Nonorganic Materials*, 1980, *16*, 559-560.

[481] Kazatchenko,E.; Markov, M.; Mousatov, A. Simulation of acoustic velocities, electrical and thermal conductivities using unified pore-structure model of double-porosity carbonate rocks, *J. Appl. Geophys*. 2006, *59*, 16-35.

[482] Sugawara, A. The precise determination of thermal conductivity of pure fused quartz. *J. Appl. Phys.* 1968, *39*, 5994-5997.

[483] Lichtencker, K. Die Dielektrizitätskonstante natürlicher und künstlicher Mischkörper. *Phys. Z.* 1926, *27*, 115-158.

[484] Touloukian, Y.S., ed., *Thermophysical properties of matter*, (New York: IFI/Plenum), 1978, vol. 14.

[485] Devyakonov, D.I. *Geothermic in Oil Geology*, (Moscow: Gostoptexizdat), 1958.

[486] Camirand, C. Measurement of thermal conductivity by differential scanning calorimetry. *Thermochimica Acta* 2004, *417*, 1-4.

[487] Kaye, G.W.C.; Higgins, W.F. *Proc. Roy. Soc. A*. 1926, *113*, 335.

[488] Incropera, F.P.; Dewitt, D.P. *Fundamentals of Heat Transfer*, (New York: John Wiley and Sons), 1981, p.780.

[489] Schulz, B. Thermal conductivity of porous and highly porous materials. *High Temperatures- High Pressures* 1981, *13*, 649-660.

[490] Petrunin, G.I.; Popov, V.G.; Soskov, A.V. Thermal conductivity *Izv. Akad. Nauk SSSR, ser. Fizika Zemli* 1999, *1*, 47-51.

[491] Woodside, W.; Messmer, J.H. Molecular effects in heat conduction *J. Geophys. Res.* 1960, *65*, 3481-3485.

[492] Ljubimova, E.A.; Lysova, L.N.; Firsov, F.V. *Izv. Akad. Nauk SSSR, ser. Geophys*. 1964, *11*, 1622.

[493] Ljubimova, E.A.; Starikova G.N.; Shuspanov A.P. Thermohysical studies of rocks. In: *Geothermal Studies*, (Moscow: Nauka), 1964, p.115.

[494] Lipaev, A.A.; Khisamov, R.S.; Chugunov, V.A. *Thermophysics of Rocks at Oil Fields*, (Moscow: Nedra), 2003.

[495] Emirov, S.N.; Gadzhiev, G.G.; Ramazanova, E.N.; Abdulagatova, Z.Z.; Magomedmirzoev, E.M. Study of the thermal conductivity of SiC-BeO ceramics at high pressures and temperatures. *Izv. Russ. Akademii Nauk, ser. Fizika* 2007, *71*, 290-292.

[496] Boyce, R.E. Electrical resistivity, sound velocity, thermal conductivity, density-porosity, and temperature, obtained by laboratory techniques and well logs: Site 462 in

the Nauru Basin of the Pacific Ocean. In*: Initial Reports of the Deep Sea Drilling Project, Leg 61 of the Cruises of the Drilling Vessel Glomar Challenger: Apra, Guam to Majuro Atoll, Marshall Islands: May–July, 1978*, eds., R.L. Larson et al., (College Station, Texas, United States: Texas AandM University, Ocean Drilling Program), 1981, vol. 61, pp.743–61.

[497] Brigaud, F., Vasseur, G. and Caillet, G. Use of well log data for predicting detailed *in situ* thermal conductivity profiles at well sites and estimation of lateral changes in main sedimentary units at basin scale. In: *Rock at Great Depth: Rock Mechanics and Rock Physics at Great Depth, Proc. Meeting: Int. Symp. Rock Mechanics, Pau, France, August 28–31, 1989.)*, eds., V. Maury and D. Fourmaintraux, (Rotterdam, Netherlands: A.A.Balkema), 1989, vol. 1, pp.403–409.

[498] Chapman, D.S.; Yi, Z.; Deming, D.; Brigaud, F.; Powell, W.G. Determining heat flow from oil/gas well data: Examples from the Utah-Wyoming thrust belt. In: *EOS, Transactions, American Geophysical Union, Abstracts of Meeting: American Geophysical Union, 1989 Spring Meeting, Baltimore, Maryland, United States, May 7–12*, (Washington DC: American Geophysical Union), *1989,* vol. 70(15), p.463.

[499] Griffiths, C.M. Thermal conductivity prediction from petrophysical data: A case study. In: *The Log Analyst, (Abstracts of Meeting: Geological Applications of Wireline Logs II (GAWL II), London, United Kingdom, June 10–11.)*, eds., A. Hurst, C. Griffiths and P.F. Worthington, (Houston, Texas, United States: Society of Professional Well Log Analysts), 1991, vol. 32(4), p.380.

[500] Griffiths, C.M.; Brereton, N.R.; Beausillon, R.; Castillo, D. Thermal conductivity prediction from petrophysical data: A case study. In: *Geological Society Special Publications, Geological Applications of Wireline Logs: II*, eds., A. Hurst, C.M. Griffiths and P.F. Worthington, (London, United Kingdom: Geological Society of London), 1992, vol. 65, pp.299–315.

[501] Hagedon, D.N. *The Calculation of Synthetic Thermal Conductivity Logs from Conventional Geophysical Well Logs*. Master's Dissertation, (Dallas: Southern Methodist University, Texas, United States), 1985.

[502] Huang, Z.; Williamson, M.A. Estimation of *in situ* thermal conductivity in the Jeanne d'Arc Basin. In: *Abstracts of Meeting: American Association of Petroleum Geologists, Annual Convention, Calgary, Alberta, Canada, June 22–24, 1992*, ed., G. Eynon, (Tulsa, Oklahoma, United States: American Association of Petroleum Geologists and Society of Economic Paleontologists and Mineralogists), 1992, p.57.

[503] Huang, Z.; Williamson, M.A. The thermal conductivity and heat flow density of the Jeanne d'Arc Basin, offshore Eastern Canada. *Tectonophysics* 1994, *233*, 177–192.

[504] Molnar, P.S. *Correlation of Thermal Conductivity with Physical Properties Obtained from Geophysical Well Logs.* Master's Dissertation, (New York: The State University of New York at Buffalo, United States), 1982.

[505] Molnar, P.S.; Hodge, D. Correlation of thermal conductivity with physical properties obtained from geophysical well logs. In: *AAPG Bulletin, Proc. Meeting: AAPG Annual Convention with Divisions SEPM/EMD/DPA, Calgary, AB, Canada, June 27–30,* (Tulsa, Oklahoma: American Association of Petroleum Geologists), 1982, vol. 66(5), p.608.

[506] Vacquier, V. Calculation of terrestrial heat flow solely from oil well logging records. In: *Abstracts of Meeting: Int. Association of Seismology and Physics of the Earth's*

Interior Regional Assembly, Hyderabad, India, October 31–November 7, eds., J.N. Brune and S.J. Singh, (Hyderabad, India: National Geophysical Research Institute), 1984, p.150.

[507] Wang, L.; Xiong, Z.; Guo, S.; Li, C.; Shi, X.; Wang, J. Estimating the original geothermal conductivity of sedimentary rock in a hydrocarbon-bearing basin by using geophysical logging data. *Oil Geophysical Prospecting* 1999, *34*, 526–531.

[508] Zoth, G. Thermolog: A new method for log display of geothermal parameters. *Scientific Drilling* 1992, *3*, 76–82.

[509] Blackwell, D.D.; Steele, J.L. Thermal conductivity of sedimentary rocks: Measurement and significance. In: *Thermal History of Sedimentary Basins*, eds., N.D. Naeser and T.H. McCulloch, (New York: Springer-Verlag), 1989, pp. 45–96

[510] Blackwell, D.D.; Steele, J.L. Heat flow and geothermal potential of Kansas. In: *Kansas Geological Survey Bulletin, Geophysics in Kansas*, ed., D.W. Steeples Wichita, (Kansas: Kansas Geological Survey), 1989, vol. 226, pp. 267–296.

[511] Blackwell, D.D., Wisian, K.W. and Beardsmore, G.R. Application of temperature logging technology to increasing the accuracy of basin thermal models. In: *Applications of Emerging Technologies: Unconventional Methods in Exploration for Petroleum and Natural Gas*, eds., R.J. Kruizenga et al., (Dallas, Texas: Institute for the Study of Earth and Man, Southern Methodist University), 1997, vol.5, pp. 41–65.

[512] Midttømme, K.; Roaldset, E.; Aagaard, P. Thermal conductivity of selected claystones and mudstones from England. In: *Clay Minerals, Proc. Meeting: Clay Minerals in the Modern Society, Rosenqvist Symposium, University of Oslo, Norway, May 21*, ed., D.C. Bain, (London, United Kingdom: Mineralogical Society), 1998, vol. 33(1), pp.131–145.

[513] Seipold, U.; Gutzeit, W. Measurements of the thermal properties of rocks under extreme conditions. In: *Physics of the Earth and Planetary Interiors, Proc. Meeting: High Pressure Physics and Core and Mantle Dynamics, Potsdam, Federal Republic of Germany, October 9–14, 1978*, eds., H. Stiller, V.N. Zharkov and H. Vollstaedt, (Amsterdam, Netherlands: Elsevier), 1980, vol. 22(3/4), pp.272–6.

[514] Beziat, A.; Dardaine, M.; Gabis, V. Effect of compaction pressure and water content on the thermal conductivity of some natural clays. *Clays and Clay Minerals* 1988, *36*, 462–466.

[515] Morgan, P. Estimates of mantle thermal conductivity based on high-temperature measurements of mantle xenolith conductivities. *EOS, Transactions, American Geophysical Union,* 1993, 74 (43, Suppl.), p. 597.

[516] Birch, F. Thermal conductivity, climatic variation and heat flow near Calumet, Michigan. *Amer. J. Sci.* 1954, *252*, 1-25.

[517] Mossop, S.C.; Gafner, G. The thermal constants of some rocks from the Orange Free State. *J. Chem. Met. Min. Soc. S. Afr.* 1951, *52*, 61–73.

[518] Bullard, E.C.; Niblett, E.R. Terrestrial heat flow in England. *Monthly Notices Roy. Astr. Soc., London, Geophys. Suppl.* 1951, *6*, 222-238.

[519] Dortman, N.B. *Physical Properties of Rocks and Minerals,* (Moscow: Nedra), 1984, p.84.

[520] Kobranova, V.N. *Petrophysics*, (Moscow: Mir), 1989, 375 p.

[521] Kappelmeyer, O.; Haenel R. Geothermic with special reference applications, (Geopublication Associates, Geoexploration Monograph), 1974, ser. 1.4.

[522] Ramazanova, A.E.; Emirov, S.N. Study of the heat transfer phenomena in rock materials. Monitoring, *Science and Technology* 2010, *2*, 93-95.

[523] Hammerschmidt, U. Thermal transport properties of water and ice from one single experiment. *Int. J. Thermophys.* 2002, *23*, 975-996.

[524] Hammerschmidt, U; Meier, V. New transient hot-bridge sensor to measure thermal conductivity, thermal diffusivity and volumetric specific heat. *Int. J. Thermophys.* 2006, *27*, 840-865.

[525] Gustafsson, E.; Karawacki, E.; Khan, M.N. Transient-hot strip method for simultaneously measuring thermal conductivity and thermal diffusivity of solids and fluids. *J. Appl. Phys.* 1979, *12*, 1411-1421.

[526] Abid, M. *Thermophysical Properties of Moist Porous Material*, Ph.D. Thesis, Technischen Universitat Carolo-Wilhelmina zu Braunschweig, 2011.

[527] Zeb, A.; Firdous, T.; Maqsood, A. Thermophysical properties of dunite rocks as a function of temperature along with the prediction of effective thermal conductivity. *Natural Science* 2010, *2*, 626-630.

[528] Maqsood, A.; Rehman, M.A. Measurements of thermal transport properties with an improved transient plane source technique. *Int. J. Thermophys.* 2003, *24*, 867-883.

[529] Wakeham, W.A.; Nagashima, A.;Sengers, J.V. (eds.) *Experimental Thermodynamics. III. Measurements of the Transport Properties of Fluids.* Blackwell Scientific Publications, Oxford, 1991.

[530] Wakeham, W.A.; Assael, M.J., *Chapter 5. Thermal Conductivity*, in *Handbook of Experimental Fluid Mechanics*, C. Tropea, J. Foss, A. Yarin (eds), Springer –Verlag, 2010.

[531] Abdulagatov, I.M.; Assael, M.J. *Chapter 5. Thermal Conductivity, in Hydrothermal Properties of Materials*, V.M. Valyshko (ed.), Wiley, 2008, pp.227-249.

[532] Hammerschmidt, U.; Sabuga, W. Transient hot strip (THS) method: Uncertainty assessment. *Int. J. Thermophys*. 2000, *21*, 217-248.

[533] Kubičár, L. *Thermal Analysis, Part E. Pulse method of measuring basic thermophysical parameters*. Vol. XII, Elseveir, New York, 1990.

[534] Maglic, K.D.; Taylor, R.E. *The apparatus for thermal diffusivity measurements by the laser pulse method. Compendium of Thermophysical Property Measurements Methods 2*, New York, 1992, pp. 281-314.

[535] Lozar, L. Uncertainty of the thermal diffusivity measurement using the laser flash method. *Tech. Dig. 15 Symp. Thermophys. Properties*, 2003.

[536] Hammerschmidt, U. A linear procedure for analyzing transient hot strip signals. *Proc. 24th Int. Thermal Conductivity Conference*, 1999, pp. 123-134.

[537] Czichos, H.; Saito, T; Smith, L. Springer handbook of materials measurements methods. Springer, 2006, pp.402-408.

[538] Assael, M.J.; Antoniadis, K.D.; Kakosimos, K.E. An improved application of the transient hot-wire technique for the absolute accurate measurement of the thermal conductiviet of Pyroceram up to 420 K. *Int. J. Thermophys.* 2008, *29*, 445-456.

[539] Assael, M.J.; Antoniadis, K.D.; Wu, J. New measurements of the thermal conductivity of PMMA, BK7 and Pyrex 7740 up to 450 K. *Int. J. Thermophys.* 2008, *29*, 1257-1266.

[540] Gustafsson, M.; Karawacki, E.; Gustafsson, S.E. Thermal conductivity, thermal diffusivity, and specific heat of thin samples from transient measuriments with hot disk sensors. *Rev. Sci. Instrum.* 1994, *65*, 3856-3859.

[541] Wulf, R.; Barth, G.; Gross, U. Intercomparison of insulation thermal conductivities measured by various methods. *Int. J. Thermophys.* 2008, *28*, 1679-1692.

[542] Hammerschmidt, U. A new pulse hot strip sensor for measuring thermal conductivity and thermal diffusivity of solids. *Int. J. Thermophys.* 2003, *24*, 675-682.

[543] Wei, G.; Zhang, X.; Yu, F. Thermal conductivity measurements on xonotlite-type calcium silicate by the transient hot-strip method. *J. Univ. Sci. Technology Beijing Miner. Mateall. Mater.* 2008, *15*, 791-795.

[544] Hammerschmidt, U. Guarded hot-plate method: Unceratinty assessment. *Int. J. Thermophys.* 2002, *23*, 1551-1570.

[545] Hamminger, W.; Jugel, R. A Guarded hot-plate apparatus for thermal conductivity measurements over the temperature range -75 to 200 °C . *Int. J. Thermophys.* 1985, *6*, 483-498.

[546] Nait-Ali, B.; Haberko, K.; Vesteghema, H.; Absi, J.; Smith, D.S. Thermal conductivity of highly porous zirconia. *J. Eur. Ceramic Soc.* 2006, *26*, 3567-3574.

[547] Goual, M.S.; Bali, A.; Queneudec, M. Effective thermal conductivity of clayey aerated concrete in the dry state: Experimental results and modeling. *J. Phys. D: Applied Phys.* 1999, *32*, 3041-3046.

[548] Rausch, M.H.; Krzeminski, K.; Leipertz, A.; Fröba, A.P. A new guarded parallel-plane instrument for the measurements of the thermal conductivity of fluids and solids. *Int. J. Heat and Mass Transfer.* 2013, *58*, 610-618.

[549] Zarr, R.R. Assesment of uncertainties for the NIST 1016 mm guarded-hot-plate apparatus: extended analysis for low-density fibrous-glass thermal insulation. *J. Res. Nat. Inst. Stand. Technol.* 2010, *115*, 23-59.

[550] Gegenhuber, N.; Schoen, J. New approaches for the relationship between compressional wave velocity and thermal conductivity. *J. Appl. Geophys.* 2012, *76*, 50-55.

[551] Osako, M.; Yoneda, A; Ito, E. Thermal diffusivity, thermal conductivity and heat capacity of serpentine (antigorite) under pressure. *Phys. Earth Planet. Inter.* 2010, *183*, 229-233.

[552] Anderson, S.; Bäckström, G. Technique for determining thermal conductivity and heat capacity under hydrostatic pressure. *Rev. Sci. Instrum.* 1986, *57*, 1633-1639.

[553] Beck, P.; Goncharov, A.F.; Struzhkin, V.V.;Militzer, B.; Mao, H.-K.; Hemley, R.J.; Measurements of thermal diffusivity at high pressures using a transient heating technique. *Appl. Phys. Lett.* 2007, *91*, 181914-181917.

[554] Novikov, S.V. Thermal properties of terrigenic rocks and saturated fluids. Ph.D. Thesis, 2009, Russian State Geological University, Moscow, 2009.

[555] Vertogradskii, V.A.; Popov, Yu. A.; Novikov, S.V. Pressure dependency of the thermal conductivity of fluid saturated high porosity sandstones. *In: Proc. VI Int. Conference "New Idea in Earth Sciences"* 2003, *3*, 133.

[556] Dos Santos, W.N. Experimental investigation of the effect of moisture on thermal conductivity and specific heat of porous ceramic materials. *J. Materials Sci.* 2000, *35*, 3977-3982.

[557] Tang, X.; Dong, J. Pressure dependence of harmonic and anharmonic lattice dynamics in MgO: A first-principles calculation and implications for lattice thermal conductivity. *Phys. Earth Planet. Inter.* 2009, *174*, 33-38.

[558] Goncharov, A.F.; Struzhkin, V.V.;Montoya, J.A.; Kharlamova, S.; Kundargi, R.; Siebert, J.; Badro, J.; Antonangeli, D.; Ryerson, F.J.; Mao, W. Effect of composition, structure, and spin state on the thermal conductivity of the Earth's lower mantle. *Phys. Earth Planet. Inter.* 2010, *180*, 148-153.

[559] Cho, W.J.; Kwon, S.; Choi, J.W. The thermal conductivity for granite with various water contents. *Engineering Geology* 2009, *107*, 167-171.

[560] Nikolaev, S.A.; Nikolaeva, N.G.; Salamatin, A.N. *Thermophysics of the Rocks*. Kazan University Pul., Kazan, 1987.

[561] Nikolaeva, N.G.; Nikolaev, S.A. Thermophysical apparatus and methods. In: *Geothermal methods in hydrogeology*. Moscow, Nedra, 1979, pp. 260-268.

[562] Veerendra, K.; Chaudhary, D.R. Indian. *J. Pure Appl. Phys.* 1980, *18*, 984.

[563] Babaev, V.V.; Budymka, V.F.; Sergeeva, T.A.; Dombrovskii, M.A. *Thermophysical Properties of rocks*. Moscow, Nedra, 1987.

[564] Blackweel, D.D.; Spaffond, R.E. Experimental methods in continental heat-flow. In: C.G. Sammis, T.L. Henyey, eds., *Geophysics field measurements. Methods of experimental physcis.* Part B., vol. 24, NY, AP, 1987, pp.189-226.

[565] Wen-Long Cheng; Yong-Hua Huang; Na Liu; Ran Ma Estimation of geological formation thermal conductivity by using stochastic approximation method based on well-log temperature data. *Energy* 2012, *38*, 21-30.

[566] Black, D.D.; Steel, J.L. Thermal conductivity of sedimentary rocks: measurement and significance. In: N.D. Naeser, T.H. McCulloh, eds., Thermal history of sedimentary basins, methods and case histories. NY, Springer, Chap. 2, 1989, pp.13-35.

[567] [Seto, A.C.; Bharatha, S. Thermal conductivity estimation from temperature logs. In: Proc. Int. Thermal Operations Symp., Bakersfield, California, SPE, 1991, pp.179-186.

[568] Smith, Jr H.D. Determination of thermal properties of a formation. Appl. No. 10/457.645, Field: Jun. 9, 2003.

[569] Holmberg, H.; Naess, E.; Evensen, J.E. Thermal modeling in the Oslo Rift, Norway. Proc. 37th Workshop on Geothermal Reservoir Engineering, Stanford University, Stanford, California, January 30-February 1, 2012, SGP-TR-194.

[570] Clauser, C. Geothermal Energy, In: K. Heinloth, ed.., Landolt-Bornstein, Group VIII: Advanced Materials and Technologies, Energy Technlogies, Subvol. C: Renewable Energies, Springer Verlag, Heidelberg-Berlin, 2006, Vo. 3, pp. 493-604.

[571] Hofmeister, A.M. Thermal diffusivity of alkali and silver halide crystals as a function of temperature. *J. Appl. Phys.* 2011, *109*, 033516-1-033516-20.

[572] Pertermann, M.; Hofmeister, A.M. Thermal diffusivity of olivine-group minerals. *American Mineralogist*, 2006, *91*, 1747-1760.

[573] Branlund, J.M.; Hofmeister, A.M. Factors affecting heat transfer in SiO_2 solids. *American Mineralogist*, 2008, *93*, 1620-1629.

[574] Friend, E. Thermal Conductivity. R.P. Tye, ed., Academic Press, New York, Vol. 2, 1969, p. 253.

[575] Miller, V.S. Contact Heat Exchange in the Elements of the High-Temperature Machinery, Naukova Dumka, Kiev, 1966, 162 pp. (in Russian).

[576] Alishaev, M.G.; Abdulagatov, I.M.; Abdulagatova, Z.Z. Effective thermal conductivity of fluid - saturated rocks. Experiment and modeling. *Engineering Geology* 2012, *135-136*, 24-39.

[577] Abid, M, Hammerschmidt, U. Thermophysical properties of Sander sandstone using transient hot-bridge sensor. Arbeitskreis Thermophysik, Karlsruhe, Germany, 4-5 March 2010.

[578] Ali, A.Z.; Gurmani, S.F.; Maqsood, A. Simultaneous measurement of thermal conductivity, thermal diffusivity and prediction of effective thermal conductivity of porous consolidated igneous rocks at room temperature. *J. Phys. D: Appl. Phys.* 2006, *39*, 3876-3881.

In: Sandstone
Editor: Afsoon Moatari Kazerouni

ISBN: 978-1-62948-382-5

Chapter 4

DIAGENETIC CHLORITE: EXAMPLE OF AN INTRA SANDSTONE RESERVOIR SEAL – FROM PALAEOCENE TYR MEMBER SANDSTONE, SIRI CANYON, DANISH NORTH SEA

A. M. Kazerouni[1,*], H. Friis[1,†] and J. P. V. Hansen[2,‡]
[1]Department of Earth Sciences, Aarhus University, Aarhus C, Denmark
[2]Noreco Oil Denmark, Holte, Denmark

ABSTRACT

This study demonstrates the diagenetic evolution of glaucony-rich deep water sandstones from the Rau-1A well in the Siri Canyon, Danish North Sea. The major diagenetic phases in the studied well are microquartz, large syntaxial quartz overgrowths, calcite and chlorite. Chlorite forms an intra-reservoir hydrocarbon seal, and our study demonstrates the influence of early diagenetic quartz on the formation of the chlorite seal. Early opal and microquartz are precipitated close to shale contacts and prevent the interaction between abundant detrital glaucony and pore-fluid and thus the formation of grain-coating berthierine. It also preserves porosity and permeability in marginal and isolated parts of the sandstone.In other parts of the sandstone grain coating berthierine precipitated. It was transformed to chlorite with increased depth. A second phase chlorite is distinguished by its morphology and slightly different x-ray diffraction pattern and fills the porosity and reduced permeability in the upper marginal part of the reservoir which had preserved porosity and permeability due to microquartz cementation. The pore-filling chlorite acts as an impermeable seal to hydrocarbon migration and prevents the topmost part of the reservoir to be oil filled. On basis of XRD data, it is possible to demonstrate a slight shift of actual peak position in overlapping peaks of chlorite in the studied samples. This variation indicates two phases of chlorite in different time. On basis of the SEM studies, these phases are identified as an early grain coating chlorite and a later pore filling chlorite. The change in diffraction pattern indicates that the relative amount of two

* E-mail: afsoon.moatari@geo.au.dk.
† E-mail: henrik.friis@geo.au.dk.
‡ E-mail: jens.peter.vind.hansen@noreco.com.

types of chlorite varies in relation to reservoir properties. The study demonstrates the conditions for formation of an intra-formational chlorite seal and relates its possible formation to high-porosity/high-permeability zones of the reservoir. The study suggests that the formation of an intra-sandstone seal of diagenetic chlorite relates to the distribution of early diagenetic microquartz (which preserve porosity and permeability), and early diagenetic chlorite which reduces the permeability.

Keywords: Siri Canyon, Cecilie Field, microquartz, macroquartz, pore lining chlorite, chlorite rosettes, berthierine

1. Introduction

The presence of diagenetic intra-reservoir seals may pose serious problems in hydrocarbon exploration because they may be complicated or impractical to map with geophysical methods. Diagenesis may cause significant variation in properties within a sandstone reservoir ranging from gradual changes in porosity and permeability to the development of effective flow barriers and compartmentalization of the reservoir. Effective flow barriers may be produced by massive cementation by quartz or calcite as for example oil-water-contact calcite cementation reported from the Siri Canyon by Stokkendal et al. (2009), or by contrasting cementation pattern in the oil and in the water zone of a reservoir (Stokkendal et al. 2009; Weibel et al., 2010).

These cementation patterns may complicate the production from the field because the connection between the oil- and the water-zone is disrupted. If flow barriers are produced earlier than hydrocarbon migration they may form the top-seal of the reservoir and restrict the distribution of oil. The distribution of a diagenetic intra-reservoir seal may have significant influence on the evaluation of an oil show (Ketzeret et al. 2005), and it is important to investigate the factors, which control the formation and location of the intra-reservoir seal.

This study explores the formation of a diagenetic chlorite seal in the Rau-1A well, Danish North Sea (Figure 1). Chlorite cement dominates most of the sandstone units. The presence of chlorite strongly influenced the physical properties of the reservoir by reducing the permeability and porosity; however, the uppermost 1 m of the reservoir sandstone is so tightly cemented by chlorite that oil did not migrate further into it.

The main aim of this study is to determine the evolutionary history of the chlorite cement and to relate it to the development of reservoir properties, especially the formation of an intra-reservoir seal of diagenetic chlorite. The example may be helpful in understanding the development of flow barriers in hydrocarbon reservoirs.

2. Geological Setting

The Palaeocene Siri Canyon extends for more than 120 km from the Stavanger Platform to the Tail End Graben along the Danish-Norwegian North Sea boundary (Figure 1).

Initial formation of the canyon system took place in the Early Palaeocene (Danian) and was related to major, submarine sliding and erosion in the uppermost chalk section caused by uplift of the Scandinavian hinterlands (Hamberg et al., 2005).

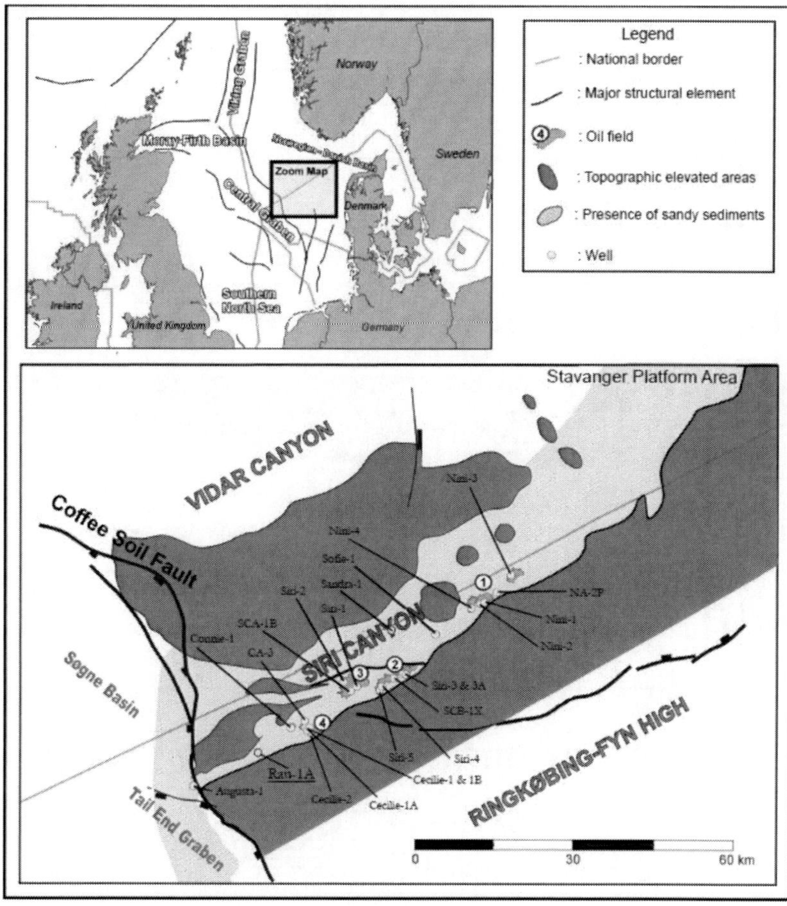

Figure 1. Structural map of Siri Canyon. The well utilised in this investigation is underlined.

The trend of the canyon partly follows salt structures along the southern edge of the Norwegian-Danish Basin (Stokkendal et al., 2009). The Rau-1A well is located farther down-canyon than the producing Palaeocene oil fields of the Siri Canyon (Figure 1).

The canyon fill consists of deep marine, hemipelagic and turbidite marls and mudstones interbedded with highly glauconitic sandstones (Figure 2; Hamberg et al., 2007). As for many Palaeocene deep-marine sandstones in the North Sea, the sandstones of the Siri Canyon are massive and blocky. Injected sandstone sills/dykes are common and post-deposition remobilization is common (Hamberg et al., 2005; Hamberg et al., 2007; Poulsen et al., 2007).

All sandstones are essentially free of detrital clays and most likely formed from non-cohesive, concentrated density flows (Hamberg et al., 2005). The sand flows were confined by the canyon and the depositional pattern indicates a complex interaction of turbidity currents and gravity flows, salt-induced seafloor topography and differential compaction. Release of sandy flows may have been caused by seismic shocks resulting in a collapse of unstable shelf sands concentrated on the Stavanger Platform (Hamberg et al., 2005).

The oil fields in the Siri Canyon rely on long distance migration of hydrocarbons, up to 75 km from mature Jurassic source rocks in the Central Graben. This was facilitated by the elongate and confined geometry of the sandstones forming a well-connected system for up-dip migration, (Hamberg et al., 2005, Ohm et al., 2006).

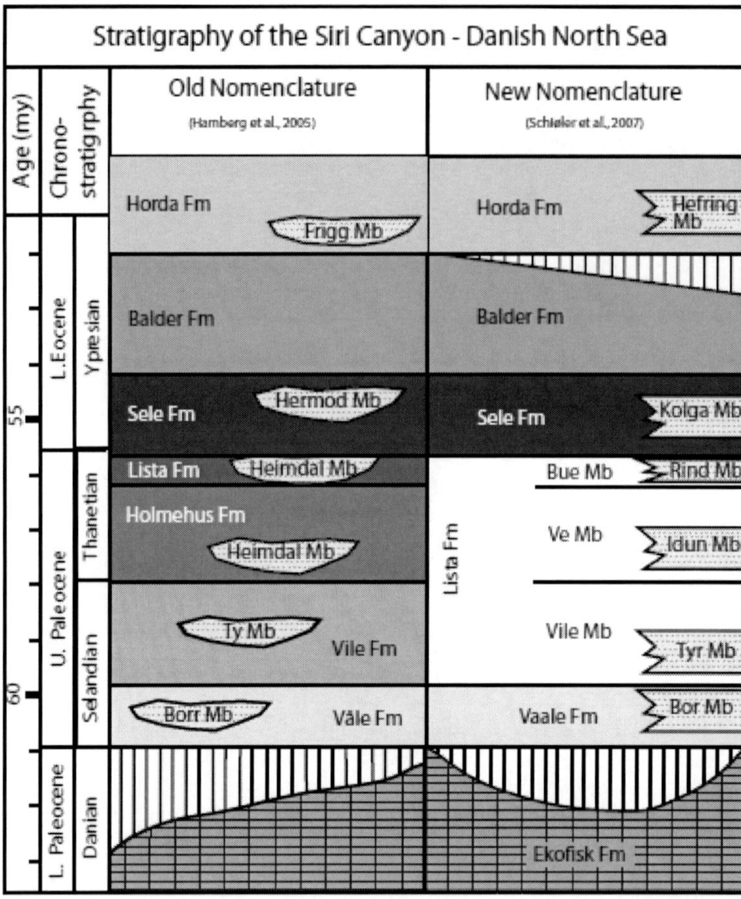

Figure 2. Schematic lithostratigraphy of the Palaeocene-Early Eocene infill of the Siri Canyon.

2.1. Tyr Member

The Tyr Member consists of glaucony-rich sandstones that are laterally equivalent to and commonly underlain and overlain by mudstones of the Vile Member (Figure 2). The Tyr Member has only been encountered in the Siri Canyon, and it may be restricted to that area (Schiøler et al., 2007).

The thickness is variable. Hamberg et al. (2007) have reported units up to 60-80 m thick which include bodies of remobilized sandstone. The Tyr Member is characterised by thick beds of olive-green to greenish grey, very fine-grained to fine-grained and well-sorted sandstone. Rounded and translucent quartz grains dominate, but the content of glaucony grains is high (15–20%), hence the greenish colour of the sandstones. Mica and small pyrite concretions are present in small amounts throughout the member.

Angular chalk and claystone clasts occur locally in the sandstones. The sandstones are partly calcite-cemented. Intrusive sandstones are common, particularly towards the top of the member where they may be several meters thick. Subordinate interbedded dark grey noncalcareous mudstones resemble those of the laterally equivalent Vile Member (Schiøler et al., 2007).

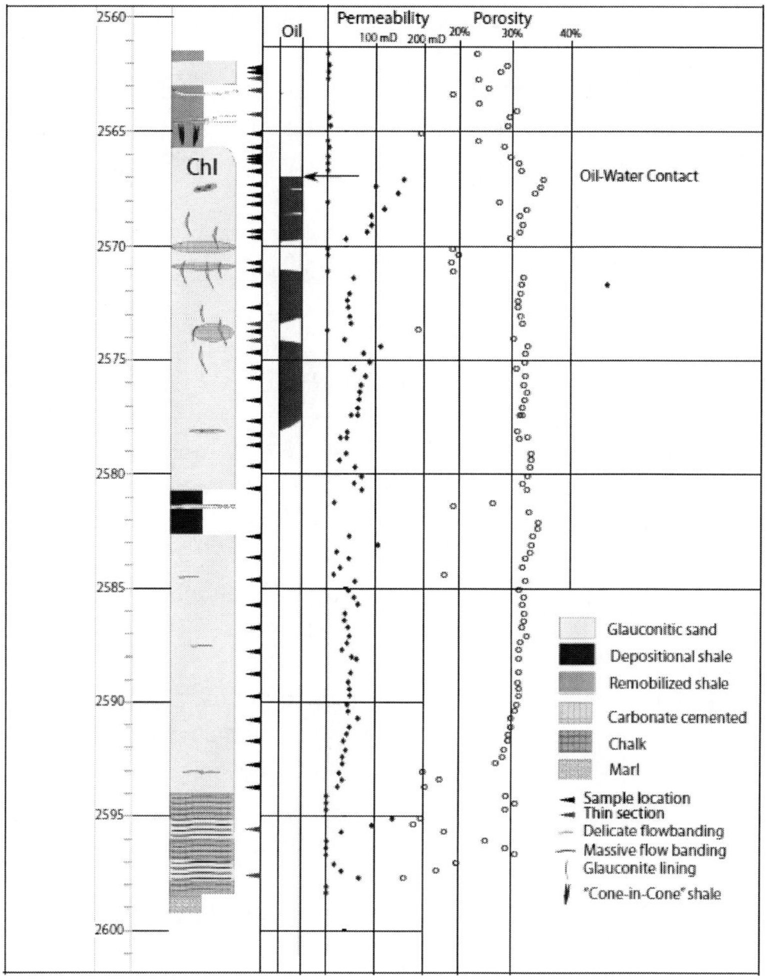

Figure 3. Lithostratigraphy and sample depths in Rau 1A well.

In the Rau-1A well the Tyr member has a total thickness of 32 m. It contains 2 main sand units and a number of thin sands (Figure 3). The interbedded Vile Shale contains layers which are rich in microquartz, presumably recrystallized diatomite.

3. Sample Material and Methods

The study of the Palaeocene Tyr Member sands in Rau-1A has been carried out on 47 core samples. Polished thin sections were prepared using blue epoxy to facilitate estimation of porosity. Standard petrographic examination was performed on 8 polished thin sections. The mineralogical composition was determined by point counting (500 points).

26 samples were examined by electron microscope (SEM). A JEOL JSM-5800 LV scanning electron microscope fitted with an Oxford energy dispersive X-ray spectroscope (EDXS) was used. Polished thin sections were carbon-coated and studied in backscatter mode (SEM-BSE) while intact sample chips were gold-coated and studied in secondary electron

mode (SEM-SEI). 12 selected samples were analyzed by XRD to determine the clay mineral cement.

The clay mineral fraction (<2μm fraction) was separated mechanically from the sandstone using repeated ultrasonic treatment, tumbling and finally decantation. The applied technique has released both diagenetic clay minerals and a variable amount of disintegrated detrital glaucony (together with some diagenetic micron-sized quartz). Analyses were performed using a Phillips PW3710 X'pert diffractometer system with CuKα1-monocromized radiation. The samples were run as air-dried, ethylene glycollated and heated to 500 °C.

4. Petrography

Detrital components: The studied samples are fine to very fine-grained sandstones. Quartz is the most common detrital component (monocrystalline quartz 18-46%; polycrystalline quartz 3-11%), glaucony constitutes 6-20%, feldspar 1-3%, and mica 1-5%. Heavy minerals (rutile, apatite, tourmaline, zircon, garnet, magnetite, monazite, amphibole, pyroxene and opaque minerals) constitute 1-2%. Mudstone intraclasts (<1%) are occasionally present in most samples. Fe-Ti oxides are altered to leucoxene in variable degree. Percentages are of whole rock volume.

Diagenesis: The diagenetic minerals are microquartz (<1%), calcite (both pre- and post compactional (0-48%), chlorite (0-26%), and large syntaxial quartz overgrowth (0-2%). There are minor amounts of diagenetic K-feldspar. Kaolinite occurs sporadically (0-1%). Pyrite (trace-1%) occurs as framboids in most samples. Percentages are of whole rock volume.

5. Diagenetic Sequences of Rau-1A

The diagenetic history of the studied well can be generally described by the following events: precipitation of microquartz, chlorite, feldspar overgrowth, quartz overgrowth, precipitation of calcite cement, dissolution of feldspar, and oil migration.

Some of the events are characterised by multiple precipitations and some overlap each other.

5.1. Quartz Cements

Microquartz is recognized as the first growth stage of quartz. It has been identified in all samples although its recognition in many samples is complicated by intensive chlorite precipitation and later quartz overgrowths. Figure 4A and B show quartz coatings consisting of small euhedral crystals (2 to 5 μm) developed in primary pores. Although microquartz is found in most samples, only one sample is dominated by microquartz cement (Figures 4B; 4C). Individual microquartz crystals may also be found within the chlorite coating.

The outer margin of the microquartz coating is generally characterized by larger crystals, which are gradually merging into macroquartz. The growth pattern of microquartz is generally "ordered" according to the description of Weibel et al. (2010).

Microquartz has been observed during point counting. As this phase is difficult to identify in transmitted light microscope, the point-count value is probably strongly underestimated.

In SEM-BSE it is seen, that microquarz is growing on most detrital quartz grains but also on other detrital grains.

Secondary porosity after dissolution of framework grains which were coated by microquartz is only moderately compacted (Figure 4C).

Microquartz is partly intergrown with grain coating chlorite or forms a dense mixture of scattered microquartz and small chlorite rosettes (Figures 4D; 5B). Microquartz is also observed beneath macro-quartz (Figure 4A), calcite (Figure 5A), and pore-filling chlorite (Figures 5B; 5D).

Macroquartz is found as large individual overgrowths on detrital quartz reflecting the crystallographic orientation of detrital grain. They cover earlier microquartz coatings (Figures 4A; 5B; 5C).

Macroquartz is found in the oil zone and deeper. It is more abundant and is represented by larger crystal in the water-saturated zone compared to the oil-saturated zone.

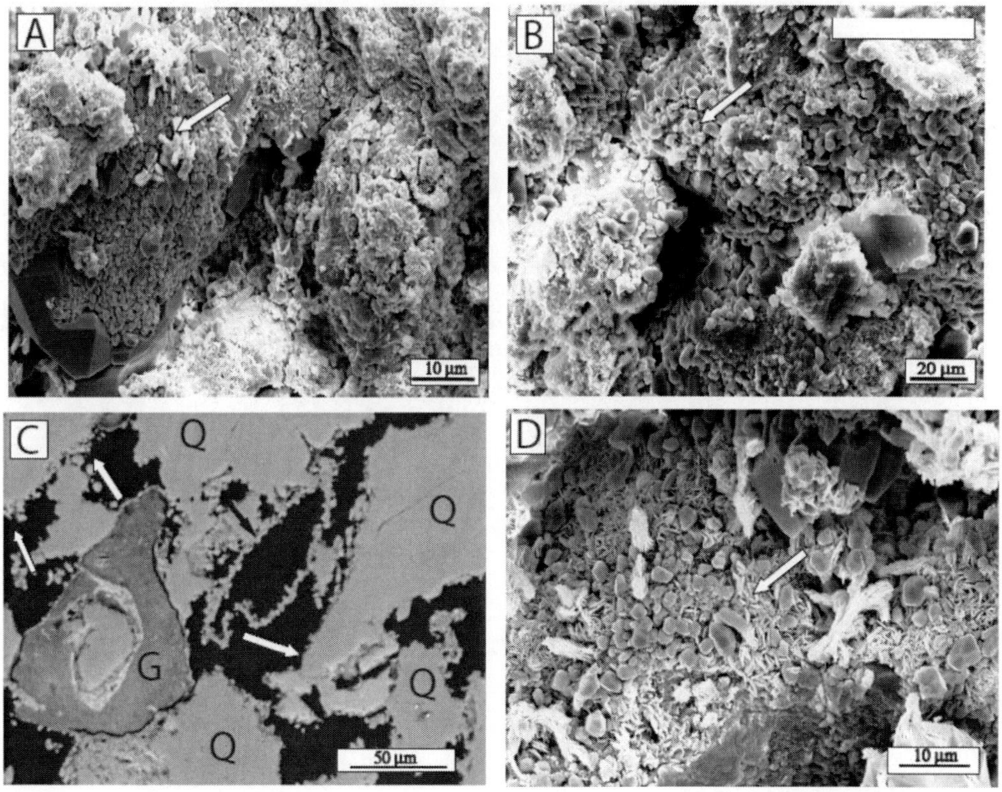

Figure 4. A. Early cementation with microquartz. Chlorite coating on detrital quartz grain (arrow). Overgrown by euhedral macroquartz cement. Rau-1A, 2567.30 m. B. All detrital grains are covered by thick coating of microquartz (arrows). Rau-1A, 2563.30 m. C. SEM-BSE micrograph showing microquartz covering most detrital grains. Moldic porosity after dissoloution of detrital grain which was coated by microquartz (black arrow). The mold is moderately deformed by compaction. Q is detrital quartz; G is detrital glaucony. Rau-1A, 2563.30 m. D. Grain coating intergrown microquartz and chlorite (arrow). Rau-1A, 2567.30 m.

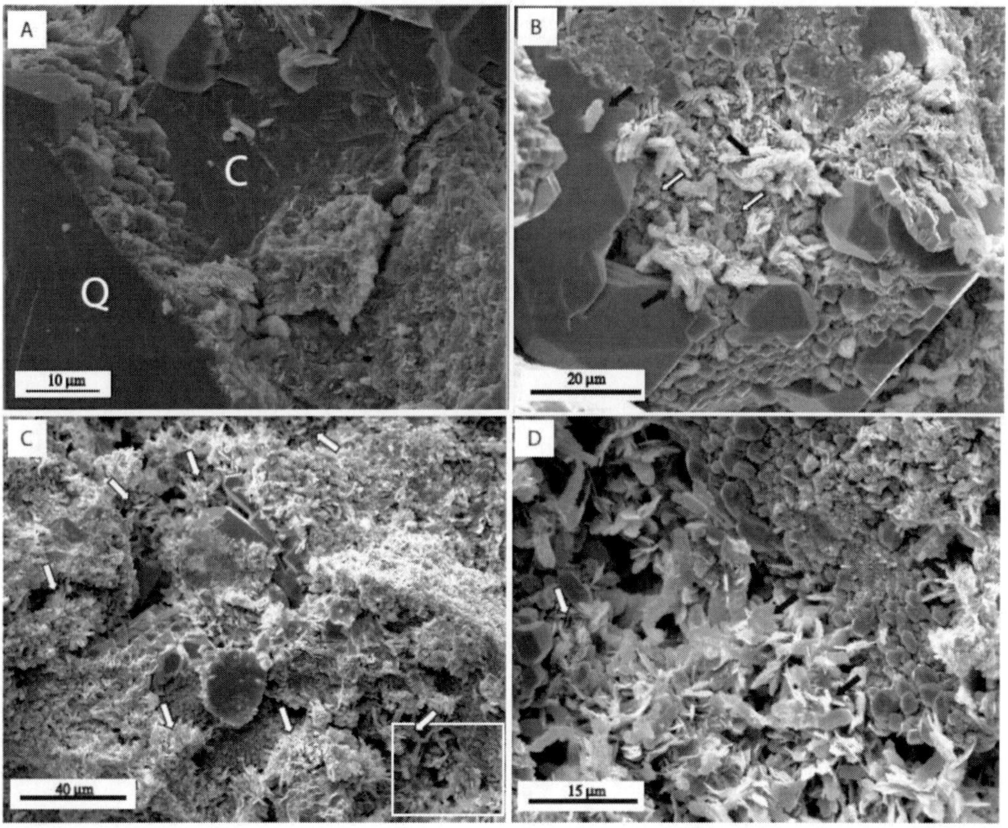

Figure 5. A. Calcite cement fills remnant porosity after growth of quartz. Q is detrital quartz. C is calcite. Open. Rau-1A, 2569.80 m.B. Macroquartz cement partly fills remnant porosity after microquartz and grain coating chlorite (open arrows).Pore-filling chlorite (filled arrows) is partly included in macroquartz. Rau-1A, 2567.30 m. C. Strongly cemented sandstone with very low macroporosity. Remnant porosity after early cementation with microquartz is filled with macroquartz and pore-filling chlorite (arrows). White box is upper left part of D. Rau-1A, 2566.20 m.D. Pore-filling chlorite (filled arrows). Open arrow points to grain-coating chlorite and microquartz. Rau-1A, 2566.20 m.

5.2. Chlorite

Chlorite is present in most samples, although only in traceable amounts in sample Rau-1-2563.30 which is dominated by microquartz cement (Figure 4B and 4C).

There seems to be two chlorite phases: The first phase occurs as bladed or small rosettes in a grain coating growth pattern (Figs. 4D; 5B). It is partly intergrown with microquartz or forms a dense mixture of small chlorite rosettes and scattered microquartz.

The pore-lining chlorite is partly pre-compactional and grows into fractures in glaucony grains. Pore-lining chlorite is formed prior to macroquartz and post-compactional calcite; whereas early calcite cemented samples did not evolve chlorite.

The second phase is characteristic of the very tightly cemented part above the oil zone in the upper reservoir unit. Above the oil zone, the sandstone is cemented by abundant grain coating microquartz cement, similar to what is seen in most studied samples, but after cementation with microquartz, the remnant porosity was filled by irregular, bladed chlorite cement, which seems to totally block the permeability (Figs. 3; 5C; 5D).

The pore-filling chlorite can be distinguished from the pore-lining chlorite by its morphology (Figure 5B). The irregular bladed chlorite is known from other wells in the deeper part of the Siri Canyon, the Cecilie Field and Augusta-1. It dominates in the uppermost part of the upper sand in Rau-1A, but this type is not important in the thin isolated sand above the reservoir unit. Chlorite is also abundant in the water zone.

The chlorite is identified by XRD by its distinct reflections which do not react on treatment with ethylene glycol and decrease after heating (Figure 6; the mixed-layer smectite/illite represent the disintegrated detrital glaucony). The large equal-order reflections compared to the odd-order reflections indicate that the chlorite is iron-rich. . The diffraction pattern also reveals the presence of abundant mixed-layer clay minerals. Compared with data from Stokkendal et al. (2009) the mixed-layer minerals are assumed to be the main component of detrital glaucony. Towards the top of the reservoir unit (between samples 2566.40 m and 2567.30 m) there is a shift to slightly larger d-spacings of the (002) and (004) chlorite peaks and the peaks are broader and slightly asymmetrical (Illustrated for the (004) reflection in Figure 7).

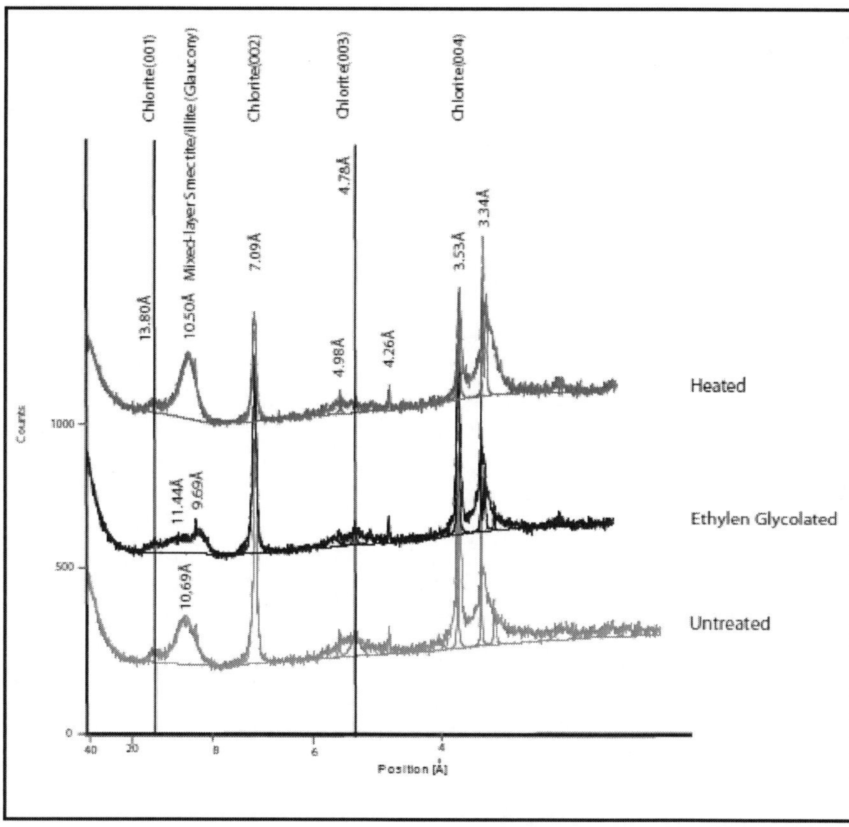

Figure 6. Typical X-ray diffractograms of untreated, ethlynglycolated, and heated sample (2567.30m). The peaks in the 12-9Å interval are from disintegrated glaucony with a mixed layer smectite/illite structure.

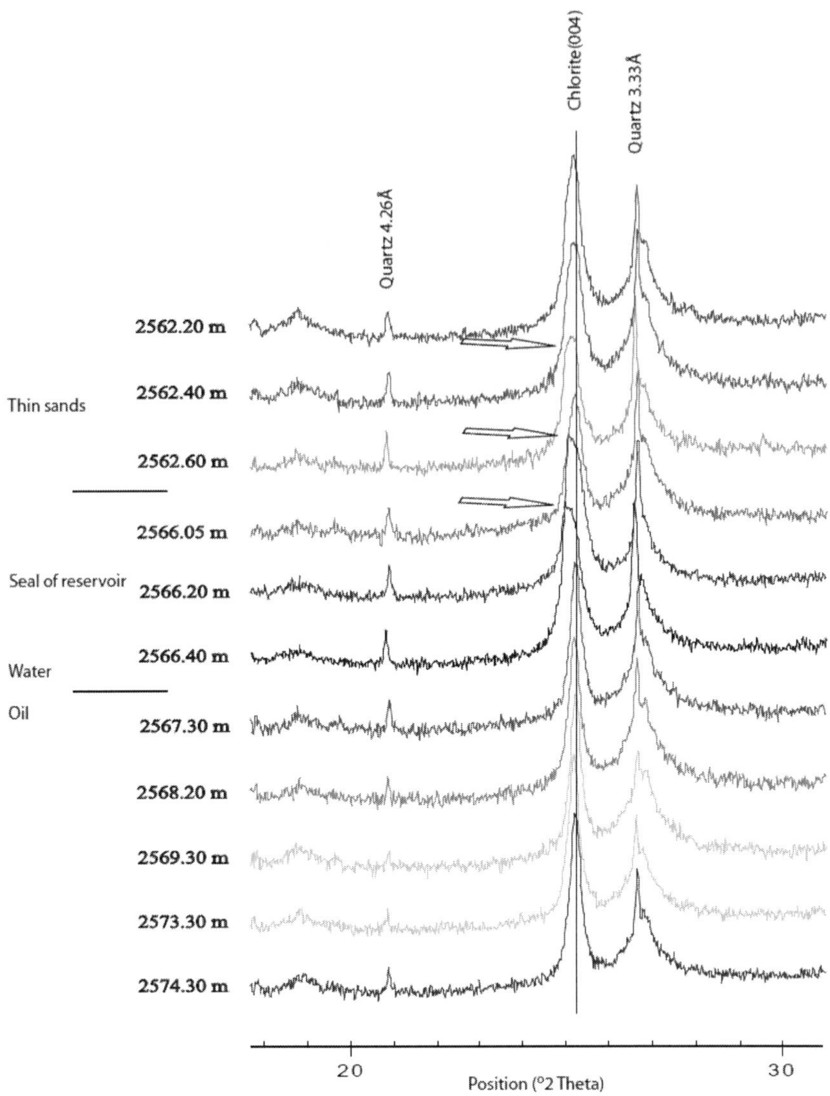

Figure 7. XRD difractograms of all samples showing the slight shift in position of chlorite (004) peak towards the top of the reservoir unit –arrow point the broadening and slight migration to the left of the peak itself (corresponding to slightly larger d-spacings of (004) chlorite peak). This variation indicates two different facies of chlorite, and on basis of the SEM studies an early, grain coating chlorite and a late, pore filling chlorite can be identified. The distribution of these two types of chlorite in the well indicate that the slight shift in position of the (002) and (004) peaks relates to variation in the relative amounts of the two types of chlorite. The peaks are very closely spaced and overlapping. The shift to larger d-spacings correlates with increased proportions of the late, pore-filling chlorite.

The shift is not visible in the (001) and (003) peaks which are small and and more poorly defined than the (002) and (004) peaks. This variation indicates the presence of two different types of chlorite. On basis of the SEM studies an early, grain coating chlorite and a late, pore filling chlorite can be identified. The distribution of these two types of chlorite in the well indicate that the slight shift in position of the (002) and (004) peaks relates to variation in the relative amounts of the two types of chlorite. The peaks are very closely spaced and overlapping. The shift to larger d-spacing correlates with increased proportions of the late, pore-filling chlorite in the low-permeability zone at the top of the reservoir (Figure 3).

The chlorite cement is more abundant and better developed in the lower parts of sandstone units, whereas it may be scarce and poorly developed in the upper part of the oil zone.

5.3. Calcite

There are two phases of calcite cementation, an early phase, which fills an almost uncompacted initial porosity (Figure 8A), and a later phase, which occur after substantial compaction (Figure 8B) and cementation with quartz and/or chlorite The calcite cementation may be patchy and both cements may occur in one single sample and within one thin section. Some fully cemented samples have calcite contents above 40% - representing intergranular volume close to the depositional porosity (Figure 8A). However, the early calcite cement also includes a minor volume of replaced detrital grains.The distribution of calcite cement in the sandstone ranges from patchy to uniform. The early calcite cement is concentrated within the bottom 4 m of the lower sandstone (Figure 3), and around rafted clasts of marl in the upper sand unit.

Post-compactional calcite cement was precipitated in minor volume (trace to 2%) after growth of chlorite (Figure 8B). It is generally patchy, leaving some porosity open. Some calcite precipitated after growth of macroquartz (Figure 8B).

5.4. Pyrite

Pyrite occurs as framboidal aggregates and as single cubic pyrite; both are engulfed by calcite. Pyrite cements comprise less than 0-0.5% of the bulk rock components.

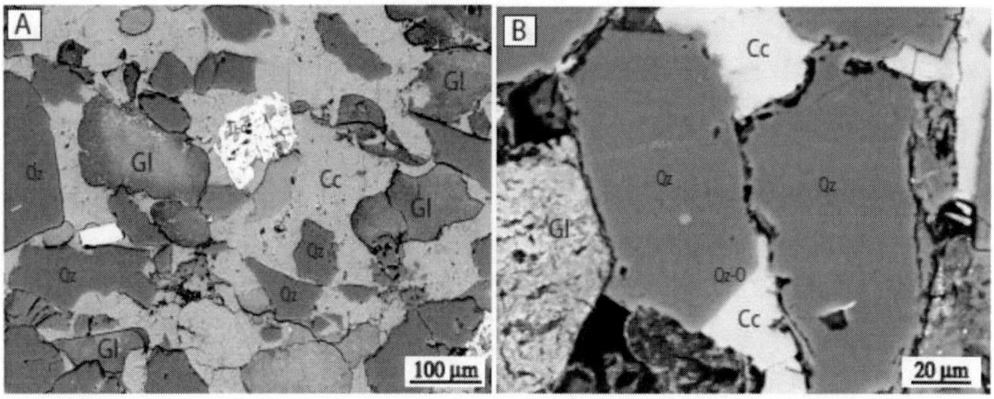

Figure 8. A. Early calcite cement after minor compaction. Fractures in glauconite has been filled with calcite.Cc: Calcite cement; Qz: detrital quartz; Gl: detrital glaucony. SEM_BSE. Rau-1A, 2595.75 m. B. Calcite cemented part, strongly compacted, and cemented by chlorite and quartz before calcite cementation. Cc: calcite cement; Qz-O: quartz overgrowth; Qz: detrital quartz; Gl: detrital glaucony. Note that detrital quartz is separated from calcite cement by grain coating chlorite. Quartz overgrowth has tight contact to calcite cement. SEM-BSE, Rau-1A 2597.75 m.

6. Discussion

6.1. Calcite

Except for framboidal pyrite, which occurs locally embedded in calcite cement, the earliest cement is calcite. The first phase of calcite cement occupies a large volume (up to 48%), although it varies on a thin section scale.

It precipitated before any significant compaction as evidenced by the high proportion of cement and the open framework of detrital grains. As seen from the Figure 8A, some quartz grains have been pressed into softer glaucony grains, and glaucony grains may be deformed at the contact point to other glaucony grains. Muscovite grains may be slightly bent. The degree of compaction is low compared to neighbouring areas with no calcite cement.

6.2. Microquartz

Microquartz is a significant early precipitate as also decribed from other Palaeocene sandstones from the Siri Canyon (Stokkendal et al., 2009; Weibel et al., 2010). It occurs in all samples although it may be difficult to recognize because of successive cement (chlorite and macroquartz).

Thick rims of grain-coating microquartz may have reduced compaction due to the formation of a strong skeleton of ingrown grain coatings (Figure 4C) and resulted in preservation of primary porosity as described by Hendry and Trewin (1995) and Stokkendal et al. (2009). Aase et al. (1996) found that high-porosity zones in Upper Jurassic sandstones from the Norwegian continental shelf correlate well with the presence of grain-coating microquartz, and therefore proposed that microquartz inhibits the precipitation of normal macrocrystalline quartz cement. In the Rau 1-A well grain coating microquartz is present in most samples, but only a few samples in the topmost part of the Tyr Formation have retained their high porosity. Most sandstones are further cemented by macroquartz or chlorite (Figure 4-5). Continued growth on individual microquartz crystals gradually resulted in the formation of macroquartz, and there does not seem to be an inhibiting effect of microquartz on further quartz cementation as proposed by Aase et al. (1996). Although all quartz crystal faces seem clean and without nucleation of chlorite (Figs. 5 B; D), abundant porefilling chlorite is present in most samples and is also found intergrown between macroquartz crystals (Figure 5B). This indicates that the apparent inhibition of chlorite nucleation on quartz crystal faces may not be efficient.

Siliceous microfossils, such as sponge spicules have been considered important sources for microquartz development and opal-CT precipitation (Hendry and Trewin, 1995; Aase et al., 1996). Remnants of microfossils (such as partly dissolved diatoms) have been identified in very small amount Tyr sandstone in Rau-1A and in other Siri Canyon sandstones (Stokkendal et al., 2009; Weibel et al., 2010) suggesting that biogenic silica embedded in the sandstone was not a sufficient source for the microquartz cement. Stokkendal et al. (2009) concluded that abundant cementation by microquartz is mainly expected close to upper or lower margin of sandstone units and is sourced from interbedded shale.

The Paleocene Shale including the Vile Formation in the Siri Canyon (Figs. 1; 2) in the northern North Sea contains several levels which are rich in diatoms and volcanic ash (Mudge and Copestake, 1992; Mudge and Bujak, 1996; Weibel; 2010) and it probably had a large potential for export of dissolved silica during early phases of burial due to dissolution and transformation of biogenic silica and volcanic ash. An external fluid generated in the underlying Ekofisk Formation (or down dip in Central Graben) is another possible silica source. Silica microfossils are also well known from the chalk, where they are commonly dissolved and re-precipitated as chert (Madsen, 2010).

6.3. Macroquartz

This phase reflects a slow growth of quartz from fluids saturated with silica. This may reflect exhaustion of the first silica sources, which resulted in microquartz precipitation.

Silica liberation will typically continue during increased burial, due several diagenetic alterations, for example dissolution of feldspar, diagenetic transformation of clay and pressure solution of quartz.

These diagenetic reactions will liberate silica at a slower rate to the pore fluid, which may be in equilibrium with quartz rather than amorphous silica.

The macroquartz has developed by gradual enlargement and merging of individual microquartz crystals. This growth pattern is related to the presence of ordered microquartz crystals (Weibel et al. 2010) and may result in poor connection between the detrital quartz nucleus and the large quartz overgrowth because of the intermixed grain coating microquartz and chlorite (Figure 5B). As macroquartz is more abundant and is represented by larger crystal in the water-saturated zone compared to the oil-saturated zone, it is assumed that quartz continued to precipitate within the water zone after hydrocarbon filled the upper part of the sandstone. The hydrocarbon may thus have stopped or retarded quartz precipitation. The source of continued quartz precipitation may have been fluids generated in the Central Graben and migrated up-canyon together with hydrocarbon.

6.4. Chlorite

Authigenic chlorite has been reported from many Paleocene and Eocene deepwater sandstones in the North Sea area and in the Faroe-Shetland Basin (Tonkin and Fraser, 1991; Wills, 1991; Chandler and Dickinson, 2003; Kunka et al., 2003; Sullivan et al., 1999; Friis et al., 2007). It is generally reported that the abundance is too low to have influenced reservoir properties significantly. However, Sullivan et al. (1999) report the presence of thick pore-lining chlorite which may block the pore throats and thus reduce permeability. On the other hand it is also suggested that abundant pore-lining chlorite has preserved porosity by preventing or delaying quartz cementation with depth (Tonkin and Fraser, 1991; Sullivan et al., 1999).

In Rau-1A sandstones the authigenic chlorite has a major impact on permeability (Figure 3) and may reduce permeability to almost zero values. The two phases of chlorite have a different distribution and consequently influence the reservoir properties differently. The pore-lining chlorite (Figs. 4D; 5B) did not reduce permeability strongly as also indicated from other studies of Paleocene North Sea sandstones (Tonkin and Fraser, 1991; Wills, 1991; Chandler and Dickinson, 2003; Kunka et al., 2003). Its tight location on grain surfaces causes a moderate porosity reduction but may block pore throats.

Pore-lining chlorite is often interpreted to have preserved porosity by retarding or preventing growth of other diagenetic minerals, especially quartz (Tonkin and Fraser 1991, Ehrenberg 1993, Sullivan et al., 1999, Aagaard et al. 2000). However, in the Rau-1A sandstones, the pore-lining chlorite is closely intergrown with microquartz (Figs. 4D; 5B), which may form nuclei for the further growth of quartz. Therefore, the cementation by late stage macroquartz may be independent of the presence of chlorite linings and only correlate with the orientation of microquartz as discussed by Weibel et al. (2010).

Pore-lining chlorite in deep-water sandstones has been interpreted as a maturation product of berthierine which may partly have been present as a depositional or very early diagenetic grain-coating (Jahren et al. 2000; Gould et al. 2010).

Grain coating berthierine has been reported from a shallower oil field in the Siri Canyon (the Stine Field (Figure 1), Stokkendal et al. 2009) and it also occurs in the shallower Nini Field (Figure 1). Berthierine is reported to convert to chlorite at temperatures above 70-100 °C (Aagaard et al., 2000, Jahren et al. 2000, Worden and Morad 2003). Downward the canyon the temperature increases from 50-60 °C in the Nini field and 65 °C in the Stine Field to a level of approximately 90°C in the Cecilie Field, 100 °C in Augusta-1 (Figure 1). The down-canyon change in mineralogy from berthierine to chlorite thus seems to take place at temperatures from 80-90 °C corresponding quite closely to the suggestion of Jahren et al. (2000).

This indicates that the grain coating chlorite in Rau-1A sandstones may also have a berthierine precursor and that berthierine precipitated locally as a grain coating cement in all the Palaeocene sandstones in the Siri Canyon while they were buried at shallow depth, corresponding to the model proposed by Stokkendal et al. (2009) for the Stine and Nini Fields, and that berthierine eventually transformed into chlorite with increasing burial depth and temperature.

The second phase pore-filling chlorite (Figs. 5C; 5D) is known mainly from the deeper part of the Siri Canyon - the Cecilie Field and deeper. In the Rau-1A well the pore-filling chlorite is seen in all depths, but generally in small amounts (Figure 5B). It is a major constituent in the uppermost part of sand unit-1 (above the oil -zone) where permeability has been lost due to dense, pore-filling chlorite cement (Figure 3; Figs. 5C; 5D). The same morphological type is known from the Cecilie Field and from Augusta-1. It is assumed to be related to higher temperature, but still formed before oil migration. Macroquartz encloses the pore-filling chlorite and was thus precipitated later, or the precipitation of quartz continued after precipitation of chlorite. The distribution pattern of the two chlorite phases may indicate their different sources.

The grain-coating chlorite is present throughout the sandstone although in minor amounts than described by Stokkendal et al. (2009) from the Stine Field. The characteristic distribution reported by Stokkendal et al. (2009) seems to indicate an interaction between formation of thick rims of opal/microquartz, formation of early diagenetic berthierine and an internal, detrital source. As in the Rau-1A well, the marginal parts of the sand layers (or thin, isolated sands) have a thick coating of microquartz and very little grain-coating chlorite. The pore-lining microquartz seems to have prevented the formation or the grain-coating berthierine/chlorite. We suggest that the formation of the grain-coating berthierine/chlorite was controlled by chemical reactions in the detrital glaucony, which thus acted as a chemical reservoir. The glaucony appears to consist of a mixed layer smectite/illite which may transform to a more illitic composition with depth and release ions for the formation of berthierine. Where glaucony grains are fully enveloped in thick opal/microquartz coatings, this reaction may not release components to the pore fluid, thus not promoting berthierine to intergrow or overgrow the microquartz (as is seen in the upper part of the reservoir unit and the thin isolated sands (Figs. 4B; 4C). Thus, the pore-lining berthierine/chlorite may have been sourced internally by clay mineral transformations in detrital glaucony grains and its distribution must be determined by the earlier precipitation of opal/microquartz. Alternatively, miroquartz may form a poor substrate for nucleation of berthierine/chlorite, although we have no support for this suggestion.

The second phase chlorite has its major occurrence in the top-most part of the reservoir unit (Figure 7); there is an increase in the amount of pore-filling chlorite relative to the amount of pore-lining chlorite in these samples. The pore-filling chlorite acts as an impermeable seal to hydrocarbon migration and prevents the top-most part of the reservoir to be oil filled. This part of the reservoir appears to have been more porous and permeable before the second phase of chlorite precipitation. This suggests that the formation of the intra-sandstone seal of diagenetic chlorite depends on variations in the flow pattern of the sandstone and relates to the distribution of early diagenetic microquartz (which preserves porosity and permeability), and early diagenetic berthierine/chlorite which reduces the permeability. The high permeability zone in the topmost part of the reservoir sand was therefore more heavily cemented by the second phase chlorite and suffered a significant porosity and permeability loss.

Therefore, it must be expected that similar early cementation patterns (dense coatings which preserve porosity and permeability) may evolve intraformational seals as a response to late chlorite cementation.

Even though chlorite may locally inhibit quartz precipitation in the investigated Palaeocene sands it has not served as an important porosity preserving constituent as described from some reservoir sandstones (Ehrenberg 1993, Aagaard et al. 2000).

Conclusion

The diagenetic evolution of Tyr sanstone in the Rau-1A well is mainly controlled by precipitation of quartz and berthierine/chlorite.

Thick microquartz coating were sourced from adjacent shale and formed in marginal parts of the sandstone and thin isolated beds. The microquartz coating prevented reaction with detrital glaucony to form grain-coating berthierine.

Thin coatings of microquartz allowed reaction with detrital glaucony and grain coating berthierine formed.

The berthierine transformed to chlorite with depth and increasing temperature.

A second phase of pore-filling chlorite formed in porous sandstone with minor or no grain-coating chlorite.

The pore-filling chlorite forms an intraformational seal within the upper one meter of the reservoir due to preserved porosity/permeability within the microquartz cemented interval.

We suggest that similar early cementation patterns (dense coatings which preserve porosity and permeability) may evolve intraformational seals as a response to late chlorite cementation.

ACKNOWLEDGMENTS

This study was supported by and published with the permission of by Noreco, RWE, Danish North Sea Fund, and State participation in oil and Gas Licences.

REFERENCES

Aagaard, P., Jahren, J., Harstad, A.O., Nilsen, O., Ramm, M., 2000. Formation of graincoating chlorite in sandstones; laboratory synthesized vs. natural occurrences. *Clay Minerals* 35, 261–269.

Aase, N.E., Bjørkum, P. A., Nadeau, P., 1996. The effect of grain coating microcrystalline quartz on preservation of reservoir porosity. *Bulletin of the American Association of Petroleum Geologists*, 80, 1654-1673.

Chandler, P.M., Dickinson, B., 2003. The Moira Field, Block 16/29a, UK Central North Sea. In: United Kingdom Oil Gas Fields Commemorative Millenium Volume (Ed. by J. G. Gluyas and H. M. Hichens), Geological Society of London Memoirs 20, 603-609.- Ehrenberg, S.N., 1993. Preservation of anomalous high porosity in deeply buried sandstones by grain-coating chlorite: examples from the Norwegian continental Shelf. *Bulletin of American Association Petroleum Geologists* 77, 1260–1286.

Friis, H., Poulsen, M.L.K., Svendsen, J.B., Hamberg, L., 2007. Discrimination of density flow deposits using elemental geochemistry-implications for subtle provenance differentiation in a narrow submarine canyon, Palaeogene, Danish North Sea. *Marine and Petroleum Geology* 24, 221–235.

Hamberg, L., Dam, G., Wilhelmson, C., Ottesen, T., 2005. Palaeocene deep-marine sandstone plays in the Siri Canyon, offshore Denmark–southern Norway. In: Dore´, A.G., Vining, B.A. (Eds.), Petroleum Geology: North-West Europe and Global Perspectives, *Proceedings of the 6th Petroleum Geology Conference.* Geological Society of London, pp. 1185–1198.

Hamberg, L., Jepsen, A. M., Ter Borch, N., Engkilde, M.K. & Svendsen, J.B., 2007. Mounded structures of injected sandstones in deep-marine Paleocene reservoirs, Cecilie Field, Denmark. In: Sand Injectites: Implications for Hydrocarbon Exploration and Production (Ed. by A. Hurst & J. Cartwright), *AAPG Memoir*, 87, 69-79.

Hendry, J.P., Trewin, N.H., 1995. Authigenic quartz microfabrics in Cretaceous turbidites: evidence for silica transformation processes in sandstones. *Journal of Sedimentary Petrology* A65, 380–392.

Kunka, J.M., Williams, G., Cullen, B., Boyd-Gorst, J., Dyer, G.R., Garnham, J.A., Warnock, A., Wardell, J., Davies, A., Lynes, P., 2003. The Nelson Field, Blocks 22/11, 22/61, 22/7, 22/12a, UK North Sea. In: United Kingdom Oil Gas Fields Commemorative Millenium Volume (Ed. by J. G. Gluyas and H. M. Hichens), Geological Society of London

Memoirs 20, 617-646.-Madsen, H.B., Silica diagenesis and its effect on porosity of upper Maastrichtian chalk- an example from the Eldfisk Field, the North Sea., 2010. *Geological Survey of Denmark and Greenland Bulletin* 20, 2010, 47-50.

Mudge, D.C,. Copestake., 1992. Lower Paleogene stratigraphy of the northern North-Sea. *Marine and Petroleum Geology* 9, 287-301.

Mudge, D.C., Bujak, J.P., 1996. Palaeocene biostratigraphy and sequence stratigraphy of the UK central North Sea. *Marine and Petroleum Geology* 13, 295-312.

Ohm, S.E., Karlsen, D.A., Roberts, A., Johannessen, E., Høiland, O., 2006. The Paleocene sandy Siri Fairway: an efficient "pipeline" draining the profilic Central Graben. *Journal of Petroleum Geology* 29 (1), 53–82.

Poulsen, M.L.K., Friis, H., Svendsen, J.B., Jensen, C.B., Bruhn, R., 2007. The application of bulk geochemistry to reveal heavy mineral sorting and flow units in thick massive gravity flow deposits, Siri Canyon Palaeocene sandstones, Danish North Sea. In: Mange, M., Wright, D. (Eds.), Heavy Minerals In Use. *Developments in Sedimentology*. Elsevier, Amsterdam, pp. 1099–1121.

Schiøler, P., Andsbjerg, J., Clausen, O. R. Dam, G., Dybkjær, K., Hamberg, L., Heilmann-Clausen, C., Kristensen, L. E., Prince. I., and Rasmussen, J. A, 2007. *A revised lithostratigraphy for the Palaeogene-lower Neogene of the Danish North Sea.*

Stokkendal, J., Friis, H., Svendsen, J.B., Poulsen, M.L.K., Hamberg, L., 2009. Predictive permeability variations in a Hermod sand reservoir, Stine Segments, Siri Field, Danish North Sea Marine and Petroleum Geology, Volume 26, Issue 3, Pages 397-415.-Tonkin, P.C., Fraser, A.R. 1991. The Balmoral Field, Block 16/21, UK North Sea. In: United Kingdom Oil and Gas Fields 25 Years Commemorative Volume (Ed. by I.L. Abbotts), *Geological Society of London Memoirs* 14, 237-244.

Weibel, R., Friis, H., Kazerouni, A.M., Johan B. Svendsen, J.B., Stokkendal., J., Poulsen.M.L.K., 2010. Development of early diagenetic silica and quartz morphologies- Examples from the Siri Canyon, Danish North Sea, *Sedimentary Geology*.

Wills, J.M., 1991. The Forties Field, Block 21/10, 22/6a, UK North Sea. In: United Kingdom Oil and Gas Fields 25 Years Commemorative Volume (Ed. by I.L. Abbotts), *Geological Society of London Memoirs* 14, 301-308.

In: Sandstone
Editor: Afsoon Moatari Kazerouni

ISBN: 978-1-62948-382-5

Chapter 5

THE EFFECT OF BURIAL ON CLAY MINERAL DIAGENESIS AND QUARTZ CEMENTATION IN MUDSTONES: EXAMPLES FROM THE SIRI CANYON, DANISH NORTH SEA

A. M. Kazerouni[1,*], H. Friis[1,†], J. B. Svendsen[2,‡] and O. B. Nielsen[1]
[1]Department of Geoscience, Aarhus University, Aarhus, Denmark
[2]DONG Energy, Hørsholm, Denmark

ABSTRACT

The Palaeocene sequence of the Siri Canyon fill consists of hemipelagic and turbidite marl and claystones interbedded with massive and blocky glauconitic sandstones deposited from sandy mass flows and sandy turbidites. The Palaeogene sediments in the Danish area are rich in siliceous microfossils as well as late Palaeocene-early Eocene volcanic ash layers. We have examined the mineralogical composition of the Sele Formation (late Palaeocene) in 49 core samples from five wells and depths ranging from 1700 to 2900 m. Based on previous studies it has been suggested that the diagenesis of the shales and the interbedded sandstones is closely related and that during diagenesis the shales acted as a significant source for diagenetic quartz in the sandstones. It has been shown that the sandstones were silica cemented in two major phases - an early opal/microquartz phase and a late stage massive cementation by macroquartz. This study explores the potential sources of SiO_2 in the shale, and the timing of their active phases. The main diagenetic phases recorded in the Sele Formation are: Smectite, heulandite, opal-CT, microcrystalline quartz and chlorite. Pyrite is abundant in all samples. Altered volcanic ash-layers in the Sele Formation consist of almost pure smectite, heulandite and plagioclase, but most samples have a minor detrital component of illite as well as partially altered smectite. Altered biogenic silica is a major component of some samples.

With-depth alteration of the various components results in successive stages of silica-release. In shallow samples the alteration of volcanic ash has already been

* E-mail: afsoon.moatari@geo.au.dk.
† E-mail: henrik.friis@geo.au.dk.
‡ E-mail: johbs@dongenergy.dk.

completed. Released silica was partly consumed for the precipitation of smectite and heulandite. Also a major part of the biogenic silica has been transformed into opal-CT and partly to microcrystalline quartz. The microcrystalline quartz is an internal sink for dissolved silica, but the shale was also an active silica exporter during this transition.

With deeper depth of burial, opal-CT is fully transformed to microcrystalline quartz. During this phase, silica has been partly mobile and depending on the rate of dissolution compared to the rate of precipitation, silica has been lost to sandstone cementation. Heulandite was completely dissolved. The heulandite dissolution may also have mobilized silica and activated the shale as silica exporter.

At deep burial, iron-rich chlorite has replaced a minor part of smectite. Therefore, a third phase of silica mobility was active. The Sele Formation in the Siri Canyon has not been influenced by illite formation; even at burial depth of approximately 3000 m. Potassium feldspar is not recorded, even at the shallowest depth, suggesting that the lack of available potassium sources has delayed the smectite to illite transformation.

The major phases of silica mobilization occurs at shallow depths (related to the transformation of biogenic opal and volcanic ash), and at intermediate depths (related to the dissolution of opal-CT and heulandite). During these phases, the shale may have exported silica. Reported sandstone cementation patterns indicate that the early silica phases are related to major silica import from the adjacent shales, whereas the shale itself was the major consumer during the intermediate diagenesis. Late phase mobilization of silica in the shale (due to smectite to chlorite transformation) is at an initial stage and has not contributed significantly to the massive late-stage quartz cementation of the interbedded sandstones.

Keywords: Burial, clay mineral, diagenesis, quartz cementation, opal-CT, heulandite, smectite, detrital composition, mobilization

1. INTRODUCTION

A major issue in the study of sandstone diagenesis is the question of active silica sources for quartz cementation and the timing of their active phases. Some of the suggested sources are internal, such as dissolution of feldspar (Fothergill, 1955; Hawkins, 1978) or other silicate framework grain, or pressure solution at grain contacts in sandstones, pressure solution at stylolites in sandstones (Harris, 1992; Weibel et al., 2010); others are external sources supplied by large-scale flow systems (Archer et al., 2004) or supply from the adjacent shales by compactional flow or diffusion (Thyne, 2001; McBride et al., 1989; Hesse, 1989). There are may be several silica sources in shale and they might be dynamic at different times during burial. Commonly cited silica sources are dissolution of quartz grains (Molenaar et al., 2007), opaline fossils (diatoms, radiolaria, sponge spicules, Siever, 1957), hydration of volcanic glass, conversion of smectite to illite during burial diagenesis (Hower, 1983), and decomposition of feldspars (Siever, 1957). Early cementation by opal and microquartz depends on the supply of dissolved silica from a readily dissolvable source. Silica release in shale does not necessarily source sandstone diagenesis but re-precipitate as opal, quartz or other mineral phases inside the shale itself (Foster and Custard, 1980; Hower et al., 1976; Yeh and Savin, 1977; Small, 1994; Peltonen et al., 2009; Thyberg et al., 2010) although many studies have concluded that silica released as a result of clay mineral reactions in mudstones and shales is transported and precipitated as quartz cement in associated sandstones (e.g.,

Towe, 1962; Boles and Franks, 1979; Lynch et al., 1997; Land and Milliken, 2000; Van der Kamp, 2008). The deep marine sandstones in the Siri Canyon (Figure1) have been reported to import significant amounts of dissolved silica from adjacent Palaeocene shales during early diagenesis (Stokkendal et al, 2009; Weibel et al., 2010).

We have studied the diagenesis of one of these shales, the Sele Formation shale, to document the diagenetic steps which release silica and to and quantify the possible timing of silica export from the shale into neighboring sandstones.

2. Geological Setting of the Palaeocene Sequence

The Palaeocene Siri Canyon extends for more than 120 km from the Stavanger Platform to the Tail End Graben along the Danish-Norwegian North Sea boundary (Figure 1). Initial formation of the canyon system took place in the Early Palaeocene (Danian) and has been explained as the result of major, submarine sliding by Hamberg et al. (2005).

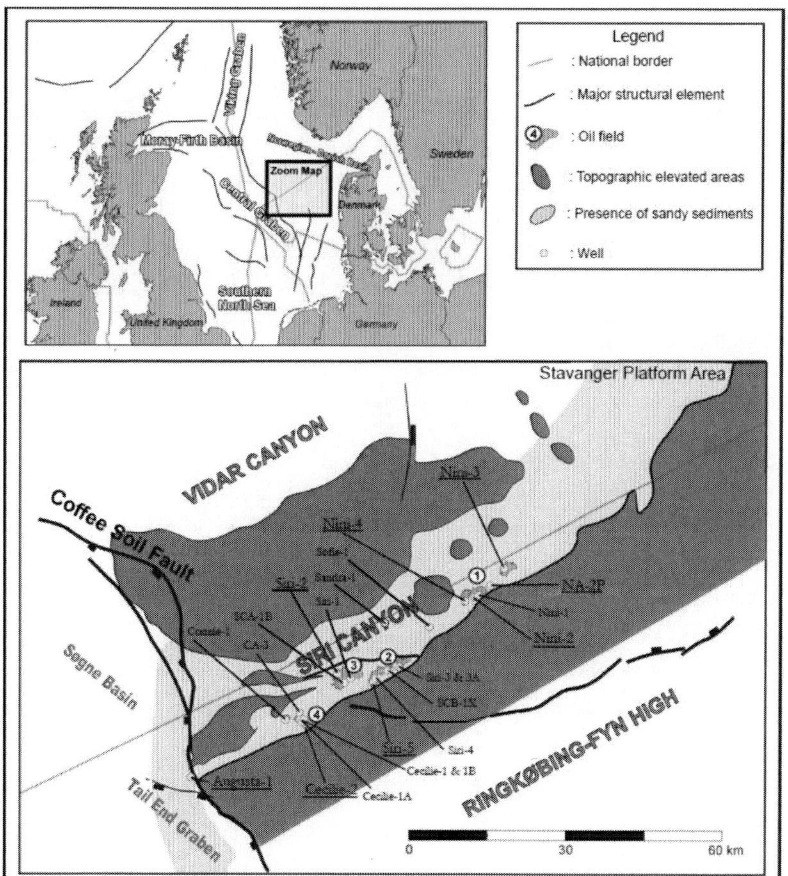

(Modified from Friis et al., 2007).

Figure 1. Location map of the Siri Canyon. (1) Nini Field; (2) Stine Field; (3) Siri Field and (4) Cecilie Field; wells applied in this study are underlined.

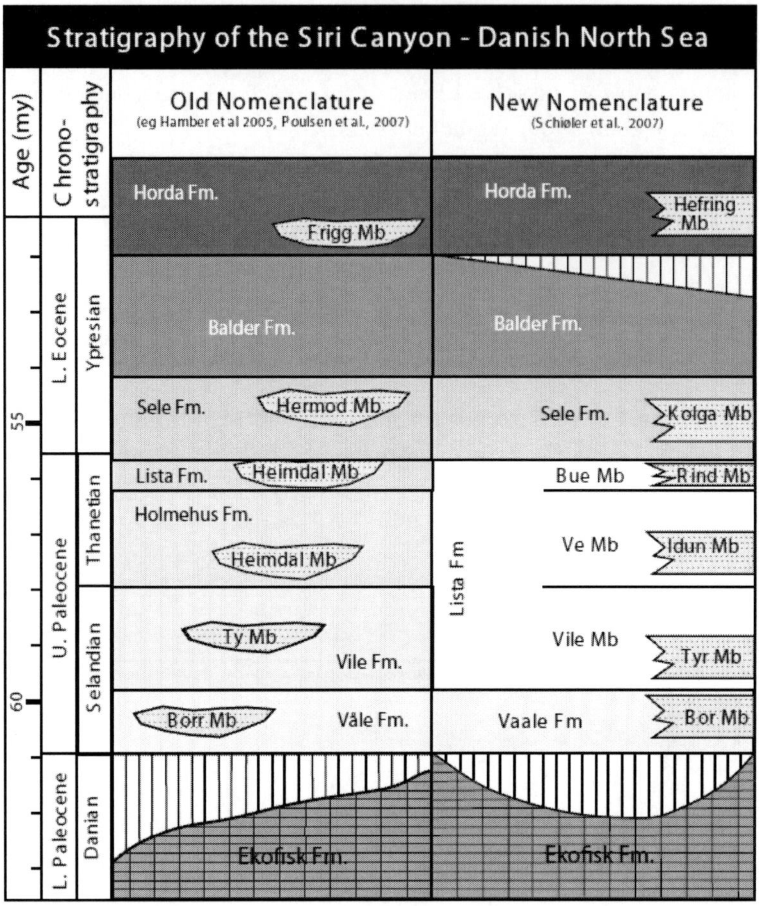

(Modified from Weibel et al., 2010).

Figure 2. Schematic lithostratigraphy of the Palaeocene-Early Eocene infill of the Siri Canyon.

The canyon fill consists of deep-marine, highly glauconitic sandstones interbedded with hemipelagic and turbidite marls and mudstones (Figure 2; Schiøler et al. 2007).

All sandstones are free of detrital clays and were most likely formed from noncohesive, concentrated density flows (Hamberg et al. 2005; Poulsen et al. 2007).

The hydrocarbon reservoirs in the Siri Canyon rely on long distance migration of hydrocarbons, up to 75 km from mature Jurassic source rocks in the Central Graben. This was facilitated by the elongate and confined geometry of the sandstones forming a well-connected system for updip migration (Hamberg et al. 2005; Ohm et al. 2006).

2.1. Sele Formation

The Sele Formation was established by Deegan & Scull, 1977. It comprises the dark grey to greenish grey, laminated shales and siltstones that overlie the non-laminated Lista Formation or local sandy sediments belonging to a variety of different units in other areas. It is rich in smectite, volcanic ash and organic carbon. The thickness varies from 5 to 54 m. Lateral equivalents onshore Denmark are diatomite of the lower part of the Fur Formation

(Pedersen and Surlyk, 1983 and laminated clay of the lower part of the Ølst Formation (Heilmann-Clausen et al., 1985). The Sele Formation is characterised by high gamma-ray readings throughout, with a number of gamma-ray peaks. The base of the Sele Formation is generally marked by an upward shift to consistently higher gamma-ray readings than those of the underlying Bue Member (of the Lista Formation).

Knox and Holloway (1992) suggested an informal threefold subdivision, units S1–3, of the Sele Formation based on the gamma-ray log signature. The same subdivision was recognised in Augusta-1, Siri-3, and Nini-3 (Schiøler et al., 2007). The mudstones of the Sele Formation represent a mixture of pelagic and hemipelagic fallout. The well-laminated character of the sediment, the high content of organic material and uranium (Schiøler et al., 2007), and the general lack of trace fossils and benthic foraminifers indicate starved sedimentation under dysoxic to anoxic bottom conditions. Common diatoms indicate a high nutrient level in the water mass (Schiøler et al., 2007). Based on microfossils Mitlehner (1996) suggested that the palaeoenvironment represent an upper bathyal setting with a palaeodepth estimate of around 300 m. The tuffs of the Sele Formation are evidence of extensive volcanism in the region and further are used to distinguish different units (Schiøler et al., 2007). The amount of ash beds identified in individual Siri Canyon wells varies significantly, partly due to poor preservation of randomly encountered thin ash layers.

In the studied wells, the Sele Formation is interbedded with thick sandstones or it grades upwards into a succession of thinly interbedded sandstones and mudstones.

Thin sandstone intrusions occur, but only in the lower part of the formation. They are referred to the Kolga Member by Schiøler et al. (2007).

3. Material and Methods

Forty-nine core samples from eight wells (Nini-2, -3, and -4; Na-2P; Siri-2 and -5; Cecilie-2 and Augusta-1) were analyzed for whole-rock (bulk) and clay fraction (<2μm) mineralogy using X-Ray Diffraction (XRD) analysis. The shale samples are dark grey to black and laminated. No bioturbation was observed. Most samples have fine white to light gray partings of diatomite or volcanic ask. A few samples contain distinct thin ash layer. The samples were selected to provide a good coverage of stratigraphy, burial depth (1700 m to 2900 m) and temperature (approximately 50 to 100 °C, Table 1). One sample is a one cm thick graded ash-layers (Cecilie-2, 2225, 60m), and one is grey, massive, very fine grained ash (Nini-4, 1786, 50 m).

Table 1. Burial depth (1700 m to 2900 m) and temperature (approximately 52 to 102 °C) of representative wells. Temperatures are bottom-hole temperatures and slightly higher than those of the sampled intervals

Well Name	Temperature (°C)	Depth (m)
Nini-3	*52*	1808
Siri-4	65	2192
Cecilie-1A	86	2389
Augusta-1	102	2968

3.1. Bulk Rock Composition

The mineralogy of bulk mudstone samples was determined from powdered samples using standard X-ray powder diffraction (XRD) analysis. One part of the cleaned core samples was powdered in a wolfram carbide mortar and mounted in an alumina sample-holder for bulk rock XRD. Analyses were performed using a Phillips PW3710 X'pert diffractometer system with CuKα1-monocromized radiation and a scanning range from 2° to 65° 2θ. The minerals present in each sample were identified by reference to the ICDD (The International Centre for Diffraction Data) powder diffraction database.

Semi quantitative estimates of mineralogical composition are based on calculation of the integrated peak area of a selected peak for each identified mineral applying in-house standards. These are based on comparison to known mixtures of two or more minerals, or to samples with successively increasing amounts of admixed quartz, and occasionally controlled by Rietveld analyses, and by chemical analyses of carbonate and sulfide phases. The bulk composition of the sediments obtained from X-ray diffraction analysis is shown in Table 2. Typical diffractograms are presented in Figure 3.

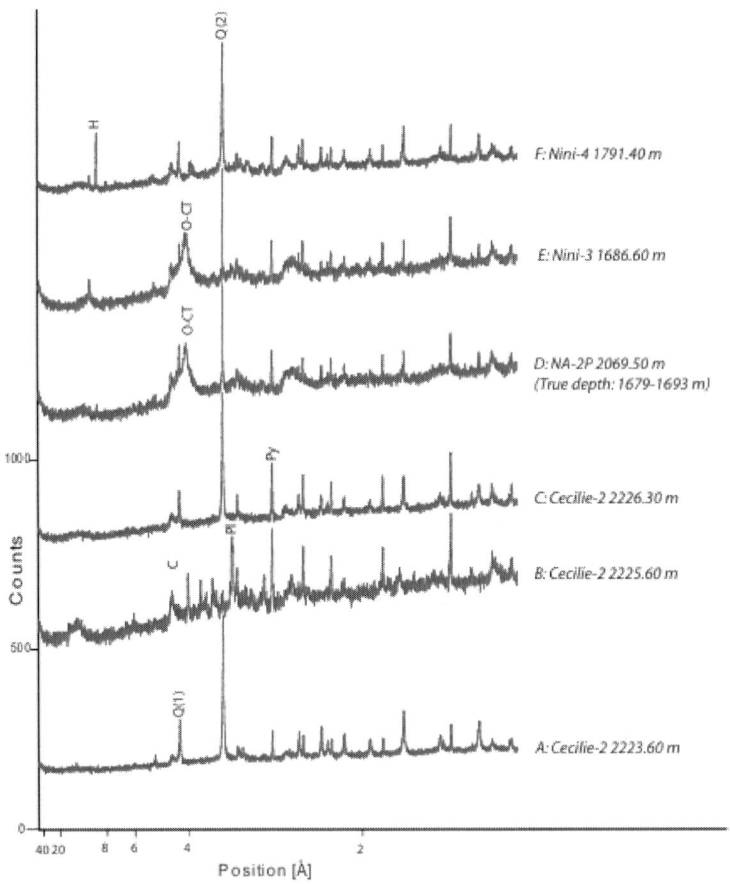

Figure 3. Typical X-ray diffractograms of bulk samples. A: Cecilie-2 (2223.60 m); B: Cecilie-2 (2225.60 m); C: Cecilie-2 (2226.30 m); D: NA2P (2069.50m – 1679-1693 m true depth); E: Nini-3 (1686.60m); F: Nini-4 (1791.40 m). H: Heulandite; C: Clay minerals; Q: Quartz; O-CT: Opal-CT; Pl: Plagioclase; Py: Pyrite.

Table 2. Bulk rock mineralogy of studied samples. *Sele Fm. interval in NA-2P situated between 1679-1693 m true depths

		Quartz	K-Feldspar	Plagioclase	Calcite	Pyrite	Clay minerals	Heulandite	Cristobalite	Jarosite
Well	Depth	%	%	%	%	%	%	%	%	%
Nini-2	1697.40	34	0	0	0	6	38	12	0	11
Nini-2	1699.60	33	0	0	0	8	46	4	0	9
Nini-3	1686.60	3	0	3	0	1	24	0	65	2
Nini-3	1688.40	4	0	3	0	2	40	0	47	4
Nini-3	1692.30	7	0	5	0	5	39	4	28	8
Nini-3	1715.50	19	0	13	0	12	51	6	0	0
Nini-4	1786.50	0	0	6	0	2	80	0	0	12
Nini-4	1791.40	24	0	0	0	8	32	23	0	13
Nini-4	1792.40	16	0	0	0	10	50	11	0	13
Nini-4	1794.70	14	0	4	0	4	77	0	0	0
Na2P *	2069.50	4	0	3	0	1	17	1	70	3
Na2P *	2073.50	4	0	0	0	3	40	1	48	5
Na2P *	2078.70	7	4	3	0	5	68	4	0	10
Na2P *	2084.80	37	6	6	0	5	43	4	0	0
Na2P *	2085.50	18	0	3	0	3	73	2	0	0
Siri-2	2092.40	50	0	0	0	5	36	0	0	9
Siri-2	2094.50	28	0	0	0	10	50	0	0	13
Siri-2	2096.70	15	0	9	0	12	51	0	0	13
Siri-5	2065.40	39	0	11	0	5	34	0	0	10
Siri-5	2067.30	15	0	0	0	4	81	0	0	0
Siri-5	2073.90	16	0	0	0	14	62	0	0	9
Siri-5	2074.60	24	0	4	0	3	63	0	0	0
Siri-5	2075.50	37	0	7	0	8	49	0	0	0
Siri-5	2088.70	30	0	6	0	7	57	0	0	0
Siri-5	2089.90	28	0	0	0	13	63	0	0	2
Siri-5	2090.30	24	0	8	4	19	43	0	0	3
Siri-5	2099.80	27	0	9	7	12	36	0	0	9
Siri-5	2100.60	34	0	0	0	6	52	0	0	8
Siri-5	2101.90	37	2	2	0	10	38	2	0	10
Siri-5	2102.20	34	0	0	0	6	52	0	0	8
Cecilie-2	2212.40	27	0	0	0	6	58	0	0	9
Cecilie-2	2213.40	31	0	3	0	3	60	0	0	4
Cecilie-2	2214.30	27	0	0	0	2	71	0	0	0
Cecilie-2	2215.80	19	0	6	0	3	70	0	0	4
Cecilie-2	2216.60	29	0	0	0	4	52	0	0	15
Cecilie-2	2217.40	41	0	3	0	7	36	0	0	13
Cecilie-2	2219.20	30	0	0	0	17	45	0	0	7

Table 2. (Continued)

		Quartz	K-Feldspar	Plagioclase	Calcite	Pyrite	Clay minerals	Heulandite	Cristobalite	Jarosite
Cecilie-2	2220.50	46	0	0	0	7	30	0	0	18
Cecilie-2	2221.60	39	0	0	0	5	50	0	0	6
Cecilie-2	2222.22	29	0	4	0	4	52	0	2	9
Cecilie-2	2223.60	38	0	0	0	4	47	0	0	5
Cecilie-2	2224.60	22	0	0	0	2	76	0	0	0
Cecilie-2	2225.60	0	0	8	0	7	77	0	0	8
Cecilie-2	2226.30	19	0	0	0	9	72	0	0	0
Cecilie-2	2227.30	12	0	0	0	18	59	0	0	12
Augusta-1	2888.75	64	0	0	0	10	21	0	0	6
Augusta-1	2890.60	15	0	14	0	5	66	0	0	0
Augusta-1	2892.60	15	0	5	0	1	79	0	0	0

3.2. Clay Fraction Composition

One part of the cleaned core sample was disintegrated and suspended by repeated tumbling in demineralised water in plastic bottles to disintegrate the sample, and finally the <2 μm-fraction was separated by repeated decantation. A few ml of the established <2 μm-fraction was smeared onto a glass-plate using the pipette-on-glass-slide technique (Gibbs, 1971), and dried at room-temperature. The plate-shaped clay minerals will thus be oriented (with 001) parallel to the glass-plate. These preparations are X-rayed 3 times:

1) As is ("untreated")
2) After being exposed to ethylene glycol vapour in exsiccator for 24 hours at 60ºC
3) After heating to 500ºC for 1 hour

Analyses were performed using a Phillips PW3710 X'pert diffractometer system with CuKα1-monocromized radiation and a scanning range from 2° to 65° 2θ. Scanning range for ethylene glycollated and heated samples was from 2° to 35° 2θ.

3.2.1. Smectite

Smectite is identified in all samples by a strong peak at 12-15Å in the untreated sample. It expands to about 17 Å following glycolation, and collapses to about 10 Å after heating.

3.2.2. Illite

Illite is identified by its (001) peak at 10 Å (which represents illite, mica, muscovite, Figure 4 A, D and E). The peak is unaltered after glycollation. The 10Å peak on the heated diffractograms also represents collapsed smectite.

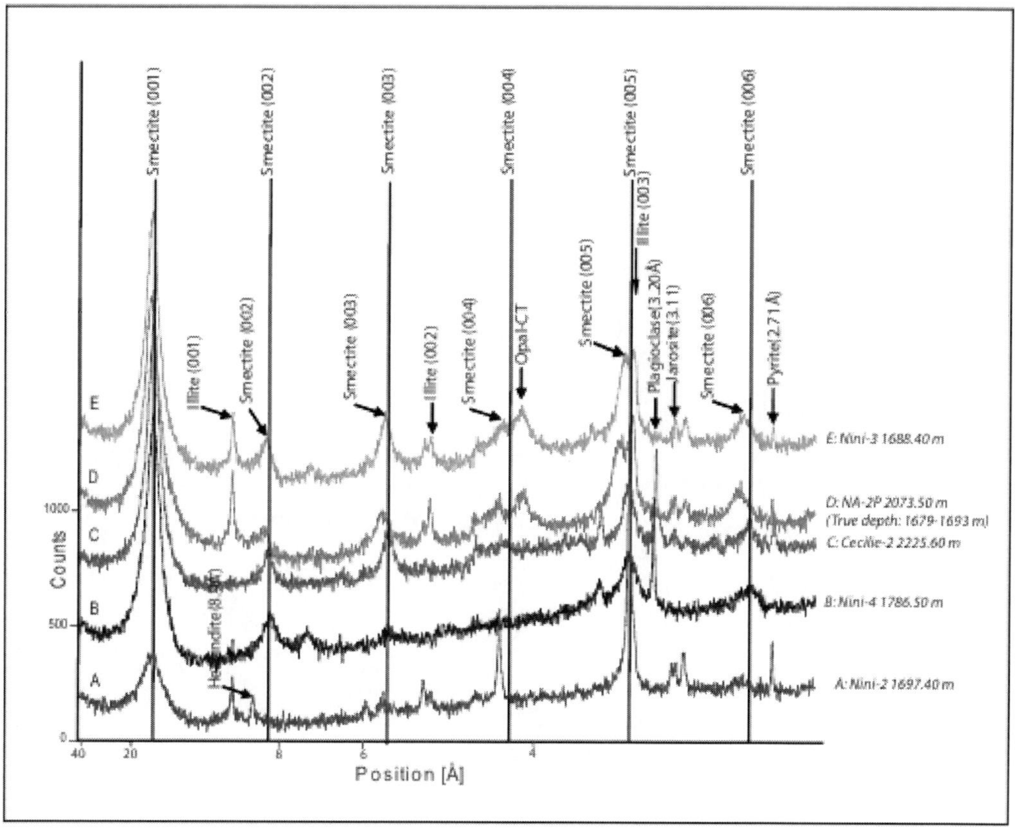

Figure 4. X-ray diffractograms of <2μ fraction (glycollated sample) illustrating smectite: A: Nini-2 (1697.40); B: Nini-4 (1786.50 m); C: Cecilie-2 (2225.60 m); D: NA-2P (2073.50 m - 1679-1693 m true depth); E: Nini-3 (1688.40 m). The position of (001), (002), (003), (004), (005) and (006) peaks are strictly equally spaced indicating that interlayering of smectite with illite is not present. The oddorder peaks are stronger than the even-order peaks and the (004) peak is poorly developed or not present. The peak position does not change significantly with the burial depth. A few samples have little or no illite present (B: Nini-4 (1786.50m), and C: Cecilie-2 (2225.60 m)). They contain plagioclase and are considered to be altered volcanic ash. Their smectite has a slightly smaller d-spacing (compare B and C (volcanic) with D and E (detrital). The lines indicate the position of smectite basal reflections for samples B and C.

3.2.3. Chlorite

Chlorite is identified by its (001), (002), (003) and (004) reflections at 14Å, 7Å, 4.7Å and 3.5Å respectively. Chlorite is a very diverse group of minerals and range from Fe-rich chamosite to Mg-rich clinochlores.

Iron-rich chlorites (chamosites have large even-order peaks compared to very small or odd-order peaks as seen in samples from Augusta-1 (Figure 5) whereas the peaks of clinochlore are more equal in intensity.

Chlorite dehydroxylate when heated to temperatures of about 500°C resulting in a marked increase of the intensity of the 14 Å peak and a marked reduction in the 7Å and 3.5Å peaks (Figure 5).

Minor amounts of chlorite were detected in samples from Siri-5, Na-2P and Augusta-1. (Figure 5; Table 3).

3.2.4. Non-Clay Minerals

Other minerals identified in the clay fraction (<2 μm) are quartz, potassium feldspar, plagioclase, opal-CT, heulandite, pyrite and jarosite (as also recorded in the bulk samples).

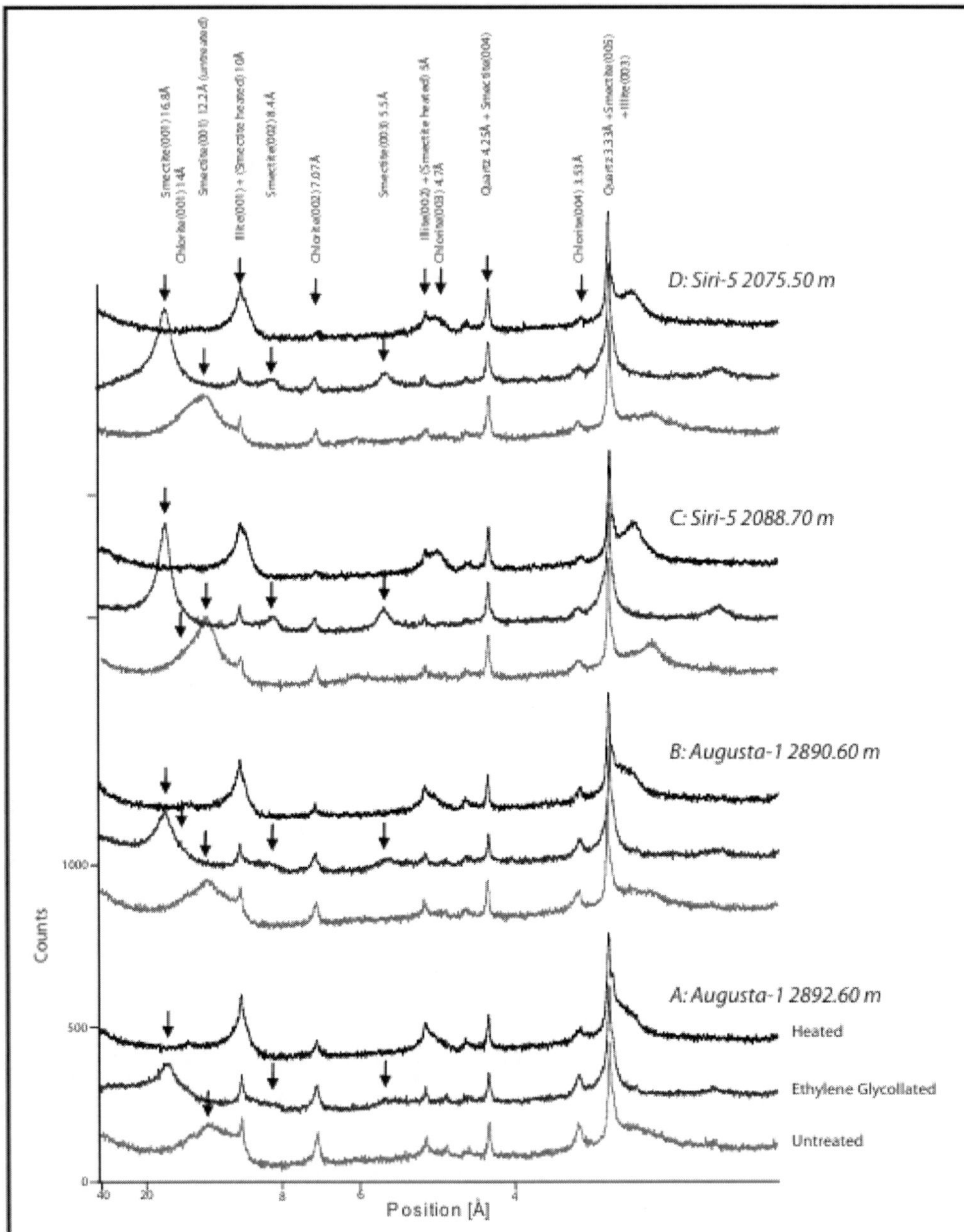

Figure 5. X-ray diffractograms of <2μ fraction, Untreated, Ethylene glycolated, and Heated samples. A: Augusta-1 (2892.60 m); B: - Augusta-1 (2890.60 m); C: Siri-5 (2088.70 m); D: Siri 5 (2075.50). *Chlorite,* is identified by its (001), (002), (003) and (004) reflections at 14Å, 7Å, 4.7Å and 3.5Å respectively. Iron-rich chlorites (chamosites) have large even-order peaks compared to very small odd-order peaks as seen in samples from Augusta-1 (A and B).

Table 3. Clay fraction mineralogy and clay mineral distribution of studied samples

		XRD-composition of <2μm fraction											Relative distribution of clay minerals			Clay mineral proportion of bulk sample		
Well	Depth	Quartz	K-feldspar	Plagioclase	Jarosite	Pyrite	Calcite	Cristobalite	Heulandite	Smectite	Illite	Chlorite	Smectite	Illite	Chlorite	Smectite	Illite	Chlorite
		%	%	%	%	%	%	%	%	%	%	%	%	%	%	%	%	%
Nini-2	1697.40	51	0	0	6	2	0	0	4	31	6	0	84	16	0.0	32	6	0
Nini-2	1699.60	22	0	0	4	1	0	0	0	64	9	0	88	12	0.0	41	6	0
Nini-3	1686.60	11	0	0	20	1	0	10	0	54	4	+	92	8	0.2	22	2	0
Nini-3	1688.40	0	0	0	8	1	0	35	0	53	3	+	93	6	0.8	37	2	+
Nini-3	1692.30	8	0	0	8	3	0	15	0	57	7	3	86	10	3.8	34	4	2
Nini-3	1715.50	13	0	0	0	2	0	0	1	77	6	0	92	7	1.1	47	3	+
Nini-4	1786.50	0	0	25	0	0	0	0	1	72	+	2	98	1	2.0	78	+	0
Nini-4	1791.40	45	0	0	8	3	0	0	1	36	7	0	84	16	0.0	27	5	0
Nini-4	1792.40	22	0	0	10	3	0	0	2	55	8	0	87	13	0.0	44	7	0
Nini-4	1794.70	9	0	+	+	1	0	0	0	82	7	1	92	8	0.2	71	6	+
Na2P *	2069.50	0	0	0	3	1	0	28	0	62	6	0	91	9	0.4	15	2	0
Na2P *	2073.50	8	0	0	3	1	0	15	0	66	7	0	90	9	0.1	36	4	+
Na2P *	2078.70	13	0	0	15	3	0	0	2	58	9	0	87	13	0.0	59	9	0
Na2P *	2084.80	43	0	2	0	1	0	0	1	48	5	+	89	10	0.6	38	4	+
Na2P *	2085.50	15	0	2	6	1	0	0	1	66	9	2	87	12	1.4	64	9	1
Siri-2	2092.40	62	0	0	9	3	0	0	0	25	2	0	93	7	0.0	33	3	0
Siri-2	2094.50	37	0	0	9	3	0	0	0	46	6	0	88	12	0.5	44	6	+
Siri-2	2096.70	22	0	0	5	3	0	0	0	44	8	0	85	15	0.0	51	5	0
Siri-5	2065.40	43	0	0	0	0	0	0	0	50	7	1	86	12	1.8	29	4	+
Siri-5	2067.30	31	0	0	0	0	0	0	0	55	5	6	80	7	13.1	64	6	11
Siri-5	2073.90	21	0	0	2	2	0	0	0	68	6	2	89	8	3.2	54	5	2
Siri-5	2074.60	51	0	0	0	0	0	0	0	43	5	2	87	10	3.3	55	6	2
Siri-5	2075.50	48	0	0	0	0	0	0	0	48	3	1	92	7	2.0	44	3	1
Siri-5	2088.70	49	0	0	0	0	0	0	0	47	3	1	92	7	1.7	53	4	1
Siri-5	2089.90	52	0	0	0	3	0	0	1	39	5	+	88	11	0.8	56	7	+
Siri-5	2090.30	58	0	0	7	4	0	0	0	29	2	+	94	5	1.0	40	2	+
Siri-5	2099.80	55	0	0	0	1	0	0	0	37	5	5	84	12	4.5	31	4	2
Siri-5	2100.60	68	0	0	0	2	0	0	0	26	3	1	87	11	1.6	45	6	1
Siri-5	2101.90	53	0	0	4	8	0	0	0	32	3	0	91	9	0.0	34	3	0
Siri-5	2102.20	48	0	0	5	8	0	0	0	35	4	0	91	9	0.0	47	5	0
Cecilie-2	2212.40	45	0	3	4	2	0	0	0	41	5	+	88	11	0.0	51	6	1
Cecilie-2	2213.40	65	0	0	2	3	0	0	0	27	2	+	92	7	0.0	56	4	1
Cecilie-2	2214.30	67	0	0	0	0	0	0	0	29	3	+	89	10	1.2	63	7	1
Cecilie-2	2215.80	38	0	0	3	0	0	0	0	55	4	+	93	6	0.7	65	4	1
Cecilie-2	2216.60	73	0	2	4	4	0	0	0	15	1	0	92	8	0.0	48	4	0
Cecilie-2	2217.40	69	0	0	3	7	0	0	0	20	1	0	91	7	2.2	33	2	0
Cecilie-2	2219.20	68	0	0	4	11	0	0	0	16	2	+	87	11	0.0	39	5	0
Cecilie-2	2220.50	73	0	0	2	5	0	0	0	17	2	0	88	12	0.0	26	4	0
Cecilie-2	2221.60	71	0	0	8	4	0	0	0	15	2	0	88	11	0.0	44	5	0
Cecilie-2	2222.22	59	0	0	4	7	0	0	0	27	2	+	94	7	0.0	49	3	0
Cecilie-2	2223.60	67	0	0	5	5	0	0	0	19	4	0	82	18	0.0	39	9	0
Cecilie-2	2224.60	58	0	0	0	1	0	0	0	41	1	+	98	2	0.0	74	1	0
Cecilie-2	2225.60	5	0	13	0	4	0	0	0	78	0	0	100	0	0.0	77	0	0
Cecilie-2	2226.30	62	0	0	6	5	0	0	0	23	3	+	89	11	0.0	63	8	0
Cecilie-2	2227.30	12	0	0	5	5	0	0	0	76	3	0	97	3	0.0	57	2	0
Augusta-1	2888.75	87	0	0	2	2	0	0	0	6	3	0	64	36	0.0	14	8	0
Augusta-1	2890.60	27	0	3	0	0	0	0	0	56	11	3	80	15	5.0	53	10	2
Augusta-1	2892.60	26	0	0	0	0	0	0	0	54	14	6	74	18	8	58	14.4	5

* Sele Fm. interval in NA-2P situated between 1679-1693 m true depths.

3.3. Quantitative Evaluation

The quantitative estimation was made using a modified version of the method of Griffin (1971). It is based on the peak areas of the XRD peaks at 17Å, 10Å and 7Å in the oriented smear slide diffractogram after glycol saturation. Peak areas (above background) were taken from those provided by the Philips APD software for each sample trace using the in-built peak search routine.

Expanding clays was quantified as smectite at the 17Å peak position. The integrated area of the 7Å (chlorite 002) peak is quantified as chlorite as no kaolinite was recorded. The quantification is based on in-house standards, which are adjusted to the specific setting of the instrument and calibrated by mixtures of pure minerals.

4. Results

4.1. Opal-CT

Opal-CT is characterized by a broad, strong (101) reflection centered at 4.09 Å (Figure 3, D and E; Figure 4, D and E). According to Mizutani (1977), Iijima et al. (1983) and Graetsch et al. (1987), opal-CT becomes more ordered during diagenesis, resulting in the formation of opal-C.

In XRD patterns it is characterized by lower d-spacings than opal-CT, with a more intense single strong (101) peak within 4.04-4.06 Å, a moderate peak at 2.49 Å and weak peaks present at ~2.85 and ~3.14 Å. In the Sele Shale, the diffraction pattern indicates opal-CT with cristobalite and tridymite stacking sequences (Flörke et al., 1991). It is abundant in Nini-field samples but is not recorded in deeper wells except for one Cecilie-2 sample (Tables 2 and 3).

4.2. Potassium Feldspar

Alkali-feldspar is present in few samples (Tables 2 and 3). It is identified by its main reflection at 3.25Å. The recorded amount is very small and no other peaks were recorded. K-feldspar is only recorded in samples which also contain plagioclase (Table 2 and 3). Plagioclase has a peak close to the diagnostic alkalifeldspar peak at 3.25Å and in these cases it is difficult to safely identify the alkalifeldspar.

4.3. Plagioclase

The presence of plagioclase is well documented by several sharp peaks in some samples, which are distinctly present in the Siri-5, Na-2P, Cecilie-2 and Augusta-1 wells (Tables 2 and 3). Three useful peaks in the plagioclase series not found in the alkali series are at 3.18 Å, 3.20 Å, and 3.21 Å (Figs. 3, B; Figure 4, B and C). The position of the main reflection varies between samples indicating a difference in composition.

4.4. Quartz

Quartz occurs in variable amount in <2-μm fractions, ranging from 1-80 w% (Table 3). Nini field samples (Nini-2, -3 and -4, NA-2P wells) generally have much lower content of quartz than the deeper samples. Very high values are only seen in deep samples. Low contents of quartz partly correlate with high contents of opal CT.

4.5. Heulandite

Heulandite is recorded from shallow samples (Tables 2 and 3), both in the bulk sample and the <2μm fraction. Some of these samples also contain plagioclase and are supposed to contain volcanic ash. Heulandite disappears with depth whereas plagioclase is found also in deeper samples (Tables 2 and 3).

4.6. Pyrite

Pyrite is present in all samples (Table 2). It is generally represented by a large number of sharp peaks. Diagnostic peaks are at 1.63 Å and 2.71 Å. In samples with much pyrite, it is generally associated with secondary jarosite (Figures 4, A and E).

4.7. Calcite

Calcite is present in a few bulk samples but is not recorded from the clay fraction. The peak position is shifted slightly towards lower values than 3.03Å indicating some Mg, Mn or Fe substitution as also reported from onshore localities by Heilmann-Clausen et al. (1985), Nielsen et al. (1986) and Nielsen (1997).

4.8. Clay Minerals

The smectite content of the bulk samples varies between 15-80 w%. The position of (001), (002), (003), (004), (005) and (006) peaks are strictly equally spaced indicating that interlayering of smectite with illite is not present. The odd-order peaks are stronger than the even-order peaks and the (004) peak is poorly developed or not present. The peak position does not change significantly with burial depth (Figure 4). The illite content varies between 1-15 w% (Table 3).

Very low illite contents are found in the samples with high smectite content. Chlorite is present in very small amounts except for few samples in the Siri 5 and Augusta 1 wells with up to 10 w% chlorite (Table 3). In the presence of chlorite, it is difficult to distinguish additional kaolinite. It may be revealed by double-peaks at 7Å and 3.5Å. In such cases, the kaolinite peak is at slightly higher Å-values than the chlorite peaks. A double-reflection at 3.5 Å (chlorite at 3, 53 Å and kaolinite at 3, 57 Å) often indicates the presence of both minerals.

In the Sele Shale samples, the chlorite peaks are single peaks at 3.52-3.53 Å, indicating that only chlorite is present.

5. DISCUSSION

5.1. Original Composition of the Sele Shale

According to Deegan and Scull (1977), the Sele Formation shales are characterized by abundant smectite and organic matter, and they are locally rich in silica fossils and volcanic ash. Heilmann-Clausen et al. (1985), Nielsen et al. (1986) and Nielsen (1997) report detrital smectite, illite and occationally kaolinite, together with detrital quartz and feldspar from the Sele Shale equivalents on-shore Denmark (Haslund Member of the Ølst Formation and Fur Formation).

Reported diagenetic minerals are pyrite (and jarosite), cristobalite, heulandite and Fe-Mn rich calcite. This is confirmed by our data (Table 2 and 3) which illustrate a very strong dominance of smectite over other clay minerals, high proportions of pyrite, and some samples being rich in opal-CT, heulandite, and plagioclase. Nielsen (1994, 1997) reported numerous ash layers from the Sele Shale level in the Harre borehole. Their XRD composition was reported to be smectite, plagioclase, quartz and heulandite. Ash layers from the Fur Formation are rich in unaltered volcanic glass and also contain feldspar (Larsen et al. 2003). The Fur Formation ashes represent a wide variation in composition, which is partly reflected by the composition of volcanic glass, partly by plagioclase composition and locally occurrence of sanidine (Larsen et al., 2003).

There seems to be a fairly distinct detrital pattern, although it is strongly overprinted by admixtures of organic matter, biogenic silica and ash fall. The present mineralogy is modified by these components and their alteration products.

The organic matter (not recorded in the XRD-mineralogical composition) is associated with early diagenetic pyrite and further with post-sampling formation of jarosite; biogenic silica is transformed into opal-CT and further into quartz; volcanic ash is recorded by unaltered plagioclase and possibly minor amounts of K-feldspar, and heulandite and smectite as alteration products. The alteration products vary with depth of burial, and it may be difficult or impossible to distinguish the amount of alteration products relative to the detrital components. Quartz as well as smectite may be partly detrital and partly diagenetic.

The samples of volcanic ash layers (Nini-4, 1786, 50 m; Cecilie-2, 2225, 60m) are dominated by smectite and only contains trace of illite or chlorite. They contain plagioclase and no quartz is recorded in the bulk sample. Thus, the admixture of volcanic ash is primarily recorded by the presence of plagioclase. Plagioclase peaks are sharp and between samples, the position of the main flection at 3.20Å is variable (Figure 4, B and C). This corresponds to differences in chemical composition as described from ash layers onshore (Larsen et al., 2003). We expect a detrital supply of plagioclase to be homogenized and therefore the variable composition of plagioclase from sample to sample represents the admixture of volcanogenic plagioclase. Approximate half of the samples contain a variable amount of plagioclase. The distribution of heulandite partly follows the occurrence of plagioclase, but heulandite is also present in many samples, which do not contain plagioclase (Tables 2 and

3). This indicates that heulandite is also related to the transformation of biogenic silica, or some volcanic ash did not supply plagioclase.

The original amount of detrital quartz is difficult to estimate because much quartz has been formed by various alteration processes. The general level of quartz in samples which appear to have had a limited potential for quartz formation is 15-25 w%. In addition, the average distribution of clay minerals can only be estimated with some uncertainty. Samples, which appear to have a minor potential for formation of additional smectite (as estimated by their content of feldspar or heulandite), have an average composition of the clay fraction of 85 w% smectite to 15 w% illite and almost no chlorite.

We therefore suggest that the average detrital composition of the Sele shale is 15-25 w% quartz, no feldspar, 85-75 w% clay minerals (85 w% smectite to 15 w% illite). This is considered the background sedimentation, which has through time been diluted by various amounts of organic matter, fossils (biogenic silica) and volcanic ash.

The main variables, with respect to mineralogy, are smectite, heulandite, quartz, opal and pyrite (Table 2 and 3).

5.2. Shallow to Deep Burial, Diagenetic Effects

5.2.1. Heulandite

Heulandite is found in a number of samples from the shallow part of the Siri Canyon (NA2P, Nini-2, -3, -4 and Siri-5), and is often associated with plagioclase (Table 2 and 3).

The association of plagioclase and heulandite is considered to demonstrate depositional volcanic ash. Heulandite is a common diagenetic product of volcanic ash (Heilmann-Clausen et al., 1985; Nielsen et al., 1986; Nielsen, 1997), and the plagioclase represents an unaltered original component of the ash fall (Larsen at al. 2003).

Heulandite is not recorded from shale intervals in deeper wells (Cecilie-2, Augusta-1 and most Siri-2 and Siri-5 samples (Tables 2 and 3). Some of these samples contain unaltered plagioclase and are expected to have formed heulandite from volcanic ash at an earlier stage of the diagenesis.

Based on the distribution of plagioclase and heulandite the transformation of volcanic glass must have produced heulandite already at shallower depth than those found in the Nini Field, and that the transformation of volcanic glass has already been completed at these depths/temperatures (1600 m; 50° C).

The heulandite survives burial to present day depths of approximately 2000 m and a temperature of 60° C. In Siri Canyon sandstones K-clinoptilolite has also been recorded down to this depth (Weibel et al., 2010).

So, in the Siri Canyon the critical depth/temperature level for heulandite dissolution appears to be approximately 2000m/60° C. Dissolution of heulandite supplies K and Si to the pore fluid and thus interfere with the smectite to illite transformation and supply silica for the precipitation of quartz. Silica may also have been exported at this stage.

5.2.2. Opal-A-Opal-CT-Quartz

Opal-CT is recorded mostly from the shallow Nini Field (NA-2P, Nini-3) but is also recorded in one sample from Cecilie-2 (Table 2 and 3).

The initial source of silica in the Sele Formation is most likely dissolution of diatoms or radiolarians. The presence of diatoms in the Sele Formation and its onshore equivalents is a well-known phenomenon (Mudge and Copestake, 1992; Danielsen and Thomsen, 1997; Pedersen and Surlyk, 1983; Heilmann-Clausen et al., 1985, Mitlehner, 1997). Thin layers rich in radiolaria have also been observed in core samples from the Siri Canyon (Weibel et al., 2010).

Biogenic silica is classified as opal-A by Jones and Segnit (1971). During burial and exposure to higher temperature and pressure the structure of the biogenic material alters, transforming it from amorphous opal-A to opal-CT and in some cases to chert.

Review articles describing opal-CT and chert formation include: Williams and Crerar (1985), Williams et al. (1985), and Knauth (1994). The chemical, mineralogical, and structural changes that occur during this process have been documented in Hurd and Theyer (1975) and Hurd (1983).

We did not record opal-A in any of the Siri Canyon samples and in accordance with the studies by Hurd and Theyer (1975), Hurd et al. (1973) and Hurd (1983) it must be assumed that the transformation from opal-A to opal-CT was completed before the Sele Formation reached its present depth of burial below 1600-1700 m and probably within a burial depth of a few hundred meters.

Williams et al., 1985 suggest that the transformation takes place as dissolution/ precipitation at a high level of Si-concentration. As a result of dissolution of opal-A and precipitation of opal-CT, dissolved silica may have been lost by diffusion along the concentration gradient into interbedded sandstones at very shallow depth.

Opal-CT is itself an unstable phase and is expected to transform into quartz at higher temperatures of greater burial depth. The transformation of opal-CT to quartz is a highly temperature-dependent reaction in which a solid-state inversion takes place converting opal-CT to microcrystalline quartz (Calvert, 1986).

There is disagreement about the order of the reactions and whether these transformations take place through solution and re-deposition or in the solid state. Carr and Fyfe (1958) concluded that amorphous silica transforms to disordered low-cristobalite and quartz by solution and precipitation. Misutany (1966), however, implied that both above mechanisms are possible for the two transformation steps but leaned toward the solid-state mechanism.

Similarly, Ernest and Calvert (1969) concluded that the opal-CT to quartz transformation takes place in the solid state. Stein and Kirkpatrick (1976) reexamined the experimental products of Ernest and Calvert (1969) in SEM and reinterpreted their results in terms of a dissolution- precipitation mechanism.

Quartz crystallizes from solutions with silica concentrations below the equilibrium solubility of cristobalite (Carr and Fyfe, 1958; Mackenzie and Gees, 1971) and, as suggested by Harder (1965) and Harder and Flehming (1970) from solutions saturated with respect to quartz in the presence of Al^{3+}, Fe^{3+}, Mg^{2+}, and Mn^{2+} hydroxides.

The presence of clay is known to speed up the replacement of opal by quartz, because clay adsorbs silica, and thus inhibits opal-CT nucleation. Thus, in the presence of abundant detrital clays, especially smectite, quartz may form directly from biogenic silica dissolution. However, a large proportion of smectite could even inhibit quartz cementation (Chang & Yortsos, 1994; Siever & Woodford, 1973; Williams et al., 1985).

Some samples from NA-2P and Nini-3 are rich in both opal-CT and quartz (Table 2, 3), possibly indicating that some transformation from opal-CT to quartz has already taken place.

The abundant occurrence of quartz in the <2μm fraction of some samples also from deeper wells (Table 3) indicates that further transformation of opal-CT to quartz has been largely completed in most samples where opal-CT was originally present.

The Sele Formation samples are rich in smectite, and, according to the suggestion of Chang and Yortsos (1994), Siever and Woodford (1973) and Williams et al. (1985), it cannot be excluded that in some samples all biogenic silica or part of it has been transformed directly to quartz without the formation of the intermediate opal-CT phase.

With deeper burial (below 2000 m and a temperature of more than 60°C) as in Cecilie-2 and Augusta-1, opal-CT is fully transformed to quartz.

During this phase, silica has been partly mobile and depending on the rate of dissolution of opal-CT compared to the rate of precipitation of quartz, silica may have been lost by diffusion to sandstone cementation. The quartz is an internal sink for dissolved silica, but the shale has been an active silica exporter during this transition.

The high amount of quartz suggests that the dissolution rate of precursor phases (i.e., biogenic silica/opal-A and/or opal-CT) was fast enough at least to sustain silica saturation at the microquartz saturation level.

5.2.3. Smectite

Smectite is by far the most abundant clay mineral in the studied Sele Fm. Shales (Table 1 and 2). The abundant occurrence of volcanic ash layers reported from the North Sea (Knox, 1997; Knox and Morton, 1983; Pedersen and Surlyk, 1983) indicates that the large amount of smectite in Palaeogene North Sea shales is partly an early alteration product of volcanic ash which was supplied by three different mechanisms:1) in situ transformation of detrital ash; 2) transformation of suspended ash in seawater; 3) weathering of volcanic ash fallen on land and subsequently reworked and incorporated in the shale.

A few samples (including the two ash layers: Nini-4, 1786, 50; Cecilie-2, 2225, 60m) have little or no illite present (Figure 4, B and C; Table 3). They are recorded with a high proportion of clay minerals and a lower than average content of quartz and they also contain plagioclase (Table 2). Thus they contain a very high proportion of authigenic smectite and represent the admixture of diagenetically altered volcanic ash. The formation of smectite from volcanic ash probably occurred at the same time as the formation of heulandite. The smectite of the ash layers is slightly different from the supposedly detrital smectite. As seen from Figure 4, samples B and C, the higher order basal reflections are consistently shifted slightly toward lower d-values in the ethylene glycollated sample compared to samples D and E. The consistent pattern indicates a slight difference in expansion between the two smectites.

The transformed ash layers do not contain quartz, and it must be assumed that the alteration of volcanic glass did not supply excess silica, compared to the consumption by smectite and heulandite. However, this depends on the composition of the volcanic ash.

The occurrence of fully transformed ash layers in the Nini field indicates that smectite formation occurred at burial depths/temperatures below 1600m/50°C, but it probably formed at even lower burial depth. Transformed ash layer are also reported from onshore Eocene deposits (Heilmann-Clausen et al. 1985).

Many studies have suggested that smectite releases significant amounts of silica by its with-depth transformation into illite (e.g., Weaver, 1959; Towe, 1962; Hower et al., 1976; Yeh and Savin, 1977; Boles and Franks, 1979; Foster and Custard, 1980; Abercrombie et al., 1994; Bjørlykke, 1998; Srodon, 1999; Van der Kamp, 2008; Peltonen et al., 2009). The released silica must be precipitated as quartz for the clay mineral reaction to proceed (Egeberg and Aagaard, 1989; Bjørlykke and Aagaard, 1992; Abercrombie et al., 1994).

The transformation of smectite is also considered to represent a major silica source for quartz cement in sandstone (Kastner and Bada, 1975).

However, the Sele Formation in the Siri Canyon has not been much influenced by illite formation. There is no systematic change in the relative amount of illite to smectite with depth (Table 3, Figure 6) although in the Augusta-1 samples (deeper than 3000 m) the recorded amount of illite to smectite is larger than average; this could represent an early phase of smectite to illite transformation, but largely the smectite has survive to this depth.

The X-ray diffraction pattern of smectite also remains stable with depth indicating that no significant change in interlayering pattern occurs (Figure 4).

Based on the amount of illite in shallow wells from the Nini Field (Table 3), we suggest that illite is representative of the depositional composition rather than diagenetic change with depth.

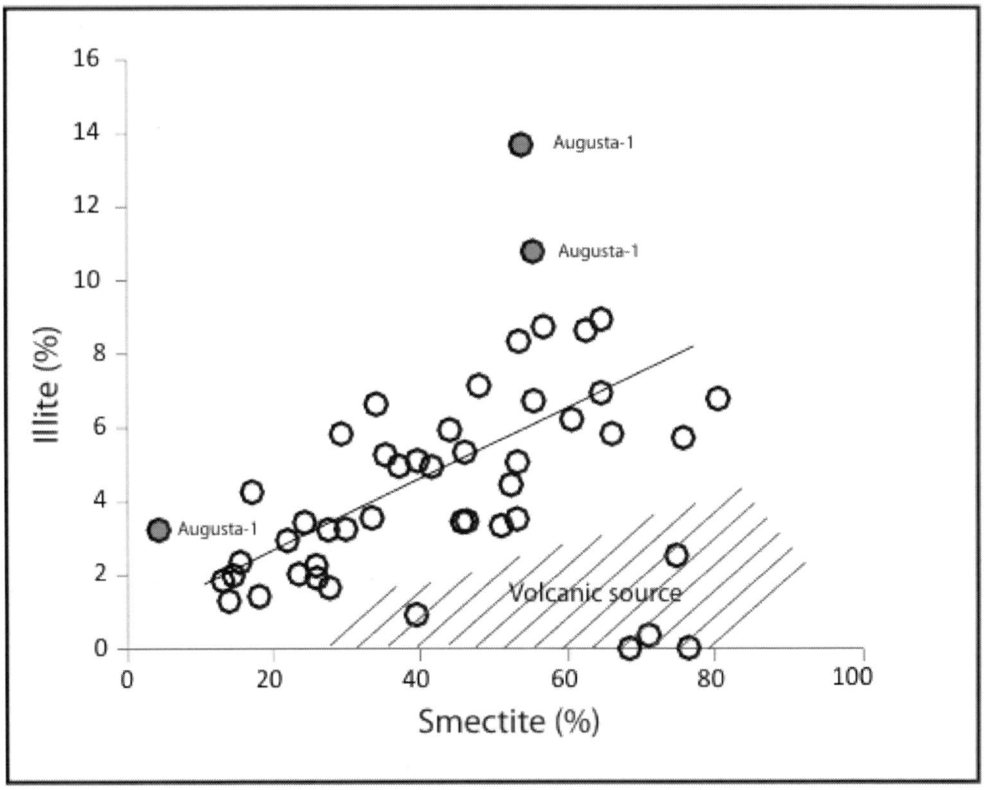

Figure 6. Relative amount of illite to smectite with depth.

The smectite to illite transformation requires an active potassium source such as potassium feldspar (Thyberg et al., in press, 2009). Potassium feldspar is generally not recorded. It occurs in a few samples together with plagioclase and (in shallow wells) with heulandite. The recorded potassium feldspar thus represents a volcanic supply. We must assume that potassium feldspar was largely not present in the detrital composition. Therefore, we suggest that the lack of dissolvable potassium sources have delayed the smectite transformation and consequently the smectite-to-illite transformation has not acted as a supplementary silica source for insitu shale diagenesis or adjacent sandstone diagenesis during deeper burial.

5.2.4. Chlorite

A small amount of chlorite is found in many samples (Table 3). There is no distinct trend in the distribution in relation to present burial depth. Only Siri-5 samples are distinguished by a consistent presence of chlorite and higher concentrations than average. It is not possible to infer any with depth chloritization, and it cannot be excluded that small amounts of chlorite were part of the detrital pattern.

6. Sele Formation Mudstones Diagenesis Compared to Siri Canyon Sandstones

The authigenic silica phases found in Palaeogene sandstones from Siri Canyon are opal, cryptocrystalline quartz, microquartz and macroquartz.

From previous SEM studies (Weibel et al., 2010) it seems that in all samples with microquartz coatings, a gritty opal coating is present below the microquartz. This opal coating can be several micrometers thick and cover all detrital grains, or it can be very thin and does not occur on other detrital grains than quartz. Very thin opal coatings on detrital quartz are usually covered by ordered microquartz, which has inherited the crystallographic orientation from the host detrital grain.

Stokkendal et al. (2009) reported that in sandstones interbedded with the Sele Formation in the wells Siri-3, Siri-3A, Siri-4 and Siri-5 thick opal rims (1-5 μm) occur abundantly in sands closest to the mudstone units (less than 1 m apart). They formed on all detrital grains and they separate the succeeding growth of microquartz from the detrital grains and prevent the formation of ordered microquartz (Weibel et al. 2010)). Because very few fossils were recorded from the sandstones, and because of a strong association of quartz morphologies with the sample proximity to the embedding mudstone, it was assumed that the major silica source was from biogenic silica within the mudstone succession (Stokkendal et al. 2009; Weibel et al., 2010).

Macroquartz overgrowths dominate in the deepest studied wells (Cecilie-2; Augusta-1). In the deepest buried sand intervals the quartz overgrowths typically completely encloses the detrital quartz grains as shown by Weibel et al. (2010).

Microquartz coatings are present beneath the quartz overgrowths in all parts of the succession but were more frequently observed in oil-filled sandstones than in water filled samples where the overgrowths are larger and cover the detrital grains more fully (Weibel et al., 2010).

The formation of early diagenetic opal, K-rich clinoptilolite and microquartz in sandstone appear to be completed in the most shallow wells of the Nini Field (Weibel et al., 2010) and is thus partly comtemporaneous with the transformation of biogenic opal and volcanic ash in the shale. Later dissolution of K-rich clinoptilolite also occurs within the same depth interval. The late stage quartz cementation in sandstones (Weibel et al., 2010) does not correspond to significant processes in the shale (Figure 7). The dissolution of heulandite and opal-CT in the shales took place at burial depths less than approximately 2000m and therefore do not relate to the supply of silica for the deep quartz cementation in the sandstones.

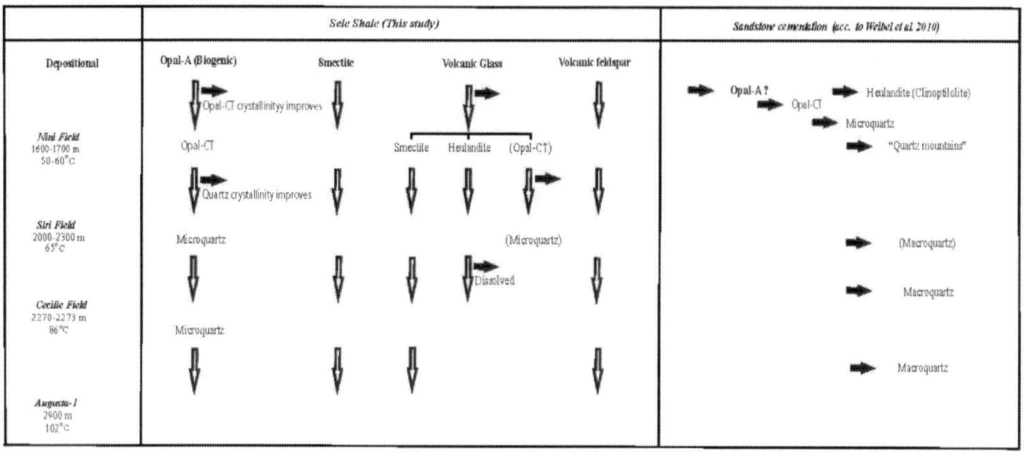

Figure 7. Schematic diagram of diagenetic zones in the Sele shale (this study) and Siri Canyon sandstones (from Weibel et al., 2010). With-depth transformation and -dissolution of detrital components (Opal-A, smectite, volcanic glass and volcanic feldspar) and their alteration products is indicated as vertical successions marked by open arrows. Related mobilization and possible export of silica is marked by filled arrows. Diagenetic silica-phases in sandstones (from Weibel et al. 2010) related to mobile silica in the sandstone – and possible import of silica (filled arrows). The phases present in the shallow Nini field samples correspond to significant mobilization of silica within shales. The macro-quartz from deep fields (Siri)-Cecilie-Augusta does not correspond to mobilization of silica in the shale and are not sourced from adjacent shales.

7. Timing of Authigenic Silica Diagenesis

Silica has been mobilised by a number of diagenetic processes in the Sele Shale. The major release appears to be related to the transformation of biogenic opal and possibly to the alteration of volcanic ash. Both these processes appear to have been early (Figure 7). They were completed in all the studied wells (at burial depth/temperatures less than 1600m/50°C) but they were probably active at much lower burial depths. The release of silica from biogenic silica resulted in precipitation of opal-CT, which occurs abundantly in shallow wells from the Nini Field. The precipitation of opal-CT requires a sustained high concentration of dissolved silica and a concentration gradient against adjacent sandstones evolved and controlled a migration of silica from shale into sandstone. Precipitation of opal-CT and possibly microquartz within the shale may have reduced the concentration gradient and resulted in a gradually lower flux rate. This evolution could control the successive precipitation of opal-CT and microquartz in the sandstones (Stokkendal et al. 2009; Weibel et al. 2010). Because the alteration of volcanic ash apparently did not result in excess silica for quartz precipitation in the shale, the volcanic ash was not an additional source of silica - except for the precipitation of heulandite.

The opal-CT was gradually replaced by quartz with depth. Dissolution of opal and precipitation of quartz indicate a second phase of major silica mobility. This phase was largely active in the depth interval down to 2000 m. However, further cementation of sandstones by quartz is minor in this interval, and the successive quartz cementation accelerates below burial depth of 2000 m (Weibel et al., 2010). Therefore, it must be assumed that silica was not exported during this transformation but was retained within the shale and precipitated as quartz.

This would also apply for the dissolution of heulandite, which occurs within this depth interval. There is no significant silica mobilization below 2000 m of burial depth. Quartz is present at all depth, and smectite is not being altered. Therefore, the possible sourcing of silica from the shales for sandstone cementation would be restricted to very shallow burial, probably a few hundred of meters and would have stopped at burial depths of 1600-1700 m (Figure6).

CONCLUSION

This study explores the potential sources of SiO_2 in the shale, and the timing of their active phases. The primary sources of silica during the diagenesis of the Sele Shale are biogenic opal and volcanic ash.

There are two major phases of silica mobilization. The first phase occurs at shallow depths (less than 1600 m but probably a few hundred meters) and is related to the transformation of both biogenic opal and volcanic ash. Diagenetic products in the shale are opal-CT, smectite, heulandite and quartz. The second phase occurs at intermediate depths (approximately 2000 m), and is related to the dissolution of opal-CT and heulandite. Below this depth opal-CT is fully transformed to quartz and heulandite was completely dissolved.

During both phases, dissolved silica has been partly mobile but only during the first of the phases dissolved silica migrated into interbedded sandstone and interacted with sandstone cementation. During the second phase all mobilized silica was internally consumed and precipitated as quartz.

The Sele Formation in the Siri Canyon has not been influenced by illite formation, even at burial depth of approximately 3000 m.

The deep cementation of sandstones (accelerating with depth below 2000 m) does not correspond to any significant transformation process within the shale, and it is concluded that the Sele shale did not source sandstone cementation at this stage.

REFERENCES

Abercrombie, H.J., Hutcheon, I.E., Bloch, J.D., de Caritat, P., 1994. Silica activity and the smectite–illite reaction. *Geology* 22 (6), 539–542.

Archer, S. G., Wycherley, H. L., Watt, G. R., Baron, M. L., Parnell, J., Chen, H., 2004. Evidence for focused hot fluid flow within the Britannia Field, offshore Scotland, UK, *Basin Research* 16, 377–395.

Bjørlykke, K., Aagaard, P., 1992. Clay Minerals in North Sea Sandstones. In: Houseknecht, D.W., Pittman, E.D. (Eds.), Origin, diagenesis, and petrophysics of clay minerals in sandstones, vol. 47. *Society for Sedimentary Geology* (SEPM), Tulsa, OK, United States, pp. 65–80.

Boles, J.R., Franks, S.G., 1979. Clay diagenesis in Wilcox sandstones of Southwest Texas; implications of smectite diagenesis on sandstone cementation. *Journal of Sedimentary Petrology* 49 (1), 55–70.

Carr, R. M. and Fyfe W. S., 1958. Some observations on the crystallization of amorphous silica. *Amer. Minwtrl.* 43. 908 -916.

Calvert, C. S. and Klimentidis R. E., 1986. The generation of silicate cements during burial diagenesis of shales. *Clay Minerals Society 23rd Annual Meeting*, 27.

Chang, J. & Yortsos, Y.C. 1994. Lamination during silica diagenesis-effects of clay content and Ostwald ripening. *Am. J. Sci.* 294:137-172.

Danielsen, M. and Thomsen, E., 1997. Paleocene/Eocene diatomite in wells in the eastern North Sea. *Aarhus Geoscience* 6, 19-24.

Deegan, C.E. & Scull B.J., 1977. A proposed standard lithographic nomenclature for the central and northern North Sea. Report of the Inst. Geol. Sci., London, 77/ 25; *Bull. Nor. Pet. Dir.* 1.

Egeberg, P.K., Aagaard, P., 1989. Origin and evolution of formation water from oil fields on the Norwegian shelf. *Applied Geochemistry* 4, 131–142.

Friis, H., Poulsen, M.L.K., Svendsen, J.B., Hamberg, L., 2007. Discrimination of density flow deposits using elemental geochemistry—implications for subtle provenance differentiation in a narrow submarine canyon, Palaeogene, Danish North Sea. *Marine and Petroleum Geology* 24, 221– 235.

Flörke, O. W., Graetsch, H., Martin, B., Röller, K. & Wirth, R., 1991. Nomenclature of microcrystalline and non-crystalline silica minerals, based on structure and microstructure. Neues Jahrbuch Für Mineralogie, *Abhandlungen* 163, 1, 19 - 42.

Foster, W.R., Custard, H.C., 1980. Smectite-illite transformation-role in generating and maintaining geopressure. *AAPG Bulletin* 64 (5), 708.

Fothergill, C., A.1955. The cementation of oil reservoir sands and its origin. *Proc. World Petrol. Cong.* 59, 301–314.

Harder, H., 1965. Experimente zur —Ausfüllung‖ der KieselsZure. *Geochirn. Cosmochim, Acta* 29, 429-442.

Harder, H. and Fleming W. 1970. Quartzsynthese bei Tiefen Temperaturen. *Geochim. Cosmochim. Acta* 34, 295-305.

Harris N.B., 1992. Burial diagenesis of Brent sandstones: a study of Statfjord, Hutton and Lyell fields. in: A.C. Morton, R.S. Haszeldine, M.R. Giles , S. Brown, (editors). Geology of the Brent Group. *Geological Society Special Publication*, 61. Geological Society of London, pp. 351-375.

Hamberg, L., Dam, G., Wilhelmson, C., Ottesen, T., 2005. Palaeocene deep-marine sandstone plays in the Siri Canyon, offshore Denmark–southern Norway. In: Dore´, A.G., Vining, B.A. (Eds.), Petroleum Geology: North-West Europe and Global Perspectives, *Proceedings of the 6th Petroleum Geology Conference*. Geological Society of London, pp. 1185–1198.

Hawkins, P.J., 1978. Relationship between diagenesis porosity reduction and oil emplacement in late Carboniferous sandstone reservoirs, Bothamsall Oilfield, E. Midlands. *J. Geol. Soc. London,* 135: 7-24.

Heilmann-Clausen, C., Nielsen, O.B. & Gersner, F. 1985: Lithostratigraphy and depositional environments in the Upper Paleocene and Eocene of Denmark. *Bulletin of the Geological Society of Denmark* 33, 287-323.

Hurd D. C., 1973. Interaction of biogenic opal, sediment and seawater in the Central Equatorial Pacific. *Geochim. Cosmochim. Actu* 37, 2257-2282.

Hurd, D. C. and Theyer F., 1975. Changes in the physical and chemical properties of biogenic silica from ihe.Central Equatorial Pacific. I. Solubility. Specific surface area and solution rate constants of acid-cleaned samples. *Advan. Chem. Ser.* :Vo. 147. 21 I -230.

Hurd, D.C. and Birdwhistell, S., 1983. On producing a more general model for biogenic silica dissolution. *Am. J. Sci.,* 283: 1-28.

Hower, J., Eslinger, E., Hower, M.E., Perry, E.A., 1976. Mechanism of burial metamorphism of argillaceous sediment; 1, mineralogical and chemical evidence. *Geological Society of American Bulletin* 87 (5), 725–737.

Iijima, A, Matsumoto, R.,Tada,.R, 1983. Mechanism of sedimentation of rhythmically bedded chert ; *Sedimentary Geology*, V41, I 2-4, 221-233.

Knauth, L. P. and Epstein, J., 1975. Hydrogen and oxygen isotope ratios in silica from the JOIDES Deep Sea Drilling Project. *Earth Planet. Sci. Lett.* 25. l-10.

Knox, R.W.O‘B., 1997. The late Palaeocene to early Eocene Ash Layers of the Danish Mo-clay (Fur Formation): stratigraphic and tectonic significance. In: Thomsen, E., Pedersen, S.A.S. (Eds.), Geology and Palaeontology of the Mo-clay, *Aarhus. Aarhus Geoscience* 6, 7–11. v.

Knox R.W.O'B. & Holloway S., 1992. Paleogene of the Central and Northern North Sea. In: Lithostratigraphic Nomenclature of the UK North 44 J. M. Huggett and R. W. O'B. Knox Sea (R.W.O'B. Knox & W.G. Cordey, editors). *British Geological Survey*, Keyworth.

Knox, R.W.O.B., Morton, A.C., 1983. Stratigraphical distribution of early Palaeogene pyroclastic deposits in the North Sea Basin. *Proceedings. Yorkshire Geological Society* 44 (25), 355–363.

Land, L.S., Milliken, K.L., 2000., Regional loss of SiO2 and CaCO3, and gain of K2O during burial diagenesis of Gulf Coast Mudrocks, USA. In: Worden, R.H., Morad, S. (Eds.), Quartz Cementation in Sandstones. *Special Publication No 29 of the International Association of Sedimentologists.* Blackwell, Oxford, International, pp. 271–280.

Larsen, L.M., Fitton, J.G., Pedersen, A.K., 2003. Paleogene volcanic ash layers in the Danish Basin: composition and source areas in the North Atlantic Igneous Province. *Lithos* 71, 47-80.

Lynch, F.L., Mack, L.E., Land, L.S., 1997. Burial diagenesis of illite/smectite in shales and the origins of authigenic quartz and secondary porosity in sandstones. *Geochimica et Cosmochimica Acta* 61 (10), 1995–2006.

McBride, E. F., 1989. Quartz cementation in sandstones: a review. *Earth Sci. Rev.* 26, 69–112.

Mackenzie F. T. and Gees R., 1971. Quartz: synthesis at Earth-surface conditions. *Science* 173. 533-534.

Mitlehner, A.G., 1996.Palaeoenvironments in the North Sea Basin around the Paleocene-Eocene boundary: evidence from diatoms and other siliceous microfossils. in: R.W.O´B. Knox, R.M. Corfield,R.E. Dunay (editors). Correlation of the Early Paleogene in Northwest Europe.. *Geological Society Special Publication* 101. Geological Society of London, pp. 255-273.

Mitlehner, A.G. 1997: Diatoms from the Ølst Formation, Denmark: Comparison with assemblages from other localities in the North Sea Basin. *Aarhus Geoscience* 6, 27-28.

Mizutani, S., 1966. Transformation of silica under hydrothermal conditions. *J. Earth Sci.* Nagoya Unio. 14. 56-88.

Mizutani, S., 1977. Progressive ordering of cristobalitic silica in the early stages of diagenesis. *Contributions to Mineralogy and Petrol0gy* 61, 129.

Molenaar, N., Cyziene, J., Sliaupa, S., 2007. Quartz cementation mechanisms and porosity variation in Baltic Cambrian sandstones. *Sedimentary Geology* 195, 135-159.

Morton, A.C., Haszeldine, R.S. Giles, M.R., Brown, S., (editors)., 1993. Geology of the Brent Group. Special Publication, 61. *Geological Society of London, and Marine and Petroleum Geology,* Volume 10, Issue 4, , Page 403.

Mudge, D.C., Copestake, P., 1992. Revised Lower Palaeogene lithostratigraphy for the Outer Moray Firth, North Sea. *Marine and Petroleum Geology* 9, 53–69.

Nielsen, O.B. 1974: Sedimentation and diagenesis of Lower Eocene sediments at Ölst, Denmark. *Sedimentary Geology* 12, 25-44.

Nielsen, O.B. 1994: Lithostratigraphy and sedimentary petrography of the Paleocene and Eocene sediments from the Harre borehole, Denmark. *Aarhus Geoscience* 1, 15-34.

Nielsen, O.B. 1997: Lateral Facies variations in the Ølst-Fur Formations in Denmark. *Aarhus Geoscience* 6, 13-18.

Nielsen, O.B., Baumann, J., Deyu, Z, Heilmann-Clausen, C & Larsen, G. 1986: Tertiary deposits in Storebælt. In: J.T. Møller (Ed.) Twentyfive Years of Geology in Aarhus. *Geoskrifter* 24, 235-253.

Ohm, S.E., Karlsen, D.A., Roberts, A., Johannessen, E., Høiland, O., 2006. The Palaeocene sandy Siri Fairway: an efficient pipeline'' draining the prolific Central Graben. *Journal of Petroleum Geology* 29, 53-82.

Pedersen T., Wangen, M., Johansen, H., 1997. Flow along fractures in sedimentary basins. In: *Fluid Flow and Transport in Rocks, Mechanisms and Effects* (eds Jamtveit B, Yardley BWD), pp. 213–33. Chapman & Hall, London.

Pedersen, G.K., Surlyk, F., 1983. The Fur Formation, a late Palaeocene ash-bearing diatomite from northern Denmark. Bulletin. *Geological Society of Denmark* 32, 43–65.

Peltonen, C., Marcussen, Ø., Bjørlykke, K., Jahren, J., 2009. Clay mineral diagenesis and quartz cementation in mudstones: the effects of smectite to illite reaction on rock properties. *Marine and Petroleum Geology* 26 (6), 887–898.

Ramm, M. and Ryseth A. E., 1996. Reservoir quality and burial diagenesis in the Statfjord Formation, *North Sea. Petrol. Geosci.* 2, 313–324.

Schiøler, P., Andsbjerg, J., Clausen,O.R., Dam,G., Dybkjær, K., Hamberg, L., Heilmann-Clausen, C., Kristensen, L.E., Prince. I., and Rasmussen, J.A, 2007. Lithostratigraphy for the Palaeogene-Lower Neogene of the Danish North Sea. *Geological Survey of Denmark and Green Bulletin* 12, 77pp.

Small, J.S., 1994. Fluid Composition, Mineralogy and Morphological Changes Associated with the Smectite-to-Illite Reaction; an Experimental Investigation of the Effect of Organic Acid Anions. *Issue Devoted to Papers Read at a Conference on Diagenesis, Overpressure and Reservoir Quality.* Mineralogical Society, London, United Kingdom, pp. 539–554.

Siever, R., 1957. The silica budget in the sedimentary cycle. *American Mineralogist* 42, 821–841.

Siever, R., & Woodford, N., 1973. Sorption of silica by clay minerals. *Geochimica Cosmochimica Acta*, 37, 1851–1880.

Srodon, J., 1999. Nature of mixed-layer clays and mechanism of their formation and alteration. *Annual Review of Earth and Planetary Sciences* 27, 19–53.

Stein, C. L. and Kirkpatrick, R. J., 1976. Experimental porcelanite recrystallization kinetics: a nucleation and growth model. *Journal of Sedimentary PetrolOgy*, 46.2, 430-435.

Stokkendal, J., Friis, H, Svendsen, J.B, Poulsen, M.L.K., Hamberg, L., 2009. Predictive permeability variations in a Hermod sand reservoir, Stine Segments, Siri Field, *Danish North Sea Marine and Petroleum Geology,* Volume 26, Issue 3, Pages 397-415.

Thyberg, B., Jahren, J., Winje, T., Bjørlykke, K., Faleide, J. I., Marcussen, Ø., 2010. Quartz cementation in Late Cretaceous mudstones, northern North Sea: Changes in rock properties due to dissolution of smectite and precipitation of micro-quartz crystals, *Marine and Petroleum Geology*, 27, 1752-1764.

Thyne, G., 2001. A model for diagenetic mass transfer between adjacent sandstone and shale. *Marine and Petroleum Geology* 18, 743–755.

Thyne, G., Boudreau, B. P., Ramm, M., and Midtbø, R. E., 2001. Simulation of potassium feldspar dissolution and illitization in the Statfjord Formation, *North Sea. A. A. P. G. Bulletin* 85, 621–635.

Towe, K.M., 1962. Clay mineral diagenesis as a possible source of silica cement in sedimentary rocks. *Journal of Sedimentary Petrology* 32 (1), 26–28.

Van der Kamp, P.C., 2008. Smectite–illite–muscovite transformations, quartz dissolution, and silica release in shales. *Clays and Clay Minerals* 56 (1), 66–81.

Weaver, C.E., 1959. The clay petrology of sediments. In: Swineford, A. (Ed.), *Clays and Clay Minerals.* Pergamon Press, Oxford, United Kingdom, pp. 154–187.

Williams, L. A., Parks, G. A. & Crerar, D. A., 1985. Silica diagenesis, I. Solubility controls. *Journal of Sedimentary Petrology* 55, 3, 301 - 311.

Weibel, R., Friis, H., Kazerouni, A.M., Svendsen, J.B., Stokkendal, J., Poulsen, M.L.K.,2010. Development of early diagenetic silica and quartz morphologies-Examples from the Siri Canyon, Danish North Sea. *Sedimentary Geology* 228, 151-170.

Yeh, H.W., Savin, S.M., 1977. Mechanism of burial metamorphism of argillaceous sediments: 3, isotope evidence. *Geological Society of American Bulletin* 88 (9), 1321–1330.

EDITOR'S CONTACT INFORMATION

Dr. Afsoon Moatari Kazerouni
Geology Researcher
Geology Department
Aarhus University
Copenhagen, Denmark
Tel: 0045-50329955
afsoon.moatari@geo.au.dk

INDEX

D

H

I

J

K

L

M

N

O

P

Q

R

S

T